Polymer Nanocomposites in Supercapacitors

Supercapacitors are energy storing devices, gaining great scientific attention due to their excellent cycling life, charge-discharge stability, energy, and power density. The central theme of this book is to review the multiple applications of polymer nanocomposites in supercapacitors in a comprehensive manner, including discussions pertaining to various unresolved issues and new challenges in the subject area. It illustrates polymer nanocomposite preparation and working mechanisms as electrodes, binders, separators, and electrolytes. This edited volume also explains different components of supercapacitors, including theory, modelling, and simulation aspects.

Features:

- Covers the synthesis and properties of polymer nanocomposites for varied usage.
- Explains roles of different types of nanofillers in polymeric systems for developing supercapacitors.
- Highlights theory, modelling, and simulation of polymeric supercapacitors.
- Gives an illustrative overview of the multiple applications of polymers and their nanocomposites.
- Includes graphene, CNT, nanoparticle, carbon, and nano-cellulose-based supercapacitors.

This book is aimed at graduate students and researchers in materials science, polymer science, polymer physics, electrochemistry, electronic materials, energy management, electronic engineering, polymer engineers, and chemical engineering.

Emerging Materials and Technologies

Series Editor:
Boris I. Kharissov

2D Monoelemental Materials (Xenes) and Related Technologies
Beyond Graphene
Zongyu Huang, Xiang Qi, Jianxin Zhong

Atomic Force Microscopy for Energy Research
Cai Shen

Self-Healing Cementitious Materials
Technologies, Evaluation Methods, and Applications
Ghasan Fahim Huseien, Iman Faridmehr, Mohammad Hajmohammadian Baghban

Thin Film Coatings
Properties, Deposition, and Applications
Fredrick Madaraka Mwema, Tien-Chien Jen, and Lin Zhu

Biosensors
Fundamentals, Emerging Technologies, and Applications
Sibel A. Ozkan, Bengi Uslu, and Mustafa Kemal Sezgintürk

Error-Tolerant Biochemical Sample Preparation with Microfluidic Lab-on-Chip
Sudip Poddar and Bhargab B. Bhattacharya

Geopolymers as Sustainable Surface Concrete Repair Materials
Ghasan Fahim Huseien, Abdul Rahman Mohd Sam, and Mahmood Md. Tahir

Nanomaterials in Manufacturing Processes
Dhiraj Sud, Anil Kumar Singla, Munish Kumar Gupta

Advanced Materials for Wastewater Treatment and Desalination
A.F. Ismail, P.S. Goh, H. Hasbullah, and F. Aziz

Green Synthesized Iron-Based Nanomaterials
Application and Potential Risk
Piyal Mondal and Mihir Kumar Purkait

Polymer Nanocomposites in Supercapacitors
Soney C George, Sam John and Sreelakshmi Rajeevan

Polymers Electrolytes and their Composites for Energy Storage/Conversion Devices
Achchhe Lal Sharma, Anil Arya and Anurag Gaur

Hybrid Polymeric Nanocomposites from Agricultural Waste
Sefiu Adekunle Bello

Photoelectrochemical Generation of Fuels
Edited by Anirban Das, Gyandeshwar Kumar Rao and Kasinath Ojha

Emergent Micro- and Nanomaterials for Optical, Infrared, and Terahertz Applications
Edited by Song Sun, Wei Tan, and Su-Huai Wei

For more information about this series, please visit:
www.routledge.com/Emerging-Materials-and-Technologies/book-series/CRCEMT

Polymer Nanocomposites in Supercapacitors

Edited by
Soney C George,
Sam John, and Sreelakshmi Rajeevan

CRC Press
Taylor & Francis Group
Boca Raton   London

CRC Press is an imprint of the
Taylor & Francis Group, an **informa** business

First edition published 2023
by CRC Press
6000 Broken Sound Parkway NW, Suite 300, Boca Raton, FL 33487–2742

and by CRC Press
4 Park Square, Milton Park, Abingdon, Oxon, OX14 4RN

CRC Press is an imprint of Taylor & Francis Group, LLC

Library of Congress Cataloging-in-Publication Data
A catalog record has been requested for this book

ISBN: 978-1-032-00545-4 (hbk)
ISBN: 978-1-032-00549-2 (pbk)
ISBN: 978-1-003-17464-6 (ebk)

DOI: 10.1201/9781003174646

Typeset in Times New Roman
by Apex CoVantage, LLC

Contents

About the Editors

Soney C George is Dean (Research) and Director, Centre for Nanoscience and Technology, Amal Jyothi College of Engineering, Kanjirappally, Kerala, India. He is a fellow of the Royal Society of Chemistry, London, and a recipient of "best researcher of the year 2018" award from APJ Abdul Kalam Technological University, Thiruvananthapuram, India. He has received other recognition for his work, such as the best faculty award from the Indian Society for Technical Education, best citation award from the International Journal of Hydrogen Energy, a fast-track award of young scientists by the Department of Science & Technology, India, and an Indian Young Scientist Award instituted by the Indian Science Congress Association. He is also a recipient of a CMI Level 5 Certificate in Management and Leadership from Dudley College of Technology, London, as part of the AICTE-UKIERI Training Programme. He did his postdoctoral studies at the University of Blaise Pascal, France, and Inha University, South Korea. He has guided eight PhD scholars and 102 student projects. He has published and presented almost 240 publications in journals and in conferences and his h-index is 27. His major research fields are polymer nanocomposites, polymer membranes, polymer tribology, pervaporation, and supercapacitors. In addition, he took leadership in organizing a sponsored international conference on engineering education in association with IEEE international chapter, as well as conferences on development of nanoscience and technology. he initiated several lecture series, including the Isaac Newton Lecture Series, the ACeNT Lecture Series, and the Lecture Series on Nobel Prize Winning Works in order to bring eminent scientists and academicians from India and abroad to the college. Scientists from the United States, Malaysia, France, Poland, and South Africa have visited the campus and interacted with students through his networking. He also initiated collaborative research among scholars at Gdansk University of Technology, Poland, Inha University, South Korea, University of Blasé Pascal, France, Centre for Nanostructures and Advanced Materials, CSIR, South Africa, and Durham University, United Kingdom.

Sam John is an assistant professor in chemistry at St. Berchmans College (Autonomous), Changanassery, affiliated with Mahatma Gandhi University, Kottayam, Kerala, India. He received his PhD degree in chemistry from the University of Calicut in 2013. His research is focused on the development of corrosion control using organic coatings and inhibitors and he has recently worked on designing and tailoring materials for energy-related applications (batteries and supercapacitors) and corrosion mitigation in metal-air batteries. Dr. John has published more than 34 papers in peer-reviewed international journals in electrochemistry and contributed five book chapters published by CRC Press and Elsevier. He has an h index of 14.

Sreelakshmi Rajeevan is a research scholar at the Centre for Nanoscience and Technology, Amal Jyothi College of Engineering, Kanjirappally, Kerala, India, affiliated with APJ Abdul Kalam Technological University, Kerala, India. She holds a master's of technology in polymer science and technology and a master's of science in chemistry from Mahatma Gandhi University, Kerala. She has two publications in international journals and three book chapters. Her major research fields are 2D nanomaterials, nanocomposites, polymer nanocomposites, electrochemistry, supercapacitor, and energy storage devices.

List of Contributors

Sanjay D. Dhole is a professor at Department of Physics, Savitribai Phule Pune University, Pune 411007 (MS), India.

Sobhi Daniel is an associate professor at T. M. Jacob Memorial Government College, Research and Postgraduate department of Chemistry, Koothattukulam-62, Kerala, India and Maharaja's College, Postgraduate and Research Department of Chemistry, Ernakulam-11, Kerala, India.

Soma Das is an assistant professor at CMR Institute of Technology, AECS layout, Kundalahalli, Bangalore-560037.

Praveena Malliyil Gopi is a researcher at Department of Physics, Maharaja's College, Ernakulam, Kerala 682011 and Department of Physics, St. Teresa's College, Ernakulam, Kerala, 682011.

Soney C George is a professor and director of the Centre for Nanoscience and Technology, Amal Jyothi Engineering College, Koovapally, Kerala, 686518, India.

K. Hareesh is an assistant professor at School of Applied Sciences (Physics), REVA University, Bengaluru 560064 (KA) India.

Anisha Joseph is an assistant professor at Department of Science and Humanities, Federal Institute of Science and Technology, Angamaly, India.

Ekta Jagtiani is a researcher at Department of Polymer and Surface Engineering, Institute of Chemical Technology Mumbai – 19, India.

Minu Joys is an assistant professor at Post Graduate and Research Department of Chemistry, Union Christian College, Aluva-683 102, Kerala, India.

Ramakant P. Joshi is an assistant professor at PDEA's Annasaheb Magar Mahavidyalaya, Hadapsar, Pune 411028 (MS), India.

Sam John is an assistant professor at St Berchmans College, Changanassery, Kerala, 686101, India.

Ashok Kumar is an associate professor at Department of Applied Sciences, National Institute of Technical Teachers Training and Research, Chandigarh, 160019, India.

Astakala Anil Kumar is an assistant professor at Nanomaterials for Photovoltaics and Biomaterials Laboratory, Godavari Institute of Engineering and Technology, Rajahmundry 533296, India.

Igor Krupa is a researcher at Center for Advanced Materials, Qatar University, P.O. Box 2713, Doha, Qatar.

Essack Mohammed Mohammed is a researcher at Department of Physics, Maharaja's College, Ernakulam, Kerala 682011.

Sella Muthulingam is an assistant professor at Sri Ramakrishna College of Arts and Science, Coimbatore, Tamilnadu, India-641006.

Ravindra U. Mene is an assistant professor at PDEA's Annasaheb Magar Mahavidyalaya, Hadapsar, Pune 411028 (MS), India.

Thanathu K. Manojkumar is a professor at School of Chemical Sciences, Kannur University, Edat, Kannur- 670327, Kerala, India and School of Digital Sciences, Kerala University of Digital Sciences, Innovation and Technology, Thiruvananthapuram-695317, Kerala, India.

Ajalesh Balachandran Nair is an assistant professor at Post Graduate and Research Department of Chemistry, Union Christian College, Aluva-683 102, Kerala, India.

Deeksha Nagpal is a researcher at Department of Physics, Chandigarh University, Gharuan, Mohali (Punjab) 140413, India.

Sona Narayanan is an assistant professor at Department of Science and Humanities, Federal Institute of Science and Technology, Angamaly, India.

Suganthi Nachimuthu is an assistant professor at PG Department of Physics, Government Arts College, Kulithalai 639 120, Tamilnadu, India.

Vijaykiran N. Narwade is a researcher at School of Physical Sciences, S.R.T.M. University Nanded 431601 (MS), India.

Ayswarya Ettuvettil Pankajakshan is an assistant professor at Department of Science and Humanities, Federal Institute of Science and Technology, Angamaly, India.

Deepalekshmi Ponnamma is a researcher at Center for Advanced Materials, Qatar University, P.O. Box 2713, Doha, Qatar.

Greeshma Kuzhipalli Perayikode is an assistant professor at Sri Ramakrishna College of Arts and Science, Coimbatore, Tamilnadu, India-641006.

Shashank Priya is a professor and Associate Vice President for Research and Director of Strategic Initiatives in the Office of the Vice President for Research (OVPR) at Materials Research Institute, Penn State University, PA 16801, USA.

Shyam Sundar Pattnaik is a professor and director at Media Engineering, National Institute of Technical Teachers Training and Research, Chandigarh-160019, India.

Vidya Thattarkudy Padmanabhan is an assistant professor at Department of Science and Humanities, Federal Institute of Science and Technology, Angamaly, India.

Kottoly Raveendran Raghi is a researcher at School of Chemical Sciences, Kannur University, Edat, Kannur, 670327, Kerala, India.

Sreelakshmi Rajeevan is a researcher at Centre for Nanoscience and Technology, Amal Jyothi Engineering College, Koovapally, Kerala, 686518, India and APJ Abdul Kalam Technological University, CET Campus, Thiruvananthapuram, Kerala, 695016, India.

Daisy Rajaian Sherin is a researcher at School of Digital Sciences, Kerala University of Digital Sciences, Innovation and Technology, Thiruvananthapuram-695317, Kerala, India.

Kala Moolepparambil Sukumaran is an assistant professor at Department of Physics, St. Teresa's College, Ernakulam, Kerala, 682011.

Pandit N. Shelke is a professor at PDEA's Annasaheb Magar Mahavidyalaya, Hadapsar, Pune 411028 (MS), India.

Shasiya Panikkaveettil Shamsudeen is a researcher at Post Graduate and Research Department of Chemistry, Union Christian College, Aluva-683 102, Kerala, India.

Ajay Vasishth is a professor at Department of Physics, Chandigarh University, Gharuan, Mohali (Punjab) 140413, India.

Neethumol Varghese is an assistant professor at Post Graduate and Research Department of Chemistry, Union Christian College, Aluva-683 102, Kerala, India.

Manishkumar D. Yadav is an assistant professor at Department of Chemical Engineering, Institute of Chemical Technology Mumbai – 19, India.

Preface

Since the industrial revolution, the societal and financial success of any country has hinged on massive consumption of fossil fuels. The search for renewable and sustainable energy production led to the development of various energy conversion technologies, such as wind, solar, and fuel cells. Numerous advanced batteries like metal-ion, metal-air, and others have been developed, but they pose problems such as lower lifecycles, less power density, higher charging times, and heating problems, and they are not environmentally safe. The best remedy is to use an alternative energy storage device like a supercapacitor. Supercapacitors possess a distinctly greater capacitance, and they combine properties of batteries and capacitors into a single device. In line with this, supercapacitors have become auspicious applied technologies in diverse fields such as hybrid electric vehicles and sensitive automation, computer chips, and portable electronic devices. Supercapacitors bridge the gap between batteries and conventional dielectric capacitors. The striking features of supercapacitors are higher power densities, rapid charge and discharge processes, and their generation of reliable power quantities. The two energy storage mechanisms of supercapacitors are double-layer capacitance and pseudocapacitance. The electric energy storage in double-layer supercapacitors is based on the electrosorption process on porous electrodes. Pseudocapacitance is the energy produced by faradaic electrochemical reactions. Today, hybrid supercapacitors attract a lot of attention because they combine the features of both conventional double-layer supercapacitors and batteries to perform as high power/high energy storage devices.

This book provides an overview of polymer nanocomposites in high-performance energy storage supercapacitors. Chemists, physicists, chemical and electrical engineers, material scientists, research scholars, and students interested in energy will benefit from this overview of the materials used in supercapacitors and its emphasis on the future potential of supercapacitors in various power-system fields.

The chapters of this book are organized as follows: Chapter 1 provides the current status of supercapacitors based on polymer nanocomposites. This chapter deals with the general introduction, mechanism, concepts, and taxonomy of supercapacitors in a straightforward and concise manner. Chapter 2 relates an overview of conducting polymer nanocomposite-based supercapacitors and the most recent research developments for addressing energy storage and harvesting issues. Chapter 3 focusses on the mechanisms of synthesis, advances in graphene/polymer nanocomposites for supercapacitors, and various measurement techniques to evaluate the performance of a graphene/polymer supercapacitor. Chapter 4 provides comprehensive and efficient coverage on the structure, properties, fabrication strategies of CNT polymer composites, their energy storage, and supercapacitance applications. Chapter 5 presents some basics of flexible and stretchable supercapacitors, a brief comparison of conventional supercapacitors and flexible supercapacitors, fabrication techniques, various applications of flexible/stretchable supercapacitors, and the future outlook including the challenges of the cost-effectiveness of flexible and stretchable supercapacitors. Chapter 6 discusses the various fluoropolymer nanocomposites containing halloysite nanotubes and their dielectric performances. Variations in dielectric stability and response with respect to the concentration, as well as modifications of the halloysite nanotubes, are discussed in detail. Chapter 7 considers some nanostructured supercapacitors, factors influencing the performance of supercapacitors, structure, morphology, and different types of carbon-based electrode materials for supercapacitors. Chapter 8 describes various aspects of bionanocomposites in supercapacitors. Chapter 9 takes a fresh look at structural design, engineering techniques, and fabrication to make nanocellulose-based energy storage devices. Chapter 10 emphasizes recent developments in the pristine and functionalized metalorganic frameworks as electrode components in supercapacitors. The analogues variation in the capacitive properties such as specific capacitance, power density, high energy density, and cycle stability are comprehensively discussed. Chapter 11 summarizes the polymeric blend nano-systems in supercapacitor applications using conducting polymers such as polyaniline, polypyrrole, and poly(3,4-ethylenedioxythiophene) etc. It also focuses on the study and development of conducting polymer-based binary and ternary composite electrodes for supercapacitor application and the future progress of conducting polymer-based supercapacitors. Chapter 12 summarizes the basic theories involved in modeling supercapacitors, such as the electrochemical model, the equivalent circuit model, the intelligent model, the transmission line model, the fractional-order model, the simplified analytical model, and the thermal model, as well as important simulation techniques. Chapter 13 describes recent advancement in the development of polymer supercapacitors regarding their design approach, configurations, and electrochemical properties for supercapacitor applications, while at the same time providing current challenges and perspectives on future energy storage applications.

All of the contributors provided an enormous amount of help and support. We gratefully acknowledge the wonderful work of all of our chapter reviewers. A big thank you to the CRC Press team for guidance and continuous encouragement on this venture. We are indebted to the support, guidance, and motivation of our management and colleagues. We hope that this book provides those who study nanocomposite-based supercapacitors with a valuable resource.

1 Polymers for Supercapacitors
An Overview

Sreelakshmi Rajeevan, Sam John, and Soney C George

CONTENTS

1.1 INTRODUCTION

Supercapacitors are powerful energy storage devices capable of controlling high power rates compared to batteries and conventional capacitors and bridge the gap between the conventional capacitor and conventional batteries. The ruination of fossil fuels and renewable energy resources is the major problem contributing to rising fuel prices, pollution, global warming, and geopolitical concern. To deal with these problems is an important goal, and it can be achieved by developing more reliable energy sources and storage technologies. Energy storage systems (ESS) such as conventional capacitors, batteries, and supercapacitors are essential for managing the intermittent nature of renewable energy sources and increasing the amount of power transferred from systems such as wind and solar to the grid. The energy storage capacity per volume of supercapacitors is ten to 100 times more than the electrolytic capacitors and three to 30 times lower than batteries. The magnitude of specific energy of supercapacitors is on the order of ten times compared to ordinary capacitors. Supercapacitors are also known as *ultracapacitors* or *supercaps* due to their high-power output, fast charging-discharging ability, and long cycle life. A comparison of the electrochemical efficiency parameters of conventional capacitors, batteries, and supercapacitors is depicted in Table 1.1.

1.1.1 CLASSIFICATION OF SUPERCAPACITORS

Supercapacitors are classified into electrical double-layer capacitors (EDLCs), pseudocapacitors, and hybrid capacitors, depending on the energy storage mechanism (1–2).

1.1.1.1 Electrostatic Double-Layer Capacitors

The EDLCs lie on the well-known electrical double layer theory put forward in the 19th century by Herman von Helmholtz, followed by Louis Georges Gouy and David Leonard Chapman, and

DOI: 10.1201/9781003174646-1

TABLE 1.1

A Comparison of the Electrochemical Efficiency Parameters of Conventional Capacitors, Batteries, and Supercapacitors

Characteristics	Capacitor	Supercapacitor	Battery
Specific energy (Wh/kg)	< 0.1	1–10	10–100
Specific power (W/kg)	> 100000	500–10000	< 1000
Discharge time	10–6 to 10–3	S to min	0.3–3 h
Charge time	10–6 to 10–3	S to min	1–5 h
Operating voltage	-20 to 65 °C	-40 to 65 °C	0 to 65 °C
Coulombic efficiency (%)	> 95	85–98	70–85
Cycle-life	Almost infinite	➢ 500000	About 1000
Storage mechanism	Physical	Physical	Chemical
Power limitation	Dielectric efficiency	Electrolyte conductivity	Reaction kinetics, mass transport
Cycle life limitations	Self-discharge	Side reactions	Mechanical stability, chemical reversibility
Charge rate	High, same as discharge	High, same as discharge	Kinetically limited
Cost	Low	high	low
Service life	10 to 15 years	Up to 5 years	5 to 10 years

Source: Reprinted from Gonzalez et al. 2016 with permission from Elsevier (2)

Otto Stern and D. C. Grahame. EDLCs consists of two electrodes, conducting electrolyte and a thin porous separator that acts as a dielectric between the electrodes. An area arises at each of the two electrode surfaces, in which the liquid electrolyte is in contact with the conductive metal surface of the electrode. This interface creates a common boundary between the two different phases of the material, the insoluble solid electrode surface and the adjacent liquid electrolyte (2). A unique phenomenon of the double-layer effect occurs in this interface. On the application of external voltage, to maintain the electric neutrality of the cell, the conducting ions (both negative and positive ions) in the electrolyte can move through the separator towards the respective electrodes (positive and negative electrodes, respectively) through coulombic force; as a result, an electrical double layer formed at the interface of electrode/electrolyte known as Helmholtz double layer (H_{DL}). The Helmholtz double layer is divided into the inner Helmholtz plane (IHP) and outer Helmholtz plane (OHP) according to the specific adsorption of ions on the surface of the electrode (3–4). The double-layer is simply two layers of charges: one electronic layer emerges from the surface of the electrode and the other from the electrolyte with opposite polarity, and each layer of charges are separated by a layer of solvent molecules. The solvent molecules are physically adsorbed on the electrode surface like a molecular dielectric, providing the inner Helmholtz plane. The rest of the charges outside the IHP are counted as the outer Helmholtz plane, which determines the amount of charge in the electrode. The capacitance of the Helmholtz double layer is the sum of the capacitance of serially connected IHPs and OHPs.

$$\frac{1}{C_{HDL}} = \frac{1}{C_{IHP}} + \frac{1}{C_{OHP}} = \frac{d_{IHP}}{\varepsilon_{IHP}\varepsilon_0} + \frac{d_{OHP}}{\varepsilon_{OHP}\varepsilon_0} \qquad (1)$$

Where C_{HDL} is the Helmholtz layer's differential capacitance, CIHP and COHP are the capacitance of the IHP and OHP, respectively, ε_{IHP} and ε_{OHP} are the relative dielectric constants of IHP and OHP respectively, and d_{IHP} and d_{OHP} represent the diameter of solvent molecules if the electrolyte is an aqueous solution. From equation 1, it is clear that the thickness of the Helmholtz layer is greatly influenced by the adsorption of specific ions. Therefore, the change in the electrode potential has a direct impact on the capacitance of the Helmholtz layer.

There is no transfer of electrons or charge occurring at the interface, so no chemical bond formation takes place. Therefore, the force that is acting behind EDLCs is purely electrostatic and non-faradaic in nature. Two double layers are generated totally at the electrode/electrolyte interface (positive electrode/electrolyte interface and negative electrode/electrolyte interface), which forms the heart of the EDLCs. These two double layers determine the electrochemical performance of the EDLCs. The fluctuations in the temperature of the electrolyte allow the remaining ions in the electrolyte to take up a scattered position within the electrochemical cell. This scattered layer of electrolytic ions with the positive charge array of electrodes constitutes the Gouy-Chapman model, and this is also called the diffuse layer. There are several factors that influence the thickness of the diffused layer, such as temperature, electrolytic concentration and dielectric constant, the conductivity of the electrolytic solution, and the charge number of the carrier ion (3). The thickness of the diffuse layer is directly proportional to the temperature and inversely related to electrolytic concentration and dielectric constant and the charge number of the carrier ion. When the temperature is low, the double layer will be thinner, and a compact Helmholtz layer is generated. The coexistence of both the Helmholtz layer and Gouy-Chapman layer together constitutes the Stern-Grahame layer. The model of the formation electrical double layer at the electrode/electrolyte interface is represented in Figure 1.1.

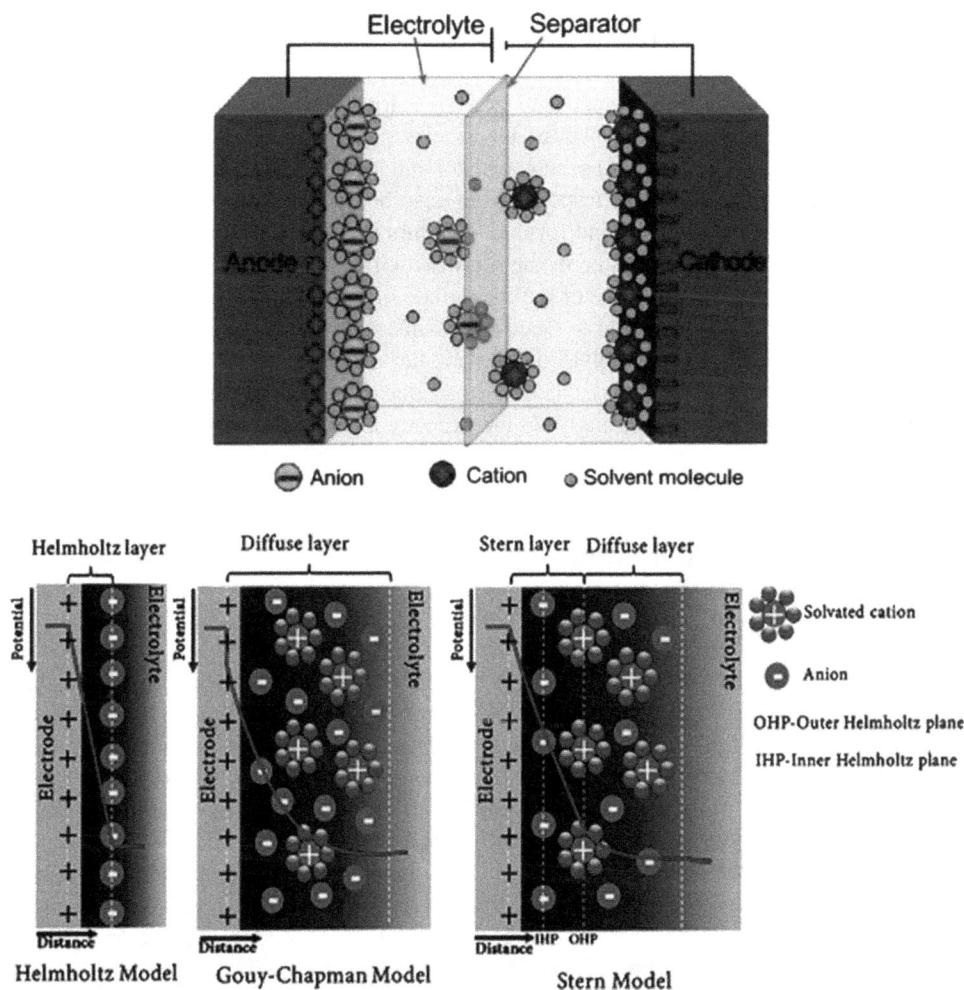

FIGURE 1.1   The model of EDLCs (a) and the energy storage mechanism of EDLCs (b) (5–6).

At the initial stage of the development of EDLCs, aqueous solutions are used as electrolytes. Later, the extension of the Helmholtz model introduced non-aqueous and ionic liquids as electrolytes for EDLCs. The charge, concentration of electrolyte, solvent, and temperature are the parameters that directly influence the electrochemical double-layer capacitance. The type and structure of the electrode material and electrolyte have no direct relation to the double-layer capacitance. In reality, differential double-layer capacitance is obtained depending on the surface structure and conductivity of both electrode and electrolyte. For a selected electrode material, the electrolyte with different sizes and morphological specialties has different interactions resulting in different adsorption strengths, and for a selected electrolyte, the mechanism is vice versa. The electrode materials generally used for the fabrication of EDLCs are carbonaceous materials such as activated carbon, graphene, carbon nanotubes, carbon-derived carbon, carbon fiber cloths, carbon organic frameworks (COFs), carbon aerogels, etc. The chosen carbon-based electrode material should be highly conductive, porous, and have a specific surface area between 500 to 4000 m²/g. The discussion regarding carbon-based EDLCs has been given in detail in the coming chapters.

1.1.1.2 Electrochemical Pseudocapacitors

The energy storage mechanism of pseudocapacitors is entirely different from the EDLCs. Pseudocapacitors store energy via a reversible faradaic redox process. In EDLCs, the charge storage mechanism depends on the formation of electrical double layers. On the application of external voltage, the specifically adsorbed or de-solvated ions in the electrolyte pervade this electrical double layer formed at the electrode/electrolyte interface, resulting in the pseudocapacitance. Pseudocapacitance is the storage of electricity in a capacitor via an electrochemical process. This can be caused by the rapid sequence of reversible faraday redox and intercalation processes on the surface of the electrode. The model of pseudocapacitor and its mechanism is depicted in Figure 1.2. The transfer of an electron or charge (especially an adsorbed ion) occurs between the electrolyte and the electrode, but no chemical bond formation occurs within the system. The formation of a static double-layer is an essential factor in the construction of pseudocapacitance. Depending on the nature and morphological structure of the magnitude of the electrode of pseudocapacitance is exceeded by a factor of 100 compared to electrical double layer capacitance for the same surface area (2–3, 7). This is because the de-solvated ion is smaller compared to the solvated ion within the electrolyte. The ability of the electrodes to generate pseudocapacitance by reversible redox reactions and intercalation of ions depends strongly on the porosity and conductivity of electrode material along with the chemical affinity of electrodes towards the adsorbed ions on the electrode surface. The materials that exhibit pseudocapacitance or redox reaction are conducting polymers such as polyaniline (PANI), polythiophene, and polypyrrole, as well as metal oxides such as RuO_2, MnO_2, or IrO_2 etc. The electrochemical performance of pseudocapacitors based on these electrode materials will be discussed in subsequent chapters.

1.1.1.3 Hybrid Supercapacitors

The combination of an electrostatic double-layer capacitor (EDLC) with an electrochemical pseudocapacitor is called a hybrid supercapacitor (HS). The properties of hybrid supercapacitors such as energy density, power density, charging-discharging ability, and cycle life are superior to those of parent supercapacitors because they are highly dependent on the electrode material and electrolyte used in the manufacture of the parent supercapacitors. The HS is made up of an equal contribution from the parent supercapacitors, i.e., one half is EDLC, and the other half is pseudocapacitor. The mechanism behind the power storage of HSs is in the formation of an electric double layer at the electrode/electrolyte interface in the EDLC and the reversible faradaic response in the pseudocapacitor. The Ragone plot for the energy storage devices shows the apex of the power density of hybrid supercapacitor well above the fuel cells and conventional battery and below the conventional capacitors. The main concern with the supercapacitor is to achieve an energy density in the range of 15–35 Wh/kg. This has played an essential role in the development of HSs (2).

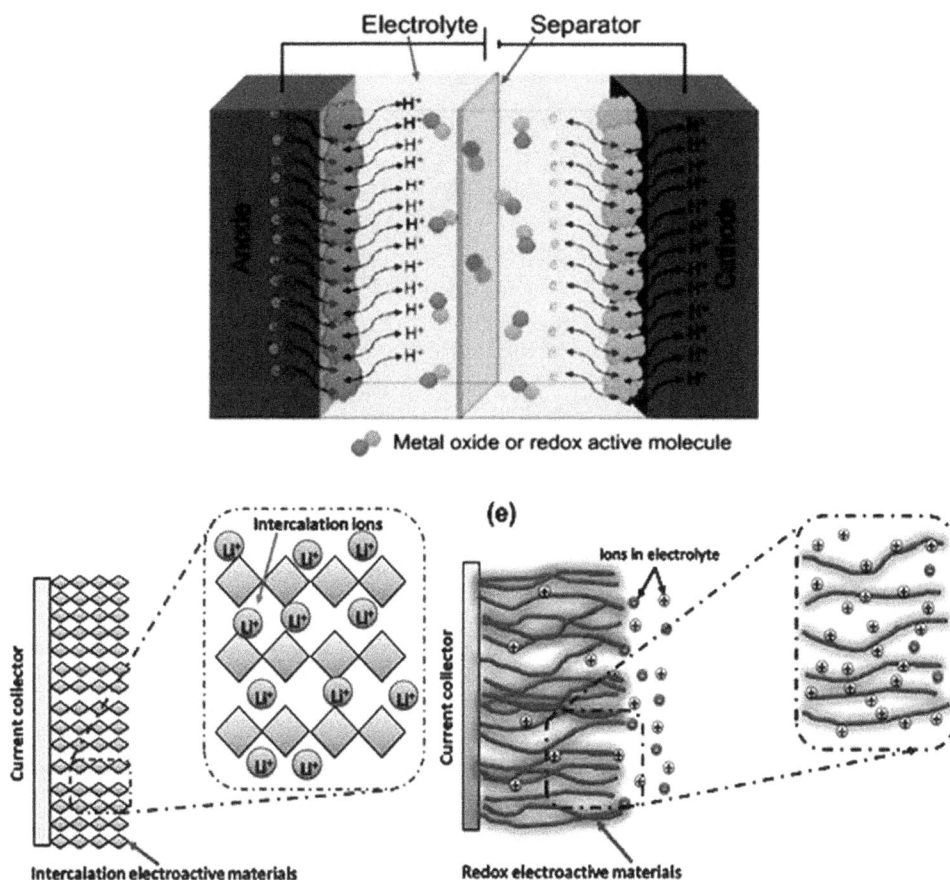

FIGURE 1.2 The model of electrochemical pseudocapacitor (a) and the energy storage mechanism of pseudocapacitors (b) (5–6).

Furthermore, the research gained in this area has found the modification and development of high porosity nano-sized active materials as an important solution to increasing the energy density of HSs. Most of the electrode materials used in the fabrication of EDLCs and pseudocapacitors can be used for the configuration of HSs, either in the original form or with some modifications. There are symmetric and asymmetric types of HSs (8–9) depending on electrode assembly, i.e., similar electrodes constitute symmetric HSs and dissimilar electrodes constitute asymmetric HSs. The asymmetric HSs show excellent electrochemical properties and maintain cycling stability compared to those prepared with symmetric ones. The widely available commercial form of HSs has an asymmetric configuration, and the most extensively used electrode material for the fabrication of these asymmetric hybrid supercapacitors are AC, MnO_2 and AC-Ni$(OII)_2$. In the case of symmetric HSs, an operational voltage up to 2.7 V comprised of binary AC electrodes inside an organic electrolyte is commercially available on the market (10). The schematic representation of the HS is represented in Figure 1.3.

Recently, the conducting polymer-based electrodes have received immense attention regarding the fabrication of symmetric HSs. This is because the mechanism of energy storage in conducting polymer-based electrodes occurs via redox reaction or doping. The doping and de-doping ions form by passing ions through conducting polymer's backbone and the electrolyte. Therefore, the charging process occurs entirely along the bulk volume of the polymer matrix, not just on the surface as in the case of carbonaceous electrodes, thereby resulting in the generation of high specific capacitance

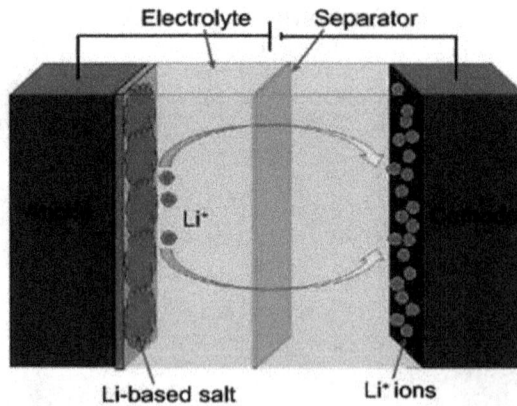

FIGURE 1.3   The model of hybrid supercapacitor and its energy storage mechanism (6).

HSs. The carbon-based electrode, in combination with conducting polymers or metal oxides, are called composite electrodes. In addition, there are battery-type electrodes (Li-ion battery electrodes in combination with carbon-based electrodes), and asymmetric electrodes (pseudocapacitive metal oxides combination with carbon-based electrodes) designed explicitly for HSs.

1.1.2   Factors Influencing Electrochemical Performance of Supercapacitors

Each of the components in the supercapacitor design has a great impact on the overall electrochemical performance of a supercapacitor. To design a well-performing supercapacitor, much attention should be given to the fabrication of its components. The electrochemical supercapacitor design consists of electrodes, a binder for the integration of electrode material, an electrolyte, a separator, and a current collector. The factors affecting the electrochemical performance of electrodes are as follows:

- The porosity, conductivity, specific surface area, and storage capacity of electrodes
- Efficiency of binders to hold the active materials together
- The thermal and electrochemical stabilities, low flammability, corrosion resistance, low volatile nature, high ionic conductance of an electrolyte
- Porosity, ionic conductivity, mechanical stability, electrochemical stability, low tortuosity of separator
- Materials and methods used for the synthesis of each component
- The ability of electrode, binder, electrolyte, and separator to withstand the applied electrochemical potential window
- Adhesion strength between the electrode and current collector
- The sealing of the assembled cell against water and other gases to prevent the cell from short-circuiting

1.1.3   Disadvantages of Supercapacitors

Besides all the advantages, supercapacitors also suffer some disadvantages. Their relatively low specific energy is one of the significant disadvantages of supercapacitors. The measure of the amount of energy stored by weight in the electrolytic supercapacitor is termed *specific energy*.

The specific energy of the supercapacitors (5 Wh/g) is very low compared to the Li-ion batteries (100–200 Wh/g). The second disadvantage is its considerably high self-discharging rate. In order to achieve higher voltage output, a series connection of the cells is required for supercapacitors. The output voltage of 50% discharge of a 2.7 V supercapacitor is about 1.35 V, which is only half of the discharge voltage of a 2.7 V conventional battery (2.7 V). Finally, the cost of fabrication of supercapacitors is relatively high compared to that of traditional batteries.

1.2 POLYMERS FOR SUPERCAPACITORS

Polymers are macromolecules composed of several repeating subunits called *monomers*. The monomers are formed by atoms or grouped atoms linked together by chemical bonds. The process of formation of polymers is called *polymerization*. Polymers play an inevitable role in fabricating energy storage devices due to their exceptional structural, morphological, mechanical, thermal, and electrochemical properties, and all these properties depend on individual constituents or monomers (11). Polymers are promising candidates for the configuration of individual components of supercapacitors or any other energy storage devices and act as a multifunctional system to achieve high performance and highly durable energy storage systems. The most fascinating features of polymers are controllable geometrical shape as per the requirement, and tunable morphology and properties by incorporating appropriate fillers, especially nanofillers. The application of polymers in the fabrication of supercapacitors can be broadly classified into four groups, including electrodes, binders, electrolytes, and separators. These are the essential components of a supercapacitor's cell design. The optimization and modification of properties of these components are significant tasks for the apprehension of achieving a high-performance energy storage system.

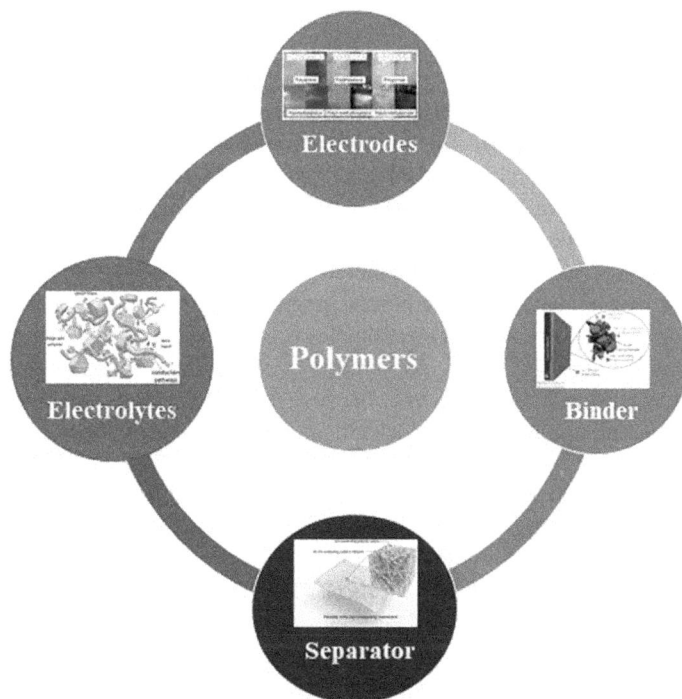

SCHEME 1.1   Role of polymers in supercapacitor fabrication.

1.2.1 POLYMERIC ELECTRODES

The electrodes constitute the heart of the supercapacitor design, and the materials used for the preparation of electrodes are called active materials. The necessary characteristics for an electrode include high electrical conductivity, porosity, high specific surface area, strong mechanical stability, long cycling stability, surface wettability, thermal conductivity towards heat fluctuations inside the cell, and a large electrochemical potential window. The conducting pathways within the electrode system enable the easy transport of electrons through the circuit; therefore, high power output is obtained. If the paths are too tortuous, then supercapacitors will suffer from low electrochemical performance (11).

When coming to the role of polymers as electrodes, as mentioned in earlier sections of this chapter, conducting polymers are used as electrodes for the fabrication of pseudocapacitors. The energy storage in pseudocapacitive material occurred via reversible redox reactions during the charge and discharge period, and this faradaic reaction occurs at the thin surface layer of the electrode. The underpotential deposition and partial electron transfer are the main reasons for the high specific capacitance of pseudocapacitors over EDLCs. The conducting polymers have proven themselves as pseudocapacitive material due to their optimistic electrochemical properties, ease of synthesis, and low cost. Conducting polymers are organic and capable of conducting electricity via the conjugated bonds present in the polymeric backbone (12–14). The most well-known conducting polymers are polyaniline (PANI) (15–16), Polypyrrole (PPy) (17), and poly-(3-4)-ethylene dioxythiophene (PEDOT) (18–19). The derivatives and nanocomposites of these polymers also receive immense interest in this area. Since the 19th century, conducting polymers have forged their own space in this field. In conducting polymers, energy storage is carried out by accumulating charge by proton doping interventions across the polymeric backbone. Subsequent chapters will discuss the mechanisms, properties, application of conductive polymers, their derivatives and nanocomposites, and their electrochemical performance as electrodes for pseudocapacitors and hybrid capacitors.

1.2.2 POLYMERIC BINDERS

Even though the active material used to fabricate supercapacitor electrodes possesses all the necessary properties, it cannot be used directly. It requires a kind of mechanical support to integrate the particles and to withstand large charge-discharge cycles. This support can be provided by polymers called *binders*. The binder holds the particles of active material together and offers a compact arrangement for the electrodes. The parameters used to evaluate a suitable binder include strong adhesion between the electrode and current collector, non-toxicity, and strong thermal and electrochemical stabilities to withstand long charge-discharge cycles. The preparation of electrodes has always depended on the ratio of the concentration of active material to the binder. In addition to that, the solvent and temperature used for the preparation of electrodes are also important. The porosity retention in the electrode is a difficult task because most of the time, the binder has a negative impact on active material, i.e., the binder blocks the pores of the active material. However, without binders, electrode preparation is a difficult task. Recently, binder-free electrodes have drawn the attention of many researchers working in this field. But the feasibility of the method of preparation suffers from many difficulties. There are natural polymeric binders such as cellulose and its derivatives, casein, alginate, starch, guar gum (GG) (20), polyvinylpyrrolidone etc. and synthetic polymeric binders such as polyvinylidene fluoride (PVDF), poly (tetrafluoroethylene) (PTFE), polyacrylic acid (PAA), and polyvinyl butyral. The choice of polymeric binders depends entirely on the active material and the researcher, as not all polymeric binders are suitable for all active materials. Each of them has its advantages and disadvantages.

Over the years, researchers have been focusing heavily on the development of eco-friendly energy devices. In itself, the development of components for supercapacitors (mainly electrodes and separators) with biodegradable polymers receives much attention. Cellulose and its derivatives are common candidates in electrode and separator preparation for supercapacitors due to its non-toxicity and strong adhesion properties. Shrestha et al. reported an active carbon-based electrode with ethylcellulose as a binder prepared via electrophoretic deposition for EDLCs. The adhesion test on the AC/ethylcellulose electrode showed a 4B level adhesion strength. The highest specific capacitance of 158.6 F/g is obtained for activated carbon with a 7% ethylcellulose concentration. The authors also proved that electrophoretic deposition was an efficient method for the preparation of electrodes (21). There is a comparative study reported in 2019 among different polymeric binders for the preparation of Si composite electrodes. They used both natural (carboxymethylcellulose (CMC), sodium alginate (SA)) and synthetic (PVDF) polymeric binders. They evaluated the effect of these binders on the cracking behavior of Si composite electrodes through this work. It was found that Si electrodes made of CMC and SA showed cracked structures in the morphology. This is due to the formation of strong chemical bonds between them, leading to high adhesion. The strong contraction and localized tensile stress ensured the opening and closing of cracks periodically at the same location in the Si electrodes. But for Si/PVDF electrodes, there was no cracked morphology. Because PVDF holds the active material via weak van der Waal's forces, this force is not enough for Si particles for creating a localized tensile stress in the electrode (22).

Generally, PVDF and PTFE are the most frequently used polymeric binders. As mentioned earlier, both these polymers have some advantages and disadvantages. Zhu et al. studied the effect of PTFE, PVDF and nafion binders on the electrochemical performance of AC electrode (23). The result of the electrochemical studies found that PTFE is a suitable binder for AC electrodes compared to PVDF and nafion. The specific capacitance obtained are 116 F/g, 124 F/g and 80 F/g for AC-Nafion, AC-PTFE, and AC-PVDF electrodes, respectively, at a current density 2 A/g. After 2000 cycles compared to AC/PVDF and AC/nafion electrodes (79.7% and 87.0%, respectively), AC/PTFE electrode attained capacitance retention above 90.8% % (23). As a binder, even though the electrochemical performance of PVDF is moderately low, it is the most widely used due to its most peculiar characteristic of compatibility with almost all types of active material. Manganese oxide is a pseudocapacitive material. Most of the binders restrict their electrochemical performance. In this sense, PVDF is an excellent choice, and it is the only polymer binder used for the preparation of MnO_2 electrodes (24).

PVDF and PTFE are fluorine-containing polymers, and both of them require toxic solvent for the preparation of electrodes. Therefore, green and water-soluble binders have always received attention from the scientific community. Polyvinylpyrrolidone (PVP) is one such greener polymer, and the preparation of AC/PVP for the first time was reported by Aslan et al. (25). The authors compared the electrochemical performance of the AC/PVP electrode AC/PVDF and AC/PVDF and found that PVP is an excellent choice for the preparation of AC electrodes for supercapacitors. Furthermore, the major advantage of using PVP as a binder is its feasibility in direct casting or spraying onto the current collectors. A similar kind of study using PVP/PVB composite was also reported in 2015 by Aslan et al. (26). The PVP/PVB binder was found to be excellent because of its solubility in ethanol and the direct casting of AC electrode slurry on the current collectors. This composite binder does not significantly affect the porosity and specific surface area of the AC electrodes.

Recently, a natural milk protein called casein was introduced as a binder for the preparation of activated carbon/conductive carbon (CC) electrodes. Casein has been a known adhesive since ancient times for the production of paints and coatings. The advantages of casein are non-toxicity, water-solubility, high adhesive strength, and excellent thermal stability up to 200 °C. It is electrochemically inert in the potential window generated by organic electrolytes, with excellent mechanical stability. The AC/CC electrode prepared with casein showed excellent rate capability and cycling stability (96.8% at 10 mA/cm after 10000 cycles) (27).

1.2.3 POLYMERIC ELECTROLYTES

Electrolytes will always play a part in the construction of a supercapacitor. The essential mechanisms that allow a supercapacitor to work are charge transfers and deposition at the electrode/electrolyte interface. An electrolyte should therefore have good ionic conductivity, low viscosity, low volatility, high density of free ions, low toxicity, high thermal stability, and good electrochemical stability with a wide potential window (28). Supercapacitor performance is heavily dependent on electrolytes, which implies that an electrolyte within these parameters can provide a high-performance supercapacitor with high specific power and specific energy, as well as low equivalent series resistance (ESR). Aqueous, non-aqueous, organic, polymer electrolytes, and ionic liquids (29) are all extensively used in the manufacture of electrochemical devices. However, with the introduction of the flexible supercapacitor concept, the importance of polymer electrolytes skyrocketed. Because of their thin-film generating capabilities, flexibility, and high ionic conductivity, polymer-based electrolytes are regarded a suitable candidate for the design of supercapacitors with a light-weight architecture. Another intriguing property of polymers is their capacity to generate various geometric designs, which may be adjusted to meet specific needs. They are more environmentally friendly and safer than other electrolytes, particularly organic electrolytes. They also aid in the prevention of liquid electrolyte leakage, which reduces electrode corrosion and supercapacitor short circuiting while also boosting the supercapacitor's reliability. Solid polymer electrolytes (SPEs) and gel polymer electrolytes (GPEs) are the most common polymer-based electrolytes utilized in the manufacture of electrochemical devices. SPEs are often high molecular weight polymers e.g., polyolefin oxides such as polyethylene oxide (PEO) and polypropylene oxide (PPO) containing anionic salts (e.g., $LiClO_4$, $LiCF_3SO_3$, NaI, KI, etc.) (30). They can be utilized as electrolytes and separators for supercapacitors due to their interoperability with metal current collectors, as well as their anti-corrosive nature, superior mechanical properties, and stability to resist extended cycles and ease of processing. The supercapacitor's whole electrochemical characteristics are improved, and internal resistance is reduced, attributable to its wide range of potential window reliability (> 4 V). In comparison to liquid electrolytes, they have a much lower ionic conductivity at ambient temperature about 10^{-8} S/cm (31).

GPEs are quasi-solid-state polymer electrolytes with substantial volumes of liquid, they differ significantly from SPEs. GPEs are made of salts and solvents that are commonly found in liquid electrolytes. The GPEs' gel consistency is the result of a synergistic impact between solid polymer cohesive capabilities and liquid electrolyte diffusive qualities. The trapping of liquid electrolytes raises the polymer's amorphous content and lowers the glass transition temperature to around -40 °C, increasing ion mobility and resulting in better ionic conductivity of the GPEs. The high ionic conductivity at room temperature (10^{-4} to 10^{-3} S/cm), which is nearly equivalent to liquid electrolytes has greatly enhanced the adoption of GPEs as electrolytes or separators. Polymers such as polyvinyl alcohol (PVA) (32), polymethylmethacrylate (PMMA) (33–34), PVDF (35), P(VDF-HFP) (36), polyacrylonitrile (PAN) (37), and polyethylene oxide (PEO) have all been used to create electrolyte systems.

PAN was primarily used in the manufacture of Li-ion batteries. Because the PAN matrix lacks oxygen functionality, lithium salts dissociate more readily, making them appropriate for Li-ion batteries. Additionally, it can generate homogeneous electrolyte films in organic solvents/salt mixture and withstand high-temperature stability (up to 300 °C). Therefore, they are suitable electrolytes for energy storage devices. The electrochemical performance of the PAN electrolyte can be enhanced by the incorporation of ionic liquid. The ionic liquid modification increases ionic conductivity as well as the thermal stability of PAN (38). Tamilarasan et al. reported a graphene EDLC with PAN/[BMIM][TFSI] electrolyte (39). The ionic conductivity of the PAN/[BMIM][TFSI] electrolyte is about 2.42 mS/cm. In this study, the incorporation of [BMIM][TFSI] facilitates the formation of a solid layer of ionic liquid on the surface pores of the PAN matrix along with liquid phase. The excellent properties of ionic liquid enhanced the cyclic stability of graphene EDLCs (39). Among polymer-based electrolytes, PVA is an attractive electrolyte due to its water solubility non-toxicity, chemical stability, and biodegradability. The peculiarity of PVA as an electrolyte is the presence of -OH groups in its structure, which can form strong hydrogen bond with water and enhances the ionic conductivity. The hydrogen bond formation also provides mechanical strength to the corresponding electrolytes (40). Karaman et al. prepared an acid blended PVA electrolyte, PVA/H_2SO_4

(41). This electrolyte is used for the fabrication of carbon nanofibers (CNF) EDLCs. The CNF based EDLC showed a specific capacitance of 134 F/g. The power and energy densities of the CNF based EDLC is 1000 W/kg and 67 Wh/kg at a current density of 1 A/g (41).

PVDF is also preferred electrolyte for second-generation devices due to the existence of a strong electron-withdrawing C-F bond in its polymer backbone and a significant dielectric constant value of roughly 8.4. It is a high molecular weight polymer that enhance the mechanical stability of gel electrolyte. Moreover, copolymerization with hexafluoropropylene (HFP) can improve the electrochemical characteristics of PVDF-based GPEs. Kumar et al. fabricated two EDLCs of AC and multiwalled CNT and evaluated their electrochemical performance in 1-ethyl-3-methylimidazolium trifluoromethanesulfonate (EMITf)/poly(vinylidene fluoride-co-hexafluoropropylene) (PVdF-HFP) as electrolyte. This electrolyte possessed high ionic conductivity in the range of 10^{-3} S/cm at room temperature. The MWCNT based EDLC and AC-based EDLC displayed a specific capacitance of 32 F/g and 157 F/g respectively. Also, rate capability and cycling stability of EDLCs are excellent (42). A novel polymer electrolyte was developed by Ortega et al. for increasing the specific capacitance of functionalized MWCNT supercapacitor (43). The electrolyte is a mixture of nanostructure silicon dioxide (SiO2) with PVDF and CNT (GPE-SiO2). The electrochemical studies of the electrolyte are carried out by comparing with GPE alone. The area of the voltammogram the GPE/SiO$_2$ electrolyte very high compared to GPE. In addition, electrochemical impedance proved that the charge transfer resistance in the GPE/SiO$_2$ low compared to GPE. The f-MWCNT supercapacitor retained 90% of specific capacitance after 2000 cycles due to the incorporation of an inorganic nano-SiO filler, which increased the ionic conductivity and thermal and mechanical stabilities of the electrolyte (43). PEO based electrolytes also promising candidates as GPEs. The Li-doped poly (ethylene oxide) (PEO)/poly(vinyl alcohol-co-acrylonitrile) (poly(VA-co-AN)) exhibits ionic conductivity in the range of 2×10^{-4} S/cm and 7×10^{-3} S/cm at the temperature 30 °C and 100 °C respectively (44).

Recently, the sulfonated polymer electrolytes have attracted attention due to their ability to improve electrochemical performance of the energy storage device. A group of researchers developed an alumina (Al2O3) incorporated ionic liquid modified sulfonated polysulfone (IL-SPSU). To withstand temperature generated inside the supercapacitor, the electrolyte should be thermally stable. The Al2O3/IL-SPSU electrolyte possess high temperature resistance up to 200 °C and high ionic conductivity in the range of 1.81×10^{-4} S/cm and 2.06×10^{-3} S/cm at 30 °C and 100 °C, respectively. The symmetric supercapacitor integrated with this electrolyte showed a specific capacitance of 144 F/g at 1 A/g. The energy density obtained for the symmetric cell is 20 Wh/kg at a power density of 1028 W/kg (45).

1.2.4 POLYMERIC SEPARATORS

The separator is used to prevent immediate contact between the electrodes and in this way guarantees a free electrochemical pathway for the progression of ions from the electrolyte to the electrodes. A separator should be strong and thin enough to give solidness to the supercapacitor and ought to forestall the movement of active materials to avoid shorting out the device. Aside from these properties, the separator ought to be steady over the working voltage window. The proper choice of separators relies upon parameters like high ionic conductance, high porosity, insulating property, low internal resistance, and wettability. Notwithstanding this load of characteristics, the separator should be chemically resistant to the impact of the electrolytes to prevent erosion and stable in ambient conditions (air, humidity etc.) (46). The choice of the separator material is likewise vital because it can make a critical commitment to the general cell series resistance and the rate capability of the supercapacitor. The separators utilized in energy devices include cellulose (paper), fiberglass, mica, silica, ceramics, and polymers. Among them, polymer-based separators have been acquiring the consideration of analysts dealing with the advancement of productive separator materials in light of their high porosity, high mechanical strength, versatility, chemical and corrosion resistances, and low cost. Both natural and synthetic polymeric separators are used for the integration of supercapacitors. Some of these include chitosan (47), cellulose (48), Polyethylene (PE), poly (ethylene oxide) (PEO), polyacrylonitrile (PAN), polyvinyl alcohol (PVA) (49), PVDF (50) and its copolymers (PVDF-co-HFP), and nylon (51).

Chitosan is a watery dissolvable natural biopolymer (the second most plentiful regular biopolymer on Earth) that has various functional groups in its structure. Due to its biodegradability and bio-viable nature, chitosan has been broadly used for biosensor, biomedical, and water treatment applications. As a separator, the specialties of chitosan include its hydrophilicity, electrochemical stability, low cost, and high ionic conductance. Raja et al. developed a activated carbon-based electrode using chitosan/poly(ethylene glycol)-ran-poly(propylene glycol) [Ch/poly(EG-ran-PG)] blend. They used this biopolymer-based blend as binder cum membrane cum separator in their study. The constructed symmetrical supercapacitor exhibited a specific capacitance of 193 F/g at 50 mV/s. The specific energy and specific power obtained for the symmetrical SCs is 4.7 W h/ kg (at 1 A/g) and 2.5 kW/kg (at 5 A/g), respectively. After 6000 cycles, about 99% capacitance retention was observed (52). Another naturally biodegradable polymer that contributed to the evolution of energy device manufacturing is cellulose. Cellulose is one of the most abundant biopolymers and has excellent properties such as biodegradability, renewability, biocompatibility, and cost-effectiveness. In the nanoscale dimension, cellulose displays unique features such as high flexibility, transparency, low thermal expansion etc. The intrinsic and extensive properties of cellulose depend on the functional groups present its structure. Moreover, it is water soluble (53). Xu et al. reported an environmentally friendly and thermally stable cellulose film separator for supercapacitors (54). The as-synthesized cellulose displayed porosity and electrolytic uptake of 74.90% and 323.68%, respectively. The supercapacitance obtained for the fabricated supercapacitor was 130 F/g.

Polyethylene is a traditional separator for energy storage devices. The separator made of PE suffered low porosity, low electrolyte uptake efficiency, low wettability, and low mechanical strength. Thickness has always been a major issue in the case of the PE separator because the thickness of the separator is directly related to the equivalent series resistance of the supercapacitor (55). The separators are an inevitable component in supercapacitor design; therefore, a lot of research has been carried out to develop the most efficient separators for second-generation devices. Electrospinning is a well-known nanofiber synthesizing technique and has always attracted scientists and researchers because of its facile and controllable method of fabricating nanofibers with auspicious properties. The nanofiberous membrane as a separator holds qualities such as high porosity, high electrolyte uptake ability, mechanical stability, electrochemical stability, etc. (56). He et al. developed porous PAN nanofiber separators via electrospinning. The resultant separator was highly porous and exhibited good mechanical, electrochemical, and thermal stabilities (57). The extensive research in this area has progressed from the development of ordinary supercapacitors to the development of self-charging supercapacitors. This long trail includes flexible, stretchable, and wearable supercapacitors. Indeed, every invention in this field is a milestone. Therefore, by using polymers in this period, researchers have found that some of them are multifunctional, e.g., PVA, PAN, and PVDF. These polymers can act as binders cum electrolytes cum separators in the fabrication of a supercapacitor. Among them, PVDF has been getting more attention due to its inherent piezo, pyro, and ferroelectric properties (24, 50, 58). The self-charging supercapacitor power cell (SCSPC) has now been added to the list of achievements of PVDF. The beta phase of PVDF is an excellent source for generating inherent electricity in its polymeric backbone under mechanical stress (24). Manoharan et al. developed a novel bifunctional phosphotungstic acid (PTA)/PVDF electrolyte cum separator (59). PTA is proton-conducting solid electrolyte, and the incorporation of PTA into the PVDF matrix enhanced the mechanical to electrochemical efficiency of the fabricated graphene based SCSPC. The areal capacitance obtained for the SCSPC is 184.94 mF/cm.

1.3 CONCLUSION

This chapter dealt with the mechanism and classification of supercapacitors as well as with the role of polymers in the fabrication of supercapacitors. Supercapacitors are powerful second-generation devices with high specific capacitance, power density, and cycling stability. The polymer is an indispensable factor in the development of energy storage devices. Now it is again paving the way for innovative development in this field. In the configuration of supercapacitor design polymers can act as electrodes, binders, electrolytes, separators, and piezoelectric generators. Moreover, the multifunctional behavior of polymers such as PVA, PAN, and PVDF help the field evolve. Green polymers also paved their path in the development of eco-friendly devices. Therefore, in the future, the implementation of eco-friendly energy devices with flexibility, stretchability, and self-powering ability will reduce the dimension of renewable energy storage and stabilize the world economy. Subsequent chapters in this book discuss in detail conducting polymers and carbonaceous materials such as activated carbon, carbon nanotubes, halloysite carbon nanotubes, and graphene electrodes as electrodes for supercapacitors. Still other chapters deal with flexible and stretchable polymers and biopolymers for supercapacitors.

ACKNOWLEDGMENT

We are grateful to the APJ Abdul Kalam Technological University (KTU), Thiruvananthapuram (KTU proceedings No. 4/1654/2019), for the financial assistance and CERD, APJ Abdul Kalam Technological University (KTU), Thiruvananthapuram (KTU/RESEARCH 4/1694/2021) for the research fellowship.

REFERENCES

1. Halper, MS, Ellenbogen, JC. *Supercapacitors: A Brief Overview* (p. 1). McLean, VA: The MITRE Corporation; 2006.
2. González A, Goikolea E, Barrena JA, Mysyk R. Review on supercapacitors: Technologies and materials. *Renewable and Sustainable Energy Reviews*. 2016 May;58:1189–206.
3. Yu, A, Chabot, V, Zhang, J. *Electrochemical Supercapacitors for Energy Storage and Delivery: Fundamentals and Applications* (p. 383). Boca Raton: Taylor & Francis; 2013.
4. Liu X, Naylor Marlow M, Cooper SJ, Song B, Chen X, Brandon NP, et al. Flexible all-fiber electrospun supercapacitor. *Journal of Power Sources*. 2018 Apr;384:264–9.
5. Balasubramaniam S, Mohanty A, Balasingam SK, Kim SJ, Ramadoss A. Comprehensive insight into the mechanism, material selection and performance evaluation of supercapatteries. *Nano-Micro Lett*. 2020 Dec;12(1):85.
6. Chen X, Paul R, Dai L. Carbon-based supercapacitors for efficient energy storage. *National Science Review*. 2017 May 1;4(3):453–89.
7. Mastragostino M, Arbizzani C, Soavi F. Polymer-based supercapacitors. *Journal of Power Sources*. 2001 Jul;97–98:812–5.
8. Choudhary N, Li C, Moore J, Nagaiah N, Zhai L, Jung Y, et al. Asymmetric supercapacitor electrodes and devices. *Advanced Materials*. 2017 Jun;29(21):1605336.
9. Kadam P, Holmukhe RM, Karandikar PB. Development of three electrode system for optimizing the parameters of hybrid capacitor. *Indian Journal of Science and Technology* [Internet]. 2016 Jul 21 [cited 2021 Oct 31];9(26). Available from: https://indjst.org/articles/development-of-three-electrode-system-for-optimizing-the-parameters-of-hybrid-capacitor
10. Muzaffar A, Ahamed MB, Deshmukh K, Thirumalai J. A review on recent advances in hybrid super-capacitors: Design, fabrication and applications. *Renewable and Sustainable Energy Reviews*. 2019 Mar;101:123–45.
11. Harris FW. Introduction to polymer chemistry. *Journal of Chemical Education*. 1981;58(11):7.
12. Shown I, Ganguly A, Chen L, Chen K. Conducting polymer-based flexible supercapacitor. *Energy Science & Engineering*. 2015 Jan;3(1):2–26.
13. Snook GA, Kao P, Best AS. Conducting-polymer-based supercapacitor devices and electrodes. *Journal of Power Sources*. 2011 Jan;196(1):1–12.

14. Meng Q, Cai K, Chen Y, Chen L. Research progress on conducting polymer based supercapacitor electrode materials. *Nano Energy.* 2017 Jun;36:268–85.

15. Eftekhari A, Li L, Yang Y. Polyaniline supercapacitors. *Journal of Power Sources.* 2017 Apr;347:86–107.

16. Chen W-C, Wen T-C, Teng H. Polyaniline-deposited porous carbon electrode for supercapacitor. *Electrochimica Acta.* 2003 Feb;48(6):641–9.

17. Jurewicz K, Delpeux S, Bertagna V, Béguin F, Frackowiak E. Supercapacitors from nanotubes/polypyrrole composites. *Chemical Physics Letters.* 2001 Oct;347(1–3):36–40.

18. Ambade RB, Ambade SB, Shrestha NK, Nah Y-C, Han S-H, Lee W, et al. Polythiophene infiltrated TiO2 nanotubes as high-performance supercapacitor electrodes. *Chemical Communications.* 2013;49(23):2308.

19. Laforgue A, Simon P, Sarrazin C, Fauvarque J-F. Polythiophene-based supercapacitors. *Journal of Power Sources.* 1999 Jul;80(1–2):142–8.

20. Ruschhaupt P, Varzi A, Passerini S. Natural polymers as green binders for high-loading supercapacitor electrodes. *ChemSusChem.* 2020 Feb 21;13(4):763–70.

21. Shrestha M, Amatya I, Wang K, Zheng B, Gu Z, Fan QH. Electrophoretic deposition of activated carbon YP-50 with ethyl cellulose binders for supercapacitor electrodes. *Journal of Energy Storage.* 2017 Oct;13:206–10.

22. Wang Y, Dang D, Li D, Hu J, Zhan X, Cheng Y-T. Effects of polymeric binders on the cracking behavior of silicon composite electrodes during electrochemical cycling. *Journal of Power Sources.* 2019 Oct;438:226938.

23. Zhu, Z, Tang, S, Yuan, J, Qin, X, Deng, Y, Qu, R, Haarberg, GM. Effects of various binders on supercapacitor performances. *International Journal of Electrochemical Science* 2016;11(10):8270–9.

24. Rajeevan S, John S, George SC. Polyvinylidene fluoride: A multifunctional polymer in supercapacitor applications. *Journal of Power Sources.* 2021 Aug;504:230037.

25. Aslan M, Weingarth D, Jäckel N, Atchison JS, Grobelsek I, Presser V. Polyvinylpyrrolidone as binder for castable supercapacitor electrodes with high electrochemical performance in organic electrolytes. *Journal of Power Sources.* 2014 Nov;266:374–83.

26. Aslan M, Weingarth D, Herbeck-Engel P, Grobelsek I, Presser V. Polyvinylpyrrolidone/polyvinyl butyral composite as a stable binder for castable supercapacitor electrodes in aqueous electrolytes. *Journal of Power Sources.* 2015 Apr;279:323–33.

27. Varzi A, Raccichini R, Marinaro M, Wohlfahrt-Mehrens M, Passerini S. Probing the characteristics of casein as green binder for non-aqueous electrochemical double layer capacitors' electrodes. *Journal of Power Sources.* 2016 Sep;326:672–9.

28. Tripathi M, Tripathi SK. Electrical studies on ionic liquid-based gel polymer electrolyte for its application in EDLCs. *Ionics.* 2017 Oct;23(10):2735–46.

29. Pandey GP, Kumar Y, Hashmi SA. Ionic liquid incorporated polymer electrolytes for supercapacitor application. *Indian Journal of Chemistry.* 2010;9.

30. Gao H, Lian K. Proton-conducting polymer electrolytes and their applications in solid supercapacitors: A review. *RSC Advances.* 2014;4(62):33091–113.

31. Tiruye GA, Muñoz-Torrero D, Palma J, Anderson M, Marcilla R. Performance of solid state supercapacitors based on polymer electrolytes containing different ionic liquids. *Journal of Power Sources.* 2016 Sep;326:560–8.

32. Alipoori S, Mazinani S, Aboutalebi SH, Sharif F. Review of PVA-based gel polymer electrolytes in flexible solid-state supercapacitors: Opportunities and challenges. *Journal of Energy Storage.* 2020 Feb;27:101072.

33. Vondrak J, Reiter J, Velicka J, Sedlaoikova M. PMMA-based aprotic gel electrolytes1. *Solid State Ionics.* 2004 May 14;170(1–2):79–82.

34. Hashmi SA, Kumar A, Tripathi SK. Investigations on electrochemical supercapacitors using polypyrrole redox electrodes and PMMA based gel electrolytes. *European Polymer Journal.* 2005 Jun;41(6):1373–9.

35. Cai J, Wang Y, Qiao X, Zhou X, Mansour AN. A novel mediator-containing polyvinylidene fluoride/lithium trifluoromethanesulfonate polymer electrolyte membrane for low-temperature solid-state supercapacitors. *Materials Letters.* 2019 Sep;251:26–9.

36. Solarajan AK, Murugadoss V, Angaiah S. Montmorillonite embedded electrospun PVdF—HFP nanocomposite membrane electrolyte for Li-ion capacitors. *Applied Materials Today.* 2016 Dec;5:33–40.

37. Hsueh M-F, Huang C-W, Wu C-A, Kuo P-L, Teng H. The synergistic effect of nitrile and ether functionalities for gel electrolytes used in supercapacitors. *Journal of Physical Chemistry C.* 2013 Aug 22;117(33):16751–8.

38. Liew C-W, Ramesh S, Arof AK. Good prospect of ionic liquid based-poly(vinyl alcohol) polymer electrolytes for supercapacitors with excellent electrical, electrochemical and thermal properties. *International Journal of Hydrogen Energy.* 2014 Feb;39(6):2953–63.

39. Tamilarasan P, Ramaprabhu S. Graphene based all-solid-state supercapacitors with ionic liquid incorporated polyacrylonitrile electrolyte. *Energy.* 2013 Mar;51:374–81.

40. Fard HN, Pour GB, Sarvi MN, Esmaili P. PVA-based supercapacitors. *Ionics.* 2019 Jul;25(7):2951–63.

41. Karaman B, Bozkurt A. Enhanced performance of supercapacitor based on boric acid doped PVA-H2SO4 gel polymer electrolyte system. *International Journal of Hydrogen Energy.* 2018 Mar;43(12):6229–37.

42. Kumar Y, Pandey GP, Hashmi SA. Gel polymer electrolyte based electrical double layer capacitors: Comparative study with multiwalled carbon nanotubes and activated carbon electrodes. *Journal of Physical Chemistry C.* 2012 Dec 20;116(50):26118–27.

43. Ortega PFR, Trigueiro JPC, Silva GG, Lavall RL. Improving supercapacitor capacitance by using a novel gel nanocomposite polymer electrolyte based on nanostructured SiO2, PVDF and imidazolium ionic liquid. *Electrochimica Acta.* 2016 Jan;188:809–17.

44. Karaman B, Çevik E, Bozkurt A. Novel flexible Li-doped PEO/copolymer electrolytes for supercapacitor application. *Ionics.* 2019 Apr;25(4):1773–81.

45. Gunday ST, Cevik E, Yusuf A, Bozkurt A. Fabrication of Al $_2$ O $_3$/IL-based nanocomposite polymer electrolytes for supercapacitor application. *ChemistrySelect.* 2019 May 24;4(19):5880–7.

46. Kar KK, editor. *Handbook of Nanocomposite Supercapacitor Materials I: Characteristics.* Cham: Springer International Publishing; 2020.

47. Roy BK, Tahmid I, Rashid TU. Chitosan-based materials for supercapacitor applications: A review. *Journal of Materials Chemistry A.* 2021;9(33):17592–642.

48. Li L, Lu F, Wang C, Zhang F, Liang W, Kuga S, et al. Flexible double-cross-linked cellulose-based hydrogel and aerogel membrane for supercapacitor separator. *Journal of Materials Chemistry A.* 2018;6(47):24468–78.

49. Ma G, Li J, Sun K, Peng H, Mu J, Lei Z. High performance solid-state supercapacitor with PVA—KOH—K3[Fe(CN)6] gel polymer as electrolyte and separator. *Journal of Power Sources.* 2014 Jun;256:281–7.

50. Jabbarnia A, Khan WS, Ghazinezami A, Asmatulu R. Investigating the thermal, mechanical, and electrochemical properties of PVdF/PVP nanofibrous membranes for supercapacitor applications. *Journal of Applied Polymer Science* [Internet]. 2016 Aug 10 [cited 2021 Oct 31];133(30). Available from: https://onlinelibrary.wiley.com/doi/10.1002/app.43707

51. Godse LS, Karkaria VN, Bhalerao MJ, Karandikar PB, Kulkarni NR. Process-based modeling of nylon separator supercapacitor. *Energy Storage* [Internet]. 2021 Feb [cited 2021 Oct 31];3(1). Available from: https://onlinelibrary.wiley.com/doi/10.1002/est2.204

52. Raja M, Sadhasivam B, Naik R J, R D, Ramanujam K. A chitosan/poly(ethylene glycol)- *ran* -poly(propylene glycol) blend as an eco-benign separator and binder for quasi-solid-state supercapacitor applications. *Sustainable Energy Fuels.* 2019;3(3):760–73.

53. Wang X, Yao C, Wang F, Li Z. Cellulose-based nanomaterials for energy applications. *Small.* 2017 Nov;13(42):1702240.

54. Xu D, Teng G, Heng Y, Chen Z, Hu D. Eco-friendly and thermally stable cellulose film prepared by phase inversion as supercapacitor separator. *Materials Chemistry and Physics.* 2020 Jul;249:122979.

55. Mandake P, Karandikar PB. Effect of separator thickness variation for supercappacitor with polythylene separator material. *IJSRSET.* 2016;2(2):967–71.

56. Li X-Y, Yan Y, Zhang B, Bai T-J, Wang Z-Z, He T-S. PAN-derived electro spun nanofibers for supercapacitor applications: Ongoing approaches and challenges. *Journal of Materials Science.* 2021 Jun;56(18):10745–81.

57. He T, Fu Y, Meng X, Yu X, Wang X. A novel strategy for the high performance supercapacitor based on polyacrylonitrile-derived porous nanofibers as electrode and separator in ionic liquid electrolyte. *Electrochimica Acta.* 2018 Aug;282:97–104.

58. Karabelli D, Leprêtre J-C, Alloin F, Sanchez J-Y. Poly(vinylidene fluoride)-based macroporous separators for supercapacitors. *Electrochimica Acta.* 2011 Dec;57:98–103.

59. Manoharan, S, Pazhamalai, P, Mariappan, VK, Murugesan, K, Subramanian, S, Krishnamoorthy, K, Kim, SJ. Proton conducting solid electrolyte-piezoelectric PVDF hybrids: Novel bifunctional separator for self-charging supercapacitor power cell. *Nano Energy.* 2021;83:105753.

2 Conducting Polymer Nanocomposites for Supercapacitors

Sona Narayanan, Ayswarya Ettuvettil Pankajakshan,
Vidya Thattarkudy Padmanabhan, and Anisha Joseph

CONTENTS

2.1   INTRODUCTION

The energy crisis is one of the important issues in the present scenario not only because of the depletion in natural energy resources but also because of the increase in world population as well as the fast growth of the global economy. This has led to the development of sustainable and renewable energy storage systems to control energy consumption and preserve natural resources. Non-conventional energy storage devices including supercapacitors, batteries, and fuel cells, have been exploited in various hybrid vehicles and portable electronic devices wherein the electrical energy is produced from chemical energy through electrochemical reactions [1]. Due to their high specific power (500–10000 W/kg), fast charging and discharging rates, high life time (> 100000 cycles), eco-friendly characteristics, and high reliability, supercapacitors or electrochemical capacitors (ECs) have drawn immense attention in the past decades [2]. A comparison of supercapacitors with various energy sources is shown in Figure 2.1 [3]. Supercapacitors can form a bridge between batteries and conventional capacitors regarding specific energy and specific power because of high specific power as that of conventional electrolytic capacitors and specific energy which is proximate to batteries. When combined with batteries, supercapacitors can enhance battery performance regarding the power density [4].

Because of their high electrical conductivity [5–6], large pseudocapacitance [8–10], high flexibility, good processability, and relatively low cost, the third most used supercapacitor materials are conducting polymers. Conducting polymers have several redox states that facilitate the control of the electrical conductivity from insulating material to metal with a pseudocapacitance mechanism [11]. Usually conducting polymers like PANI, PPy, and PEDOT show large specific capacitance of 1284 F/g [12], 480 F/g [13], and 210 F/g [14–16], respectively. Supercapacitors based on various conducting polymers are shown in Figure 2.2 [17]. Still, these polymers suffer shortcomings for application in supercapacitors owing to the poor cyclic stability in long-term charge-discharge processes and lower practical capacitance compared to theoretical predictions. This could be due to poor mechanical stability from, for example, swelling, shrinkage, cracks, or breaking of conducting

DOI: 10.1201/9781003174646-2

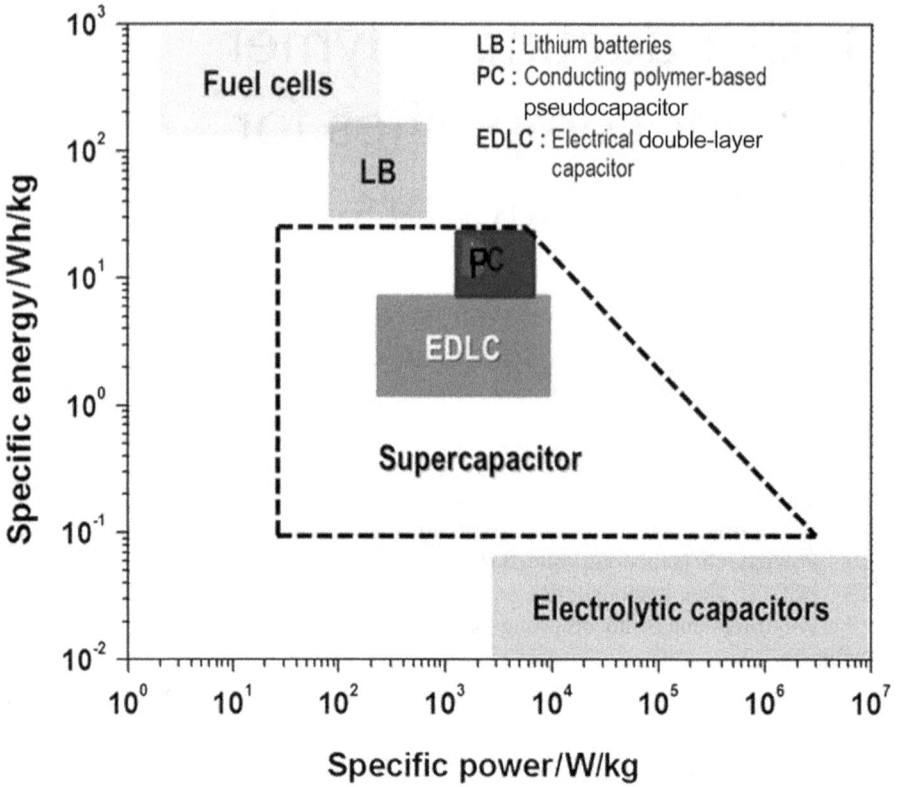

FIGURE 2.1 A Ragone plot represents the comparison of supercapacitors with various energy storage devices.

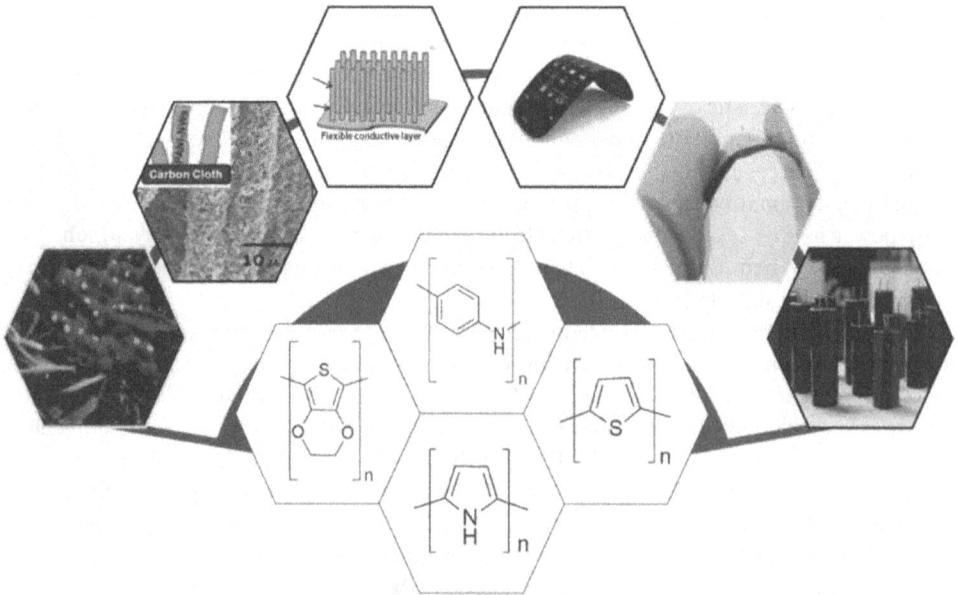

FIGURE 2.2 Supercapacitor based on various conducting polymers.

polymers owing to high intercalation, depletion of ions in charging and discharging of the polymer backbone and over oxidative degradation of conducting polymers because of the limited working potential range [18].

To overcome these shortcomings, several approaches have been explored: (1) Designing of microstructure and morphology of the conducting polymers to improve the specific capacitance (2) preparation of conducting polymer composites with different metal oxides or carbon materials to increase the mechanical stability of the conducting polymers so as to improve the long-term charge-discharge stability, (3) development of novel electrolytes with wide electrochemical windows to decrease the decomposition of the conducting polymers. Conducting polymer composites showed better electrical, mechanical, and electrochemical properties with respect to pristine conducting polymers, resulting in a large variety of applications. This chapter mainly emphasizes recent advances in conducting polymer nanocomposites as promising supercapacitor electrodes. The working mechanism of a supercapacitor is shown in Figure 2.3 [19].

Synthesis of conducting polymers can be carried out either chemically or electrochemically through oxidative polymerization of the corresponding monomers. Conducting polymers that are synthesized by chemical polymerization display good stability, high conductivity, and negligible solubility in aqueous solutions. Aniline can be chemically polymerized in an aqueous solution by using oxidizing agents like ammonium persulfate ($NH_4)S_2O_8$, ferric chloride ($FeCl_3$), potassium dichromate ($K_2Cr_2O_7$), potassium iodate (KIO_3), potassium permanganate ($KMnO_4$), potassium bromate ($KBrO_3$), and potassium chlorate ($KClO_3$) [20]. Likewise, conducting pyrrole can be polymerized by different oxidizing agents like $FeCl_3$, potassium ferricyanide ($K_3Fe(CN)_6$), ferric nitrate ($Fe(NO_3)_3$), ferric sulfate ($Fe_2(SO_4)_3$) and cupric chloride ($CuCl_2$), so as to obtain highly conducting PPy [21]. Mostly thiophene monomer and its derivatives can only be polymerized in an organic solution like chloroform, tetrahydrofuran etc. However, compared to chemical polymerization, electrochemical polymerization is better in terms of homogeneity and integrity of the sample. Also, an integrated conducting polymer film can be easily prepared by electrochemical polymerization. The synthesis

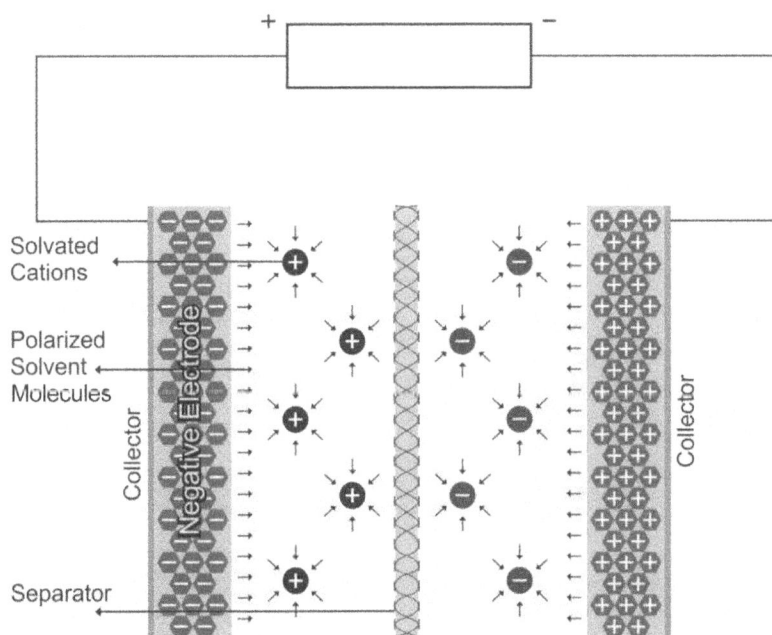

FIGURE 2.3 Supercapacitor mechanism.

of PANI and PPy is more economic and environmentally friendly compared to thiophene derivatives due to more solubility of aniline and pyrrole monomers in aqueous solutions which make them more popular among conducting polymer supercapacitors.

The conductivity of conducting polymers can be improved by doping. On doping, conducting polymer loses electrons and becomes polycations (polarons), leads to intercalation of anions (Cl- in this case in the solution) into the conducting polymer so as to maintain electro-neutrality of the polymer, resulting in pseudocapacitance in the material. The illustration of pseudocapacitance in a conducting polymer is shown in Figure 2.4 [22].

Supercapacitors are of two types, pseudocapacitors and electrical double-layer capacitors (EDLCs). A schematic diagram of conducting polymers based asymmetric supercapacitors are shown in Figure 2.5 [3]. As in the case of EDLCs, the opposite charges are stored on the surface of the two parallel plates and it is cost effective and has high-power density with tunable porosity. The disadvantage is that it possesses low energy density. Whereas, the pseudocapacitors can store charges through fast and reversible reaction, and thereby possesses high capacitance and low power density.

Nowadays, a large number of conducting polymer-based supercapacitor materials are available including composites of carbon-based materials and metal oxide/hydroxides, which combine both double-layer capacitance and pseudocapacitance in a single material. Among them, the most widely used commercial materials are carbon-based conducting polymer composites. Generally, carbon

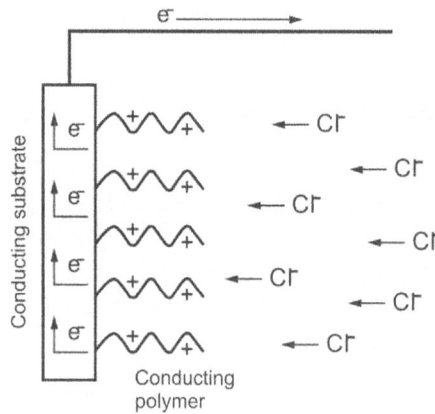

FIGURE 2.4   Illustration of pseudocapacitance in a conducting polymer.

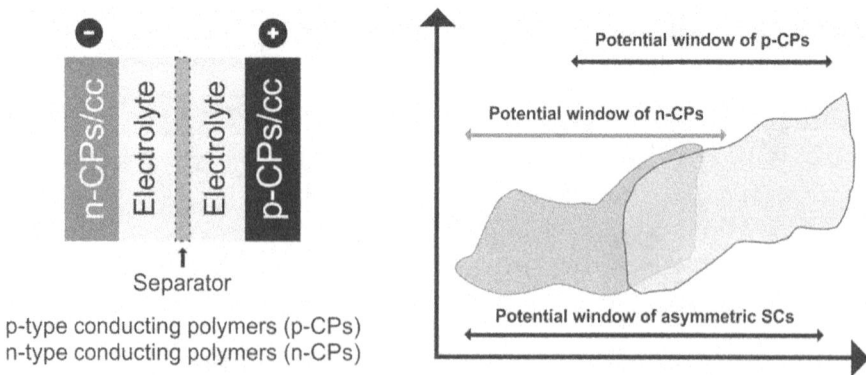

FIGURE 2.5   Schematic diagram of conducting polymers based asymmetric supercapacitors.

materials exhibit a large and stable double-layer capacitance owing to the high conductivity, cycle stability, and high specific area. Alkaline treatment with aqueous KOH or NaOH solution can introduce oxygen functional groups and increase the specific area of carbon surface [23]. Most carbon-based materials have surface oxygen, functional groups, like carboxyl, lactone, aldehyde, carboxylic, phenolic and ether groups, depending on the pre-treatment methods and source of the carbon. The functional groups present in carbon can boost the surface hydrophilicity thereby introducing redox properties that can contribute pseudocapacitance to the overall capacitance [24–25]. Another type of supercapacitor material like metal oxides or hydroxides including ruthenium, nickel, cobalt, and manganese are conducting or semi-conducting and can exhibit redox properties, resulting in pseudocapacitance [22].

2.1.1 POLYANILINE NANOCOMPOSITE-BASED SUPERCAPACITORS

PANI is a π-conjugated system that can facilitate continuous electron transport through the polymer backbone. PANI exhibits three variable oxidation states like (1) leucoemeraldine- the fully reduced form, (2) permigraniline, and (3) emeraldine base (EB) [26]. Among these, two forms are non-conductive in nature, whereas emeraldine base can be doped by protonic acid to form salt (an equal form of amine (reduced form) and quinonoid (oxidized form) so that it can show variable conductivity of 10^{-10} to 10^2 Scm^{-1} depending on the type of dopant and concentration of dopants used. Protonic acids such as HI, HCl and H_2SO_4 and organic acids like dodecylbenzene sulfonic acid (DBSA) and camphor sulfonic acid (CSA) are commonly used as doping agents. When it is doped, it gets transferred to conductive forms-polaronic and bipolaronic forms. Protonation of PANI-EB with HX acids result in a spinless structure of bipolarons. Thereafter it is rearranged to a delocalized polaron lattice called polysemiquinone radical cation site [27]. If it is doped with an organic protic acid, the negatively charged sulfonic group of the dopant gets attached with the positively charged polyaniline emeraldine salt through coulombic interactions resulting in a hydrophobic side chain which makes it soluble in organic solvents. The dopant, CSA (CSA)$_{0.5}$ showed more solubility than DBSA (DBSA)$_{0.5}$ in m-cresol owing to the availability of more interaction sites (carbonyl group present in CSA also takes part in reaction apart from sulfonic interaction) [28].

Because of excellent electrochemical properties together with high electrical conductivity, high stability, high environmental stability, [29] high theoretical capacitance (3407 Fg^{-1}), ease of synthesis, and biocompatibility, PANI has been widely used in the fabrication of supercapacitors. A schematic diagram of the CP-based flexible electrode fabrication process is shown in Figure 2.6 [3]. Like other conducting polymers, PANI suffers from short cyclic life, infusibility, and weak processability [30]. To conquer these problems, blending nanomaterial with a conducting polymer is a very attractive way not only to reinforce mechanical properties but also to introduce required electronic properties. This paves the way for the development of PANI nanocomposites with carbon-based materials and transition metals to diminish the aforementioned disadvantages. The addition of nanoparticles to the polyaniline matrix leads to factors like development of new reaction sites, an increase in surface area, short pathway of charge/mass transfer and improvement in the accommodation of strain within the electrodes during electrochemical reaction [31]. The selection of active material is crucial for supercapacitor applications for the improvement of life-cycle and energy density without losing power density. For CP nanocomposites, "pseudocapacitive" charge storage is prominent so that it can produce high energy density than EDL capacitors owing to fast surface redox reactions.

PANI shows very high theoretical capacitance up to 2000 Fg^{-1}. Nevertheless, due to the varying values of conductivity the experimental capacitance of PANI electrodes is much lower than that of theoretical capacitance. Composites of PANI and carbon materials have attracted considerable interest owing to their better electrochemical performances by integrating pseudocapacitances of PANI and the electrochemical double-layer capacitances of carbon materials. However, the performances of the PANI nanocomposites depends on several factors like morphology, the synthesis/fabrication techniques and PANI doping level of the nanocomposites.

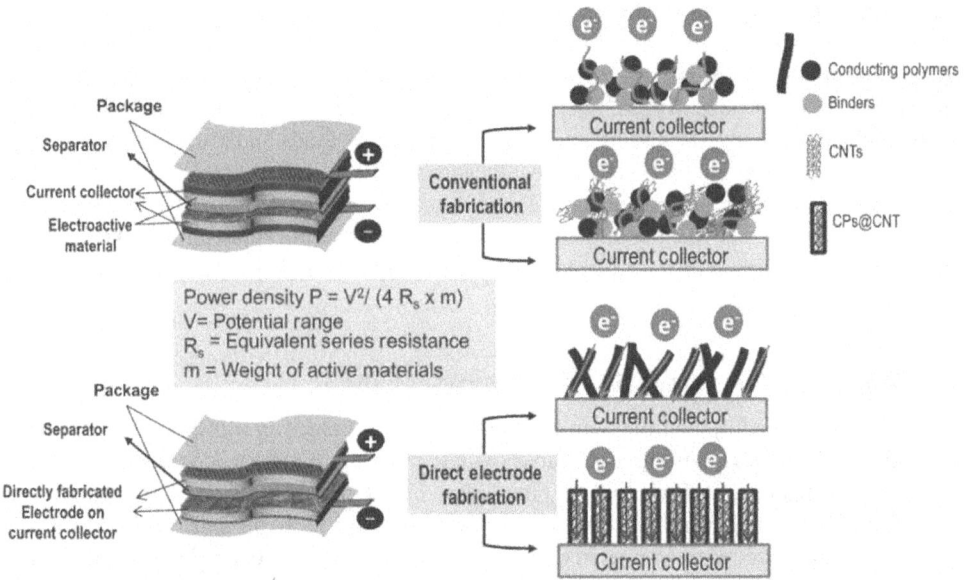

FIGURE 2.6 Schematic diagram of the CP-based flexible electrode fabrication process.

As electrode materials, carbon materials have gained more interest owing to their cost effectiveness, high mechanical properties, excellent electrical conductivity, and better thermal and chemical stability. Graphene, the parent of all graphitic structures with two-dimensional (2D) nanostructure aroused considerable interest as a good electrode candidate for supercapacitors [32–33]. Another important carbon material like carbon nanotubes (CNTs) (single-walled carbon nanotubes, SWCNT & multiwalled carbon nanotubes, MWCNT) are single or multiple atomic layers of graphite that rolled back on themselves forms nanotubes with a hollow inner cavity. Considering various carbon materials, graphene has attracted great attention owing to its large surface area, high conductivity, flexibility, excellent electrochemical stability, and superior mechanical properties [34–38]. Hence, PANI/graphene nanocomposites have become an ideal material as a promising supercapacitor. For instance, a PANI/3D graphene framework with an exceptionally high specific capacitance of 1341 Fg^{-1} has been reported [39].

PANI/graphene nanocomposites can be fabricated by different methods including simple solution mixing [40–41], layer-by-layer assembly [42], electrochemical oxidative polymerization [43–44], interfacial polymerization [45] and chemical oxidative polymerization [46–47]. Among them, chemical synthesis routes are widely used for the preparation of PANI composites [48–49] because these chemical methods can impart hydrophilic oxygen-containing functional groups on graphene oxide (GO). Positively charged in their conductive emeraldine salt form of PANI [50], and oppositely, GO carries negative charges owing to residual carboxylic groups [51], leads to better dipole-dipole interactions resulting in efficient dispersion of both PANI and graphene. Likewise, reduced graphene oxide (rGO) could impart expanded coil conformation of PANI chains owing to the basal plane of graphene and better π-π interactions between the quinoid rings of PANI [52]. This led to the enhanced electrical conductivity of PANI/r-GO nanocomposites [53–54].

Still, layered morphology and potentially more surface area of graphene make it very difficult to produce a porous electrode [55–56]. Hence, composites of graphene with CNTs, [57] conducting polymers [58], and transition metal oxide [59] have been prepared to overcome such difficulties. Electrodes fabricated from CNTs have attracted considerable interest owing to their entangled mat-like structure with a highly accessible network of mesopores. Also, the interconnecting mesopores permit a continuous charge distribution in CNTs which facilitates utilization of the entire available surface area which enable them to attain specific capacitances with respect to inactivated carbon-based supercapacitors [60]. The CNTs electrodes showed a lower equivalent series resistance (ESR) [61–62] due to the easy diffusion of electrolyte ions into the mesoporous network. The entangled mat structure of CNTs offers comparable energy densities to other carbon-based materials and the low ESR offers high power densities.

Several metal oxides have also been investigated as promising pseudocapacitive electrode materials [63–64] owing to their high electrical conductivity. Among various metal oxides, hydrous ruthenium oxide (RuO_2) has attracted great attention owing to its high specific capacitance compared to carbon-based and conducting polymer materials [65–66]. Also, the ESR of hydrous RuO_2 is much lower than that of other electrode materials, resulting in high power density and energy than that of conducting polymer pseudocapacitors. Even then, the use of RuO_2 has been limited by its high cost and toxicity. Great attention is now being paid to the oxides of iron, vanadium, copper, manganese, and nickel as potential candidates for supercapacitor electrode materials to overcome

TABLE 2.1

Specific Capacitance of Different PANI Nanocomposites

Electrode Material	Specific capacitance (Fg⁻¹)	Retention	Reference
PANI/CNT	7926	96% after 1000 cycles	69
PANI nanowire on rGO/ZrO$_2$	1360	93% after 1000 cycles	70
Stretchable isotropic buckled CNT/PANI	1147		71
PANI/GO on SS	1136	89% after 1000 cycles	72
PANI/CNY array	1030	77% after 1001 cycles	73
rGO hydrogel films embedded with PANI NFs	921	100% after 2000 cycles	74
PANI	890	89% after 1000 cycles	75
Ag/MnO$_2$/PANI	800		76
PANI/NiCo$_2$O$_4$	781	91% after 3000 cycles	77
H/Ni co-doped PANI/CNT	781	92% after 700 cycles	78
Unzipped CNT (UCNT)/PANI	762	80% after 1000 cycles	79
PANI/C/Ni	725		80
Hydrogel-assisted PANI microfiber	709		81
PANI-g-MWCNT/TiO$_2$NTs/Ti	708	88% after 1000 cycles	82
PANI/MnO$_2$/TiN	674		83
PANI nanorods on rGO sponges	662	93% after 5000 cycles	84
3D PANI on pillared rGO	652	90% after 4000 cycles	85
MoS$_2$/PANI/G	618	78% after 2000 cycles	86
PANI NFs/N-doped rGO hydrogels	610	94% after 1000 cycles	87
PANI/MoS$_2$	575	98% after 500 cycles	88
RuO$_2$/PANI/C	531		89
Cellulose/CNT/PANI	495.2	81% after 1000 cycles	90
PANI nanowhiskers	470	90% after 1000 cycles	91
CNT/CB/GNSs PANI	450	84% after 1000 cycles	92
Self-doped PANI	408		93
PANI/B-doped rGO	406	90% after 10000 cycles	94
G/PANI/MCM-41	405	91.4% after 1000 cycles	95
CoMoO$_4$/PANI	380	90% after 1000 cycles	96
Bamboo carbon/PANI	244	92% after 1000 cycles	97
PANI/OMC	146		98
CNT/PANI cotton-shaped fiber	144	84.5% after 200 cycles	99

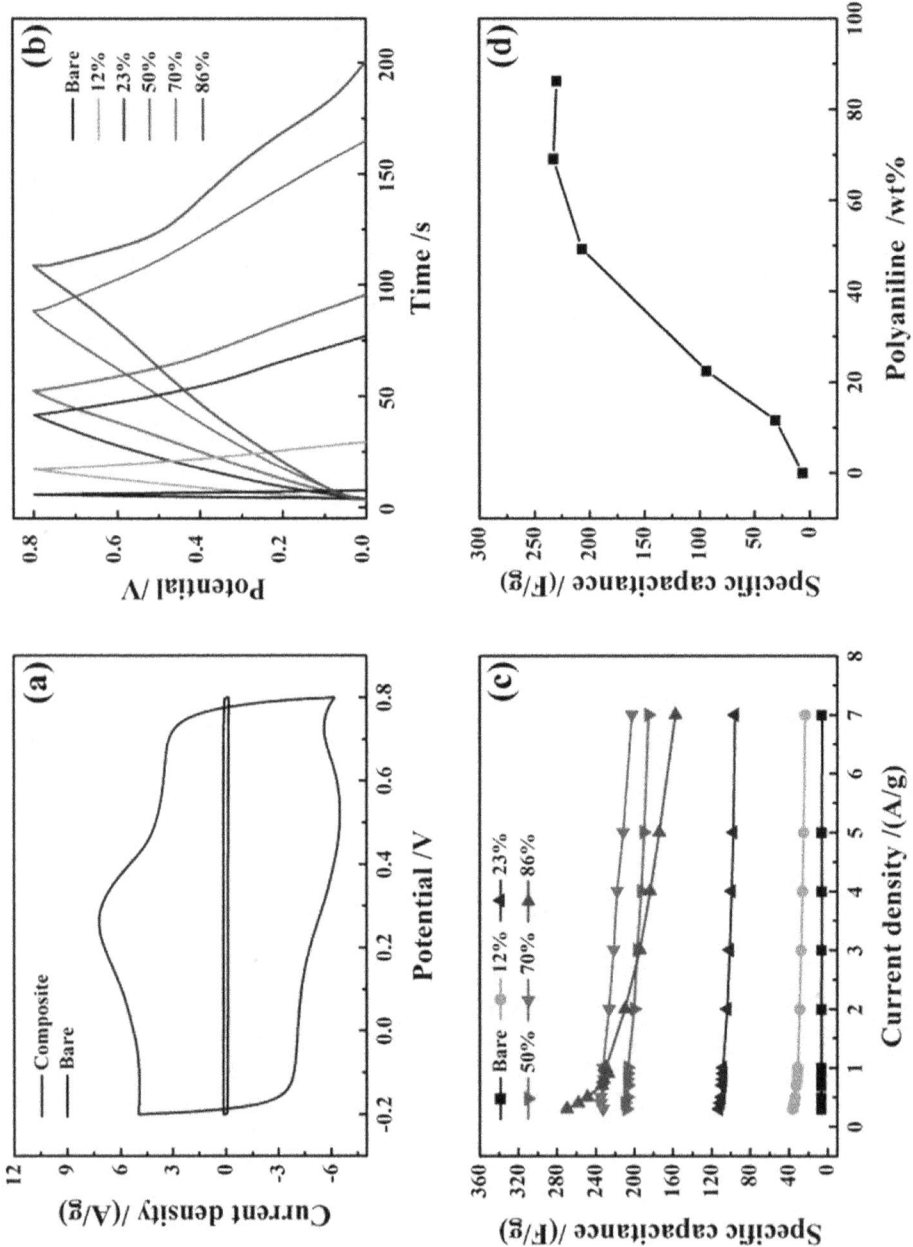

FIGURE 2.7 (a) CV curves of pristine MWCNT and composite film (70% loading) as electrodes in a potential range of 0.2 to 0.8 V at a scan rate of 50 mVs^{-1} in H$_3$PO$_4$-PVA gel electrolyte. (b) Galvanostatic charge-discharge curves of pristine MWCNT and composite films with different loading of 12, 23, 50, 70 and 86% at a current density of 1 A/g. (c) Dependence of specific capacitance on current density. (d) Dependence of specific capacitance on weight percentage of PANI.

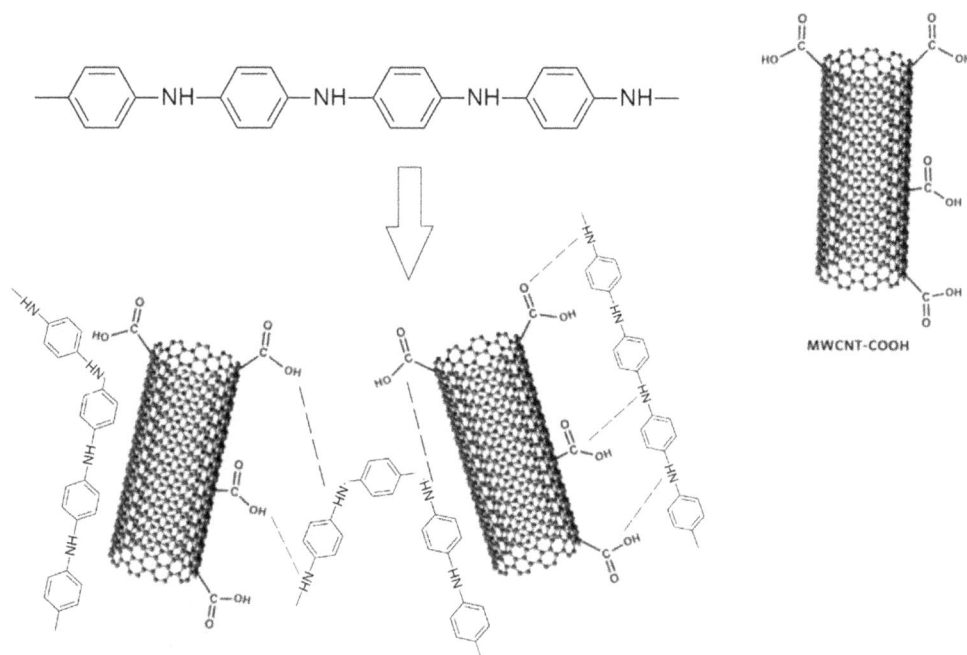

FIGURE 2.8 Schematic illustration of PANI-CNT interactions.

such difficulties. The proper mixing of metal oxide materials in the composite electrodes of carbon-based materials and conductive polymer can combine both physical and chemical charge storage mechanisms together in a single electrode material. Nowadays, most of the researchers focus on the development of cost-effective fabrication methods and efficient composite materials to make it more economic, without reducing the electrochemical performance [67–68].

Cyclic voltammetry can be used to analyze the electrochemical performance of MWCNT film and MWCNT/PANI composite films [100]. Cyclic voltammograms (CVs) of MWCNT and MWCNT/PANI composite films with PANI loading of 70% measured in two-electrode system in H_3PO_4. PVA gel electrolyte is illustrated in Figure 2.7a. From the figure it is revealed that electrochemical double layer was obtained from the pristine MWCNT film and pseudocapacitance derived from different oxidation states of PANI were observed for MWCNT/PANI films. In addition to this, it is observed that current density of composite films was much higher than that of pristine MWCNT film. Figure 2.7b describes the galvanostatic charge-discharge curves of both MWCNT/PANI composite and pristine MWCNT film electrodes between 0 and 0.8 V. The symmetry of the discharge curves indicated high reversibility between charge and discharge process.

As expected, the specific capacitance firstly increased with increasing PANI loading and reached a constant value of specific capacitance at 70% loading (Figure 2.7c and Figure 2.7d). The increment could be due to the higher pseudocapacitance of PANI and better π-π interactions of PANI and CNT (Figure 2.8). At higher loading (> 70%), the additional PANI accumulated on the surface of the films [101]. The accumulated PANI could not interact further with MWCNT, and as a result there is no change in supercapacitance at higher loading.

2.1.2 POLYPYRROLE NANOCOMPOSITE-BASED SUPERCAPACITORS

Polypyrrole (PPy) is an intrinsic conducting and relatively inexpensive polymer [102]. It possesses high energy density, high thermal stability, high conductivity, and quick charging/discharging properties. PPy is one of the most promising conducting polymers because the oxidation of pyrrole monomer can easily be done even in aqueous medium [103]. It is cost effective and commercially available. Due to the reversible electrochemical doping/de-doping and easy electrochemical processability, PPy can be used as an electrode material. Through the process of electropolymerization, PPy can enhance the properties like specific capacitance, charge storage and cycling performance and thermal stability with the use of appropriate dopants (i.e., aromatic anions) [104–106]. Structure of polypyrrole is shown in Figure 2.9.

Chemical or electrochemical polymerization methods can be used to prepare conducting polymer (CP)/CNT nanocomposite-based supercapacitors [107]. PPy/SWNT composite showed the specific capacitance value to be 134 Fg^{-1} in the presence of a binder [108]. The CNT nanofibers-PPy nanocomposites showed 192 F/g specific capacitance per gram [109]. In CNT-PPy nanocomposites, specific capacitance per area was less than 1.0 Fcm^{-2}, much greater than those results previously reported for similar electrode materials used supercapacitors [110]. The fabrication of an activated carbon nanofiber (ACNF)/PP/CNT ternary composite system could enhance the capacitance of nanocomposites [111]. As compared to ACNF, the nanocomposites had higher electrical conductivity and larger specific surface area. The ACNF/PPy/CNT composite showed specific capacitance as 333 F/g, much greater than those of activated carbon had been employed for supercapacitors [112]. The electrochemical capacitor applications based on PPy-CNT composites could be synthesized by growing CNT on ceramic fabrics using chemical vapor deposition (CVD) [113]. The energy storage capacity could be improved by the high conductivity and large surface area of CNT. PPy can act as a bridge to connect each layer of CNT to enhance capacitance.

As Ruiqiao Xu et al. report, a flexible and conductive CNT-PPy composite fiber with core-shell structure can be used for best productive supercapacitor electrode [114]. Due to high pseudocapacitance of PPy and high conductivity of CNTs, this composite fiber showed a high specific capacitance of 350 Fg^{-1}. Iurchenkova et al. synthesized PPy and MWCNT Bucky paper hybrid material by chemical and electrochemical deposition method [115]. The electrochemical behavior of these materials in 1 M H_2SO_4 revealed that the chemical deposition method was better than electrochemical deposition for increasing the specific capacitance. Chemical deposition method allowed increasing the specific capacity of the hybrid material up to ~2 times (91 Fg^{-1}) and electrochemical deposition method provided specific capacitance only up to ~1.2 (54 Fg^{-1}).

To synthesize PPy/MoS_2 nanocomposites, MoS_2 was used as nanofiller for improving electrical and ammonia sensing properties of pristine PPy [116]. MoS_2 nanoparticles have been proved to be a promising material in sensor chemistry, electronics, and optoelectronics owing to their low band gap (1.8 eV) and high chemical and thermal stability [117–119]. Niaz et al. synthesized PPy/MoS_2 nanocomposites through the in situ oxidative polymerization method [120]. The addition of MoS_2 nanoparticles into polymer matrix improved its properties such as electrical, thermal, mechanical, and optical. The PPy/MoS_2 nanocomposite electrode showed high specific capacitance of 654 Fg^{-1} and 95% retention, even after 500 cycles. A PPy/MoS_2 nanocomposite was used for the construction of electrode materials in supercapacitors [121].

Mahore et al. synthesized a ternary nanocomposite of PPy with CNT and manganese dioxide by in situ chemical oxidation polymerization method [122]. Electrochemical performance was studied in 1.0 M sodium sulphate electrolyte solution in the voltage range 0.1–0.9 V at scan rates 0.5 and 1 V/s. SC of PPy/CNT/manganese dioxide determined (in F/cm²) 5.57 F/cm² at a scan rate of 0.5 V/s, and 7.07 F/cm² at a scan rate of 1 V/s, respectively.

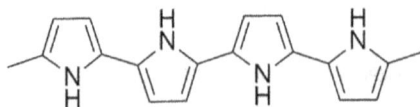

FIGURE 2.9   Structure of polypyrrole.

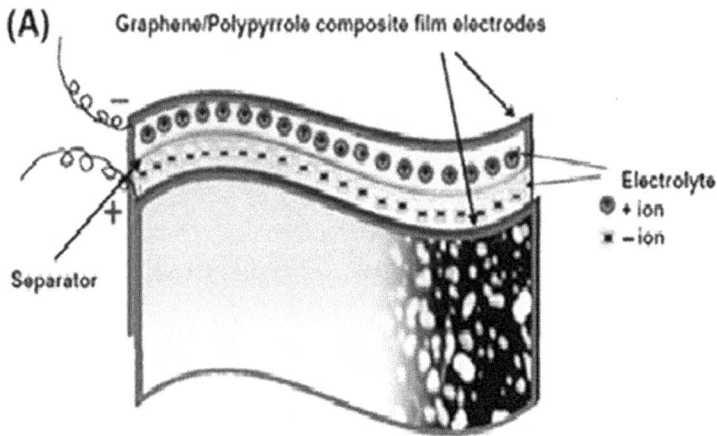

FIGURE 2.10 Graphene/polypyrrole composite film electrodes.

Adam et al. prepared polypyrrole/graphene oxide (PPy/GO) hybrid composites via a hydrothermal self-assembly process [123]. The specific capacitance of GO/PPy composites was recorded 262 Fg^{-1} at the low current density of 0.2 Ag^{-1}. PPy/GO hybrid composites attained specific energy density of 7.4 Wh kg^{-1} at a power density of 0.09 kW kg^{-1}. Y. C. Eeu et al. prepared a polypyrrole/reduced graphene oxide/iron oxide (PPy/RGO/Fe_2O_3) nanocomposite using an electrodeposition method [124]. When this scan rate is 50 mV/S in 0.1 M of KCl the specific capacitance of PPy/RGO/Fe_2O_3 was 125.7F/g. Murat Ates et al. prepared a nanocomposite of polypyrrole, reduced graphene oxide (rGO), and manganese oxide (MnO_2) (PPy/rGO/MnO_2) [125]. The Ppy/rGO/MnO_2 nanocomposites were found to have the highest specific capacitance (285.81 F/g) at 1 mV/s. A flexible, uniform thin film of PPy/graphene composite was synthesized by Davies et al. using supercapacitor electrodes (Figure 2.10) [126]. Due to appropriate nucleation of the PPy chains at defect's site in the graphene surface, this flexible film possesses high energy and power densities.

Using an in situ polymerization method, Lu et al. prepared a ternary composite of graphene/polypyrrole/carbon nanotube. Here the composite had specific capacitance of 361 Fg^{-1} and large surface of 112 m^2g^{-1} [127]. Saptarshi Dhibar et al. synthesized PPy/graphene nanocomposite decorated with on Ag nanoparticles (Ag-PPy/Gr) via in situ oxidative polymerization technique [128]. The nanocomposite gained the highest specific capacitance of 472 F/g at a 0.5 A/g current density. Also, it exhibited the better higher power density and energy density of 1548.94 W/kg and 41.95 Wh/kg respectively at 0.5 A/g current density and at a 3 A/g current density.

The flower-like Co_3O_4 (f-Co_3O_4) and ball-like (b-Co_3O_4) nanoparticles coated on carbon paper (CP) and PPy ternary composites (PPy/f-Co_3O_4/CP) were prepared by potentiodynamic electropolymerization [129]. PPy/f-Co_3O_4/CP showed specific capacitance of 398.4 F/g which is much higher than that of f-Co_3O_4/CP (40.9 F/g) and b-Co_3O_4/CP (22.0 F/g).

Jemini et al. fabricated a novel ternary composite of reduced graphene oxide (rGO)/palladium oxide/polypyrrole (PdPGO) produced by the facile electrode position technique and compared its performance with palladium/polypyrrole (PdP) [130]. The electrodeposited PdPGO composite on stainless steel exhibited better electrochemical properties with specific capacitance of 595 Fg^{-1} at 1 Ag^{-1} in 1M H_2SO_4 due to the improved electrostatic interactions at the electrode-electrolyte interface and the lower the aggregation of rGO layers in PdPGO.

Javed Iqbal et al. prepared a binary nanocomposite of polypyrrole/cobalt oxide (PPy/Co_3O_4) and ternary nanocomposites of polypyrrole (PPy)/cobalt oxide nanograin/silver nanoparticles (PPy/Ag/Co_3O_4). It was prepared using a hydrothermal method [131]. The PPy/Ag/Co_3O_4 showed high specific capacity of 355.64 Cg^{-1} as compared to its binary nanocomposite. This was due to the

TABLE 2.2

Specific Capacitance of Different PPy Nanocomposites

PPy nanocomposites	Electrolyte	Capacitance (Fg^{-1})	Conditions	Reference
PPy/carbon aerogel nanocomposites	6 M KOH	433	1 mVs^{-1}	[135]
PPy/(SG) nanocomposites	1 M aq. KCl	285	0.5 Ag^{-1}	[136]
PPy/nanoclay/graphene composite	1 M aq. KCl	347	1 Ag^{-1}	[137]
PPy/MoS$_2$ nanocomposite	1 M. H_2SO_4	400	1 Ag^{-1}	[138]
PPy/MoO$_3$ nanorods	1 M H_2SO_4	687	1 Ag^{-1}	[139]
PPy/GNS/RE3+	1 M aq. H_2SO_4	238	1 Ag^{-1}	[140]
PPy/MWCNT	1 M KCl	146.3–167.2	0.5 $mAcm^{-2}$	[141]
PPy/TiO$_2$/PANI	1 M aq. H_2SO_4	497	0.5 Ag^{-1}	[142]
GN/PPY/CNT	1 M KCl	361	0.2 Ag^{-1}	[143]
PPy cluster	0.5 M H_2SO_4	586	2 mVs^{-1}	[144]

synergetic effects of Co_3O_4 and silver nanoparticles on PPy by improving the effective charge transport mechanism through the polymer matrix.

Li-li Jiang et al. prepared flexible TiO_2/graphene/polypyrrole composite films as electrodes for supercapacitors [132]. This composite combined the advantages of both PPy and TiO_2. The electrochemical stability and pseudocapacitance of graphene-based electrodes were remarkably improved by the incorporation of TiO_2. The highest capacitance (201.8 Fg^{-1}) was obtained with 14.6% TiO_2 content. The PPy coating and incorporation of TiO_2 also enhanced the capacitance of graphene-PPy composite electrodes.

Shalini et al. constructed a novel layer-by-layer assembled nanocomposite film of polypyrrole/graphene oxide with polypyrrole/nanocrystalline cellulose (PPy/GOPPy/NCC) [133]. Polypyrrole/nanocrystalline cellulose was strongly linked to the polypyrrole/graphene oxide layer without any apparent cracks and defects. As a result, this device provided a high specific capacitance of 562.9 Fg^{-1} at 3mVs^{-1} with a maximum specific energy and specific power of 19.3 Wh kg^{-1} and 884.6 Wkg^{-1} respectively.

A ternary hybrid nanocomposite of Co_3O_4/polypyrrole/MWCNT was synthesized by Ramesh et al. The hybrid composite was prepared using a hydrothermal process [134]. The electrochemical performance of the hybrid composite displayed an excellent capacitive behaviour with a specific capacitance of 609 Fg^{-1} at an energy density 84.58 (W h kg^{-1}), a current density of 3 A g^{-1} and power density 1500 (Wkg^{-1}). It showed a 97% specific capacitance retention even after 5000 cycles.

2.1.3 PEDOT Nanocomposite-Based Supercapacitors

Poly (3,4-ethylene dioxythiophene):poly styrene sulfonate (PEDOT:PSS) is a promising conducting polymer with myriad applications. It is frequently used in thin film technology by different fabrication techniques due to its low cost, flexibility, optical properties, high work function, high electrical conductivity, and high chemical and physical stability. Thus, PEDOT:PSS has a number of applications in the field of energy storage devices [145].

PEDOT:PSS is an organic semiconductor composed of ionomers, namely, poly(3,4-ethylene dioxythiophene) (PEDOT) and poly(4-styrene sulfonate) (PSS). PEDOT can be synthesized by the oxidative polymerization of thiophene derivatives by minimizing α-β and β-β couplings [145–146].

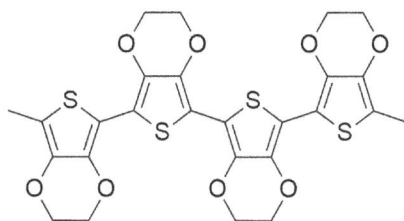

FIGURE 2.11   The structure of PEDOT.

TABLE 2.3
Carbon Modified PEDOT/ESM Supercapacitors Compared with Other PEDOT-Based Supercapacitors

Materials	Areal capacitance	Specific capacity (F/g)	Reference
PEDOT-PIDG composite	-	115.4 at 0.5 A	[153]
PEDOT/CNT/GO	150.6	-	[154]
PEDOT/SDS-GO	79.6	-	[155]
Graphite/PEDOT/MnO$_2$	316.4	195.7	[156]
MOF/GO/PEDOT	128	-	[157]
TiO$_2$ nanotube/PEDOT TiO$_2$/	255.5	393.1	[158]
PEDOT	62.3	128.7	
Vitamin C/GO/PEDOT:PSS	304.5	63.1	[159]
GO/PEDOT MOF/CNT films	30	-	[160]
Activated carbon/PEDOT:PSS	26	640	[161]
Activated carbon/PEDOT Titanium sheet	42	1183	
MWCNT/PEDOT:PSS	32.9	24.1	[162]
PEDOT/Au-PANI/Au	66	175	[163]
PEDOT/GNP	66.2	-	[164]
PEDOT:PSS	4 4.7	27.7	[165]
PEDOT/GO	25	-	[166]
PEDOT:PSS/CNT	64	85.3	[167]
PEDOT:PSS/MWCNT	20.6	-	[168]

Doping can be used to enhance the properties like electrical conductivity and solubility of PEDOT. The structure of PEDOT is shown in Figure 2.11.

Liu et al. have remarked that PEDOT:PSS can store electrical charges, as it can function as a binder as well as an anode. It has high value of capacitance and conductivity which allows the charge storage through faradaic charging [147–149]. Joseph Adekoya et al. have reported that a lithium-ion battery with a nanostructured Ge particle, composite of PEDOT:PSS exhibits electrochemical properties. These composites exhibited a reversible capacity of 405 mAhg^{-1} even after 200 cycles [150].

Zhao et al. have found that the addition of conducting nanofillers and expanding the surface area of PEDOT can improve the performance and conductivity of PEDOT based supercapacitors [151]. Choa et al. in 2020 fabricated a simple and inexpensive active layer of PEDOT-PIDG via an additive and metallic current collector-free route for the fabrication of large-scale supercapacitors and which exhibits good electrochemical performances [152].

Justino da Silva et al. have developed an organic supercapacitor by integrating carbon derivatives and PEDOT chains into the eggshell membrane which exhibits an aerial capacitance of 39.6 m

Fcm^{-2} [169]. They have proved the incorporation of carbon derivatives and PEDOT helps to develop supercapacitors with high storage capacity due to EDLC mechanisms. The modified ESM electrodes exhibit better conductivity and high mechanical resistance upto 0–1800 folds. The resulting composite device is a sustainable and environmentally friendly one. This device has better electrochemical performance with high mechanical flexibility. These capacitors exhibit better performance due to its negligible IR drop and retention characteristics [169].

Recent studies proved PEDOT:PSS can be used even in smart textiles and wearable systems as sweat can be used there as an electrolyte. This one is tested with different volumes of sweat and it exhibited an attractive performance. This opens a new way towards sustainable methods to meat power requirements in the smart textile industry [170].

2.1.4 POLYTHIOPHENE NANOCOMPOSITE-BASED SUPERCAPACITORS

Polymerized thiophenes in the firm of sulfur heterocyclic are called polythiophenes (PTs) [179]. Since Polythiophene polymers have high environmental stability, good thermal stability, and less bandgap energy, they possess a significant place in research and industrial areas. As a result, polythiophene polymers are widely used in electrochromic devices, polymer batteries, and solar cells. They are more important among conducting polymers as they possess easy polymerization and have stability in the air [180]. Compared to other conducting materials, polythiophenes and their nanocomposites are becoming more important and useful in the industries so they are widely used in different applications including display devices, light-emitting diodes, chemical/optical sensors, photovoltaic cells, rechargeable batteries, transistors, EMI shielding, supercapacitors, etc. [181–182].

TABLE 2.4
Specification table of PEDOT and its composite based Supercapacitors

Materials	Conductivity/ Resistance	Specific Capacitance	Energy Density	Power Density	Reference
PEDOT:PSS wrapped MWNT/MnO composite	-	428.2 F/g	63.8 Wh kg^{-1}	-	[171]
MWCNT-reinforced cellulose/ PEDOT:PSS film (MCPP)	low resistance of 0.45 Ω	485 Fg^{-1} at 1 Ag^{-1}	-	-	[172]
PEDOT nanowire (NW) films	1340 S cm^{-1}	667.5 mF cm^{-2} at 1 mA cm^{-2}	48.3 mWh cm^{-2} 19.1 mWh cm^{-2}	0.22 mWcm^{-2} 16.77 mWcm^{-2}	[173]

TABLE 2.5
Specifications of PEDOT and Its Composite-Based Supercapacitors

Materials	Potential limits (V)	Electrolyte	Specific capacity	Reference
PEDOT	-0.5 to 0.9 (1.4)	1 M Et$_4$NBF$_4$/ acetonitrile	103 (1000)	[174]
PEDOT-CNTs composite	composite 0 to 1.5 (1.5)	1 M H$_2$SO$_4$	127 (3000)	
Electrodeposition of PEDOT on CNTs	1.0 to 0.8 (1.8)	-0.5 M KCl (aq)	150	[175]
PEDOT-PPy	0.4 to 0.6 (1.0)	1 M KCl (aq)	270 (1000)	[176]
PEDOT:PSS-RuOx	0.2 to 1.0 (1.2)	1 M KCl (aq)	1409 (40)	[177]
PEDOT-MoO$_3$	1.3 to 0.2 (1.5) 300	1 M LiClO$_4$ in EC/DMC	300	[178]

Also, polythiophene nanocomposites exhibit excellent electro-chromic, electrical, and electronic properties along with high thermal and environmental stabilities, which makes them unique and attractive materials. The structure of polythiophene is given in Figure 2.12.

Vijeth et al. fabricated polythiophene-aluminum oxide nanocomposites by an in situ chemical polymerization method with anionic surfactant camphor sulfonic acid (CSA) [183] and their electrochemical performance was checked by using cyclic voltammetry in 1 M H_2SO_4. The polythiophene and polythiophene nanocomposites exhibited high specific capacitance of 654 and 757 F/g at a rate of 30 mVs^{-1}, respectively.

Polythiophene/graphene (GR/PT) composites can be prepared using the chemical polymerization method with the varied mass percentages of PT:GR/PT-67, -50 and -33 [184]. Compared to others, GR/PT-50 exhibited the best specific capacitance of 365 Fg^{-1}. This value is higher than that of pure GR (232 Fg^{-1}) and four times higher than that of the pure polymer PT (92 Fg^{-1}). Double-layer capacitance was observed in the cyclic voltammetry experiments. This composite material possesses good pseudocapacitance due to the synergistic effect of PT and graphene by reducing the ion diffusion length. As a result, this material can be used as an electrode material for supercapacitors.

With polythiophene/aluminum oxide (PTHA) nanocomposite as an anode and charcoal as a cathode, an asymmetric supercapacitor (ASC) can be produced (Figure 2.13) [185]. PTHA electrodes and charcoal electrodes displayed specific capacitances of 554.03 Fg^{-1} and 374.71 Fg^{-1}, respectively, at a current density (CD) of 1 Ag^{-1} and 1.4 Ag^{-1}, respectively (both measured using a three-electrode system). The specific capacitance obtained for the assembled PTHA/charcoal asymmetric supercapacitor (ASC) was 265.14 Fg^{-1} at 2 Ag^{-1}. It possessed a high energy density of 42 $Whkg^{-1}$ at a power density of 736 Wkg^{-1} and 95% capacitance retention even after 2000 cycles.

Fu et al. successfully electropolymerized polythiophene over multi-walled carbon nanotube (MWCNT) modified glassy carbon (GC) as electrodes in ionic liquid $bmimPF_6$ solution [186].

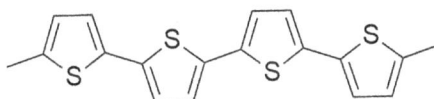

FIGURE 2.12 The structure of Polythiophene.

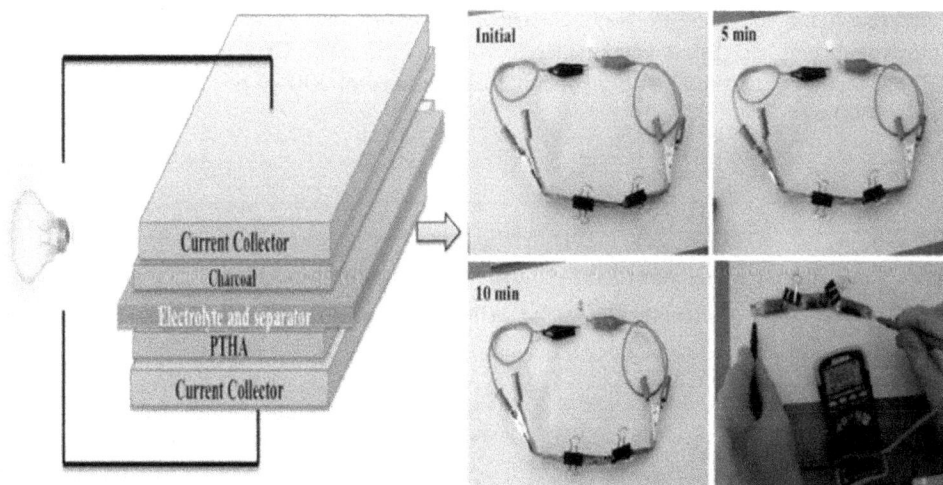

FIGURE 2.13 Pictorial representation of the fabricated solid-state ASC with PTHA anode and charcoal cathode using PVA/KOH gel electrolytes.

From the cyclic voltammograms it was found that polythiophene/MWCNT composites have better specific capacitance than pristine polymer and pure MWCNT. The specific capacitance of the composite was 110 Fg^{-1}. The stability testing of the supercapacitor showed 90% specific capacitance retention even after 1000 cycles.

Ates et al. developed ternary nanocomposites of reduced graphene oxide (rGO)/Ag nanoparticles/polythiophene (PTh), (rGO/Ag/PTh) using an in situ polymerization method [187]. Galvanostatic charge-discharge measurements revealed high specific capacitance of 904 F/g at a constant current of 10 mA. In addition to this rGO/Ag/PTh nanocomposite electrode material showed low interfacial resistance and improved conductivity because of the synergetic effects of rGO, polythiophene, and Ag nanoparticles.

Poly(methyl methacrylate) (PMMA) electrodes and hexagonal oriented holes with 0.5- to 10 μm^2 surface areas can be produced by the direct breath figure method [188]. Through the process of emulsion polymerization, polythiophene (PTh) nanoparticles, polythiophene reduced graphene oxide (PTh-G), and polythiophene-graphene oxide (PTh-GO) nanocomposites were produced. The spray-coating method was used for the deposition of electrode materials on breath figure-decorated PMMA substrates. The specific capacitance of porous PMMA electrodes coated by PTh nanoparticles, PTh-G and PTh-GO nanocomposites were observed 3.5 F/g, 28.68 F/g and 16.39 F/g, respectively at a 5 mV/s scan rate, obtained from cyclic volumetric studies.

Porous polythiophene (PTh) nanofibers were produced by surfactant aided dilute polymerization technique using FeCl$_3$ as oxidant [189]. The asymmetric supercapacitor was gathered by using carbonaceous PTh nanofibers as the anode and PTh nanofibers as the cathode in 6 M KOH electrolyte. The asymmetric supercapacitor had specific capacitance, specific energy, and specific power densities 252 F g^{-1} at the scan rate of 5 mV s^{-1}, 54.6 W h kg^{-1} and 1.7 kW kg^{-1}, respectively with good cyclic stability.

Dhibar et al. successfully produced graphene (Gr)- single-walled CNT (SWCNT)-poly (3-methyl thiophene) ternary nanocomposite by a simple and cost-effective in situ chemical oxidative polymerization method [190]. The ternary nanocomposites showed the high specific capacitance of 561 F/g at a scan rate of 5 mV/s. Also, it showed better specific capacitance of 551 F/g at a current density of 0.5 A/g with 93% specific capacitance retention after 1000 cycles. This nanocomposite gained a power density of 1579.35 W/kg and the highest energy density of 48.97 W h/kg at current densities of 3 and 0.5 A/g, respectively together with enhanced electrical conductivity of 4.68 S/cm.

The PTh/Ni nanocomposites thin films were synthesized by the electrochemical oxidative polymerization method showed good electrical properties due to the polycrystalline granular structures with high roughness and surface to volume ratio [191]. Cyclic voltammograms showed that PTh/Ni nanocomposites had high specific capacitance of 3000 F g^{-1} and a rapid discharging process of 200 s.

TABLE 2.6

Specific Capacitance of Different Polythiophene Nanocomposites

Materials	Current Density	Specific Capacitance	Electrolyte	Reference
Gr-PTh nanocomposite	1.33 A/g	154 F/g	2 M H$_2$SO$_4$	[192]
Gr-PTh	1 A/g	365 F/g	1 M KCl	[193]
GO-PTh derivative	0.3 A/g	296 F/g	2 M KCl	[194]
GR-P3MT	0.5 Ag^{-1}	332 Fg^{-1}	1 M KCl	[195]
GR-P3MT	10 mVs^{-1}	240 Fg^{-1}	1 M KCL	[196]
GNPs-P3MT	0.5 Ag^{-1}	215.5 Fg^{-1}	1 M TEABF$_4$ in PC	[197]
PTh- CNT	1 A/g	125 Fg^{-1}	1 M H$_2$SO$_4$	[198]

2.2 CONCLUSION

Sustainable and renewable energy storage devices are inevitable in fulfilling future energy demands. Supercapacitors are widely acknowledged as ideal energy storage devices for satisfying the energy requirements of sustainable and renewable energy storage for electronic applications such as hybrid electric vehicles and in electronic devices. A brief overview of conducting polymer composites for advanced electrode materials for supercapacitors has been investigated in this chapter. Composites of conducting polymers with other materials ensure enhanced energy density and excellent cycling stability by the proper integration of pseudocapacitance and electrochemical capacitance in a single material. Their high flexibility, excellent structural diversity, great durability, and superior electro-chromic properties of conducting polymers make such composites potential candidates in smart supercapacitors. Therefore, conducting polymer composite electrodes will play an important role in flexible, smart, and economical energy storage devices in the coming years.

REFERENCES

[1]. M. Rajkumar, C. T. Hsu, T. H. Wu, M. G. Chen, C. C. Hu, Advanced materials for aqueous supercapacitors in the asymmetric design, *Prog. Nat. Sci.*, 25 (2015) 527–544.

[2]. L. L. Zhang, X. S. Zhao, Carbon-based materials as supercapacitor electrodes, *Chem. Soc. Rev.*, 38 (2009) 2520–2531.

[3]. I. Shown, A. Ganguly, L. C. Chen, K. H. Chen, Conducting polymer-based flexible supercapacitor, *Energy Sci. Eng.*, 3 (2014) 2–26.

[4]. G. Boara, M. Sparpaglione, Synthesis of polyanilines with high electrical conductivity, *Synth Met.*, 72 (1995) 135–140.

[5]. S. Hung, T. Wen, A. Gopalan, Application of statistical design strategies to optimize the conductivity of electrosynthesizedpolypyrrole, *Mater Lett.*, 55 (2002) 165–170.

[6]. M. C. Morvant, J. R. Reynolds, In situ conductivity studies of poly(3,4-ethylenedioxythiophene), *Synth Met.*, 92 (1998) 57–61.

[7]. K. A. Noh, D. W. Kim, C. S Jin, Synthesis and pseudo-capacitance of chemically-prepared polypyrrole powder, *J Power Sources*, 124 (2003) 593–598.

[8]. V. Gupta, N. Miura, High performance electrochemical supercapacitor from electrochemicallysynthesized nanostructured polyaniline, *Mater Lett.*, 60 (2006) 1466–1475.

[9]. K. Lota, V. Khomenko, E. Frackowiak, Capacitance properties of poly(3,4-ethylenedioxythiophene)/carbon nanotubes composites, *J Phys Chem Solids*, 65 (2004) 295–301.

[10]. A. J. Heeger, Nobel Lecture: Semiconducting and metallic polymers: The fourth generation of polymeric materials, *Rev. Mod. Phys.*, 73 (2001) 681–700.

[11]. C. Peng, D. Hu, G. Z. Chen, Theoretical specific capacitance based on charge storage mechanisms of conducting polymers: Comment on 'Vertically oriented arrays of polyaniline nanorods and their super electrochemical properties', *Chem. Commun.*, 47 (2011) 4105–4107.

[12]. J. F. Zang, S. J. Bao, C. M. Li, H. J. Bian, X. Q. Cui, Q. L. Bao, C. Q. Sun, J. Guo, K. R. Lian, Well-aligned cone-shaped nanostructure of polypyrrole/RuO2 and its electrochemical supercapacitor, *J. Phys. Chem. C.*, 112 (2008) 14843–14847.

[13]. J. Bobacka, A. Lewenstam, A. Ivaska, Electrochemical impedance spectroscopy of oxidized poly(3,4-ethylenedioxythiophene) film electrodes in aqueous solutions, *J. Electroanal. Chem.*, 489 (2000) 17–27.

[14]. R. Liu, S. Il Cho, S. B. Lee, Poly(3,4-ethylenedioxythiophene) nanotubes as electrode materials for a high-powered supercapacitor, *Nanotechnology*, 19 (2008) 1–8.

[15]. H. Randriamahazaka, C. Plesse, D. Teyssie, C. Chevrot, Relaxation kinetics of poly(3,4-ethylenedioxythiophene) in 1-ethyl-3-methylimidazolium bis((trifluoromethyl)sulfonyl)amide ionic liquid during potential step experiments, *Electro-chim. Acta.*, 50 (2000) 1515–1522.

[16]. T. Kobayashi, H. Yoneyama, H. Tamura, Oxidative degradation pathway of polyaniline film electrodes, *J. Electroanal. Chem. Interfacial Electrochem*, 177 (1984) 293–297.

[17]. F. Shen, D. Pankratov, Q. Chi, Graphene-conducting polymer nanocomposites for enhancing electrochemical capacitive energy storage, *Curr Opin Electrochem.*, 4 (2017) 133–144.

[18]. W. J. Feast., *Hand Book of Conducting Polymer*, New York: Marcel Inc.; 1986, 1–43.

[19]. Q. Meng, K. Cai, Y. Chen, L. Chen, Research progress on conducting polymer based supercapacitor electrode materials, *Nano Energy* 36 (2017) 268–285.

[20]. H. S. Nalawa, L. R. Dalton, W. F. Schmidt, Conducting polyaniline and polypyrrole: studies of their catalytic properties. *Polymer Commun.*, 27 (1985) 240–242.

[21]. C. Peng, S. Zhang, D. Jewell, G. Z. Chen, Carbon nanotube and conducting polymer composites for supercapacitors, *Prog. Nat. Sci.*, 18 (2008) 777–788.

[22]. H. Wang, J. Lin, Z. X. Shen, Polyaniline (PANi) based electrode materials for energy storage and conversion, *J Sci-Adv Mater Dev.*, 1 (2016) 225–255.

[23]. Bleda-Martínez, M.J., Lozano-Castello, D., Morallón, E., Cazorla-Amorós, D. and Linares-Solano, A, Chemical and electrochemical characterization of porous carbon materials, *Carbon.*, 44 (2006) 2642–2651.

[24]. H. Li, H. Xi, S. Zhu, Preparation, structural characterization, and electrochemical properties of chemically modified mesoporous carbon, *Microporous Mesoporous Mater.*, 96 (2006) 357–362.

[25]. H. A. Andreas, B. E Conway, Examination of the double-layer capacitance of an high specific-area C-cloth electrode as titrated from acidic to alkaline pHs, *Electrochim Acta.*, 51 (2006) 6510–6520.

[26]. K. Dutta, P. Kumar, S. Das, P. P. Kundu, Utilization of conducting polymers in fabricating polymer electrolyte membranes for application in direct methanol fuel cells, *Polym Rev.*, 54 (2014) 1–32.

[27]. S. Thomas, P. M. Visakh, Handbook of engineering and specialty thermoplastics, *Wiley Online Laboratory*, 4 (2011) 183–210.

[28]. R. S. Norouzian, M. M. Lakouraj, E. N. Zare, Novel conductive PANI/hydrophilic thiacalix[4]arene nanocomposites: synthesis, characterization and investigation of properties, *Chin. J. Polym. Sci.*, 32 (2014) 218–229.

[29]. A. S. Aricò, P. Bruce, B. Scrosati, J. M. Tarascon, W. Schalkwijk, Nanostructured materials for advanced energy conversion and storage devices, *Nat. Mater.*, 4 (2005) 366–377.

[30]. J. Zhang, J. Jiang, H. Li, X. S. Zhao, A high-performance asymmetric supercapacitor fabricated with graphene-based electrodes, *Energy Environ. Sci.*, 4 (2011) 4009–4015.

[31]. K. S. Novoselov, A. K. Geim, S. V. Morozov, D. Jiang, M. I. Katsnelson, I. V. Grigorieva, S. V. Dubonos, A. A. Firsov, Two-dimensional gas of massless Dirac fermions in graphene, *Nature*, 438 (2005) 197.

[32]. W. Yu, L. Sisi, Y. Haiyan, L. Jie, Progress in the functional modification of graphene/graphene oxide: a review, *RSC Adv.*, 10 (2020) 15328–15345.

[33]. T. J. Booth, P. Blake, R. R. Nair, D. Jiang, E. W. Hill, U. Bangert, A. Bleloch, M. Gass, K. S. Novoselov, M. I. Katsnelson A. K. Geim, Macroscopic graphene membranes and their extraordinary stiffness, *Nano Lett.*, 8 (2008) 2442–2446.

[34]. C. Lee, X. Wei, J. W. Kysar, J. Hone, Measurement of the elastic properties and intrinsic strength of monolayer graphene, *Science* 321 (2008) 385–388.

[35]. J. Xia, F. Chen, J. Li, N. Tao, Measurement of the quantum capacitance of graphene, *Nat. Nanotechnol.*, 4 (2009) 505–509.

[36]. H. Li, L. Xu, H. Sitinamaluwa, K. Wasalathilake, C. Yan, Coating Fe_2O_3 with graphene oxide for high-performance sodium-ion battery anode, *Comp. Comm.*, 1 (2016) 48–53.

[37]. T. Wang, J. Yu, M. Wang, Y. Cao, W. Dai, D. Shen, L. Guo, Y. Wu, H. Bai, D. Dai, J. Lyu, N. Jiang, C. Pan, C.-T. Lin, Effect of different sizes of graphene on thermal transport performance of graphene paper, *Comp. Comm.*, 5 (2017) 46–53.

[38]. M. Yu, Y. Huang, C. Li, Y. Zeng, W. Wang, Y. Li, P. Fang, X. Lu, Y. Tong, Building three-dimensional graphene frameworks for energy storage and catalysis, *Adv. Funct. Mater.*, 25 (2015) 324–330.

[39]. Q. Wu, Y. Xu, Z. Yao, A. Liu, G. Shi, Supercapacitors based on flexible graphene/polyaniline nanofiber composite films, *ACS Nano.*, 4 (2010) 1963–1970.

[40]. H. Bai, Y. Xu, L. Zhao, C. Li, G. Shi, Non-covalent functionalization of graphene sheets by sulfonated polyaniline, *Chem. Commun.*, (2009) 1667–1669.

[41]. T. Lee, T. Yun, B. Park, B. Sharma, H.-K. Song, B.-S. Kim, Hybrid multilayer thin film supercapacitor of graphenenanosheets with polyaniline: importance of establishing intimate electronic contact through nanoscale blending, *J. Mater. Chem.*, 22 (2012) 21092–21099.

[42]. X. M. Feng, R. M. Li, Y. W. Ma, R. F. Chen, N. E. Shi, Q. L. Fan, W. Huang, One-step electrochemical synthesis of graphene/polyaniline composite film and its applications, *Adv. Funct. Mater.*, 21 (2011) 2989–2996.

[43]. M. Xue, F. Li, J. Zhu, H. Song, M. Zhang, T. Cao, Structure-based enhanced capacitance: In situ growth of highly ordered polyaniline nanorods on reduced graphene oxide patterns, *Adv. Funct. Mater.*, 22 (2012) 1284–1290.

[44]. Q. Hao, H. Wang, X. Yang, L. Lu, X. Wang, Morphology-controlled fabrication of sulfonated graphene/polyaniline nanocomposites by liquid/liquid interfacial polymerization and investigation of their electrochemical properties, *Nano Res.*, 4 (2011) 323–333.

[45]. K. Zhang, L. L. Zhang, X. S. Zhao, J. Wu, Graphene/polyaniline nanofiber composites as supercapacitor electrodes, *Chem. Mater.*, 22 (2010) 1392–1401.

[46]. L. Q. Xu, Y. L. Liu, K.-G. Neoh, E.-T. Kang, G. D. Fu, Reduction of graphene oxide by aniline with its concomitant oxidative polymerization, *Rapid Commun.*, 32 (2011) 684–688.

[47]. H. Y. He, J. Klinowski, M. Forster, A. Lerf, Reduction of graphene oxide by aniline with its concomitant oxidative polymerization, *Chem. Phys. Lett.*, 287 (1998) 53–56.

[48]. A. Lerf, H. He, M. Forster, J. Klinowski, Structure of graphite oxide revisited, *J. Phys. Chem. B*, 102 (1998) 4477–4482.

[49]. D. Li, R. B. Kaner, Processable stabilizer-free polyaniline nanofiber aqueous colloids, *Chem. Commun.*, (2005) 3286–3288.

[50]. D. Li, M. B. Müller, S. Gilje, R. B. Kaner, G. G. Wallace, processable aqueous dispersions of graphene nanosheets, *Nat. Nanotechnol.*, 3 (2008) 101–105.

[51]. M. Kim, C. Lee, J. Jang, Fabrication of highly flexible, scalable, and high-performance supercapacitors using polyaniline/reduced graphene oxide film with enhanced electrical conductivity and crystallinity, *Adv. Funct. Mater.*, 24 (2014) 2489–2499.

[52]. N. A. Kumar, H.-J. Choi, Y. R. Shin, D. W. Chang, L. Dai, J.-B. Baek, Polyaniline-grafted reduced graphene oxide for efficient electrochemical supercapacitors, *ACS Nano.*, 6 (2012) 1715–1723.

[53]. Ce. NeR. Rao, Ae. K. Sood, Ke. S. Subrahmanyam, A. Govindaraj, Graphene: The new two-dimensional nanomaterial, *Chem. Int. Ed.*, 48 (2009) 7752–7777.

[54]. L. Q. Xu, Y. L. Liu, K. -G. Neoh, E. -T. Kang, G. D. Fu, Reduction of graphene oxide by aniline with its concomitant oxidative polymerization, *Macromol. Rapid Commun.*, 32 (2011) 684–688.

[55]. X. M. Feng, R. M. Li, Y. W. Ma, R. F. Chen, N. E. Shi, Q. L. Fan, W. Huang, One-step electrochemical synthesis of graphene/polyaniline composite film and its applications, *Adv. Funct. Mater.*, 21 (2011) 2989–2996.

[56]. M. Xue, F. Li, J. Zhu, H. Song, M. Zhang, T. Cao, Structure-based enhanced capacitance: in situ growth of highly ordered polyaniline nanorods on reduced graphene oxide patterns, *Adv. Funct. Mater.*, 22 (2012) 1284–1290.

[57]. H. Y. He, J. Klinowski, M. Forster, A. Lerf, A new structural model for graphite oxide, *Chem. Phys. Lett.*, 287 (1998) 53–56.

[58]. Q. Hao, H. Wang, X. Yang, L. Lu, X. Wang, Morphology-controlled fabrication of sulfonated graphene/ polyaniline nanocomposites by liquid/liquid interfacial polymerization and investigation of their electrochemical properties, *Nano Res.*, 4 (2011) 323–333.

[59]. D. Li, R. B. Kaner, Processable stabilizer-free polyaniline nanofiber aqueous colloids, *Chem. Commun.*, 126 (2005) 3286–3288.

[60]. D. Li, M. B. Müller, S. Gilje, R. B. Kaner, G. G. Wallace, processable aqueous dispersions of graphene nanosheets, *Nat. Nanotechnol.*, 3 (2008) 101–105.

[61]. X. Lu, H. Dou, S. Yang, L. Hao, L. Zhang, L. Shen, F. Zhang, X. Zhang, Fabrication and electrochemical capacitance of hierarchical graphene/polyaniline/carbon nanotube ternary composite film, *Electrochim. Acta*, 56 (2011) 9224–9232.

[62]. L. Ma, L. Su, J. Zhang, D. Zhao, C. Qin, Z. Jin, K. Zhaob, A controllable morphology GO/PANI/metal hydroxide composite for supercapacitor, *J. Electroanal. Chem.*, 777 (2016) 75–84.

[63]. M. Morshed, J. Wang, M. Gao, C. Cong, Z Wang, A controllable morphology GO/PANI/metal hydroxide composite for supercapacitor, *Electrochim. Acta*, 370 (2021) 137–714.

[64]. P. Sekar, B. Anothumakkool, S. Kurungot, 3D polyaniline porous layer anchored pillared graphene sheets: enhanced interface joined with high conductivity for better charge storage applications, *ACS Appl. Mater. Interfaces.*, 7 (2015) 7661–7669.

[65]. S. Park, R. S. Ruoff, Chemical methods for the production of graphenes, *Nat. Nanotechnol.*, 4 (2009) 217–224.

[66]. K. S. Kim, I. Y. Jeon, S. N. Ahn, Y. D. Kwon, J. B. Baek, Edge-functionalized graphene-like platelets as a co-curing agent and a nanoscale additive to epoxy resin, *J. Mater. Chem.*, 21 (2011) 7337–7342.

[67]. C. Basavaraja, W. J. Kim, Y. Do Kim, Synthesis of polyaniline-gold/graphene oxide composite and microwave absorption characteristics of the composite films, *Mater. Lett.*, 65 (2011) 3120–3123.

[68]. L. Mao, K. Zhang, H. S. O. Chan, J. Wu, Surfactant-stabilized graphene/polyaniline nanofiber composites for high performance supercapacitor electrode, *J. Mater. Chem.*, 22 (2012) 80–85.

[69]. Y. Zhang, X. Cui, Lei Zu, X. Cai, Y. Liu, X. Wang, H. Lian, New supercapacitors based on the synergetic redox effect between electrode and electrolyte, *Mater.*, 9 (2016)734.

[70]. S. Giri, D. Ghosh, C. K. Das, Growth of vertically aligned tunable polyaniline on graphene/zro2 nanocomposites for supercapacitor energy-storage application, *Adv. Function. Mater.*, 2 (2004) 1312–1324.

[71]. J. Yu, W. Lu, S. Pei, K. Gong§, L. Wang, L. Meng, Y. Huang, J. P. Smith, K. S. Booksh, Q. Li, J. H. Byun, Y. Oh, Y. Yan, T. W. Chou, Omnidirectionally stretchable high-performance supercapacitor based on isotropic buckled carbon nanotube films, *ACS Nano.*, 10 (2016) 5204–5211.

[72]. Q. Zhang, Y. li, Y. Feng, W. A. Feng, Electropolymerization of graphene oxide/polyaniline composite for high-performance supercapacitor, *Electrochim. Acta.*, 90 (2013) 95–100.

[73]. H. Zhang, G. Cao, W. Wang, K. Yuan, Electropolymerization of graphene oxide/polyaniline composite for high-performance supercapacitor, *Electrochim. Acta.*, 54 (2009) 1153–1159.

[74]. N. Hu, L. Zhang, C. Yang, J. Zhao, Z. Yang, H. Wei, H. Liao, Z. Feng, A. Fisher, Y. Zhang, Z. J. Xu, Three-dimensional skeleton networks of graphene wrapped polyaniline nanofibers: an excellent structure for high-performance flexible solid-state supercapacitors, *Scientific Reports.*, 6 (2016) 19777–19787.

[75]. M. Liu, Y. Miao, C. Zhang, Hierarchical composites of polyaniline–graphene nanoribbons–carbon nanotubes as electrode materials in all-solid-state supercapacitors, *Nanoscale*, 5 (2013) 7312–7320.

[76]. J. Kim, H. Ju, I. Akbar, I. Y. Jo, J. Han, H. Kim, H. Im, Synthesis and enhanced electrochemical supercapacitor properties of Ag–MnO$_2$–polyaniline nanocomposite electrodes, *Energy.*, 70 (2014) 473–477.

[77]. N. Jabeen, Q. Xia, M. Yang, H. Xia, Unique core–shell nanorod arrays with polyaniline deposited into mesoporous nico$_2$o$_4$ support for high-performance supercapacitor electrodes, *ACS Appl Mater Interfaces.*, 8 (2016) 6093–6100.

[78]. D. Ghosh, S. Giri, A. Mandala, C. K. Das, Supercapacitor based on H$^+$ and Ni^{2+} co-doped polyaniline–MWCNTs nanocomposite: synthesis and electrochemical characterization, *RSC Adv.*, 3 (2013) 11676–11685.

[79]. M. Fathi, M. Saghafi, F. Mahboubi, S. Mohajerzadeh, Synthesis and electrochemical investigation of polyaniline/unzipped carbon nanotube composites as electrode material in supercapacitors, *Synth. Met.*, 198 (2014) 345–356.

[80]. Y. Li, Y. Fang, H. Liu, X. Wub, Y. Lu, Free-standing 3D polyaniline–CNT/Ni-fiber hybrid electrodes for high-performance supercapacitors, *Nanoscale*, 4 (2014) 2867–2869.

[81]. A. Yahya, J. Chang, S. R. Shin, R. S. Mane, S. H. Han, S. J. Kim, Hydrogel-assisted polyaniline microfiber as controllable electrochemical actuatable supercapacitor, *J. Electrochem. Soc.*, 156 (2009) 313–317.

[82]. M. Faraji, P. N. Moghadam, R. Hasanzadeh, Fabrication of binder-free polyaniline grafted multiwalled carbon nanotube/TiO2 nanotubes/Ti as a novel energy storage electrode for supercapacitor applications, *Chem. Eng. Sci.*, 304 (2016) 841–851.

[83]. S. K. Tam, K. M. Ng, High-concentration copper nanoparticles synthesis process for screen-printing conductive paste on flexible substrate, *J Nanopart Res.*, 17 (2015) 1–12.

[84]. K. H. S Lessa, Y. Zhang, G. Zhang, F. Xiao, S. Wang S, Conductive porous sponge-like ionic liquid-graphene assembly decorated with nanosized polyaniline as active electrode material for supercapacitor, *J. Power Sources*, 302 (2016) 92–97.

[85]. X. Zang, X. Li, M. Zhu, X. Li, Z. Zhen, Y. He, K. Wang, J. Wei, F. Kang, H. Zhu, Conductive porous sponge-like ionic liquid-graphene assembly decorated with nanosized polyaniline as active electrode material for supercapacitor, *Nanoscale.*, 7 (2015) 7318–7322.

[86]. C. Sha, B. Lu, H. Mao, J. Cheng, X. Pan, J. Lu, Z. Ye, 3D ternary nanocomposites of molybdenum disulfide/polyaniline/reduced graphene oxide aerogel for high performance supercapacitors, *Carbon*, 99 (2016) 26–34.

[87]. J. Luo, W. Zhong, Y. Zou, C. Xiong, W. Yang, Preparation of morphology-controllable polyaniline and polyaniline/graphene hydrogels for high performance binder-free supercapacitor electrodes, *J. Power Sources*, 319 (2016) 73–81.

[88]. S. Zhou, S. Mo, W. Zou, F. Jiang, T. Zhou, D. Yuan, Preparation of polyaniline/2-dimensional hexagonal mesoporous carbon composite for supercapacitor, *Synth. Met.*, 161(2011) 1623–1628.

[89]. D. Zhao, X. Guo, Y. Gao, F. Gao, Preparation of polyaniline/2-dimensional hexagonal mesoporous carbon composite for supercapacitor, *ACS Appl. Mater. Interfaces.*, 10 (2012) 5583–5589.

[90]. X. Shi, Y. Hu, M. Li, Y. Y. Duan, Y. Wang, L. Chen, L. Zhang, Highly specific capacitance materials constructed via in situ synthesis of polyaniline in a cellulose matrix for supercapacitors, *Cellulose*, 21 (2014) 2337–2347.

[91]. Y. Yan, Q. Cheng, G. Wang, C. Li, Growth of polyaniline nanowhiskers on mesoporous carbon for supercapacitor application, *J. Power Sources* 196 (2011) 7835–7840.

[92]. Zhou, G. Min, Wang, D. Wei, Li, Feng, Zhang, Li-li, Weng, Zhe, Cheng, H. Ming, The effect of carbon particle morphology on the electrochemical properties of nanocarbon/polyaniline composites in supercapacitors. *New Carbon Mater.*, 26 (2011) 180–186.

[93]. L. J Bian, F. Luan, S. Sha, L. X. Xia, Liu, Self-doped polyaniline on functionalized carbon cloth as electroactive materials for supercapacitor, *Electrochim. Acta.*, 64 (2012) 17–22.

[94]. J. Pedrós, J. Martínez, S. R. Gómez, L. Pérez, V. Barranco, F. Calle, Polyaniline nanofiber sponge filled graphene foam as high gravimetric and volumetric capacitance electrode, *J. Power Sources.*, 317 (2016) 35–42.

[95]. X. Feng, Z. Yan, N. Chen, Y. Zhang, X. Liu, Y Ma, X. Yang, W. Hou, Synthesis of a graphene/polyaniline/MCM-41 nanocomposite and its application as a supercapacitor, *New J. Chem.*, 37 (2013) 2203–2209.

[96]. M. Mandal, D. Ghosh, S. Giri, I. Shakir, C. K. Das, Polyaniline-wrapped 1D $CoMoO_4 \cdot 0.75H_2O$ nanorods as electrode materials for supercapacitor energy storage applications, *RSC Adv.*, 4 (2014) 30832–30839.

[97]. X. Zhou, L. Li, S. Dong, X. Chen, P. Han, H. Xu, J. Yao, C. Shang, Z. Liu, G. Cui, A renewable bamboo carbon/polyaniline composite for a high-performance supercapacitor electrode material, *J Solid State Electrochem.*, 16 (2012) 877–882.

[98]. S. W. Woo, K. Dokko, H. Nakano, K. Kanamur, Incorporation of polyaniline into macropores of three-dimensionally ordered macroporous carbon electrode for electrochemical capacitors, *J. Power Sources.*, 190 (2009) 596–600.

[99]. D. Xie, Q. Jiang, G. Fu, Y. Ding, X. Kang, W. Cao, Y. Zhao, Preparation of cotton-shaped CNT/PANI composite and its electrochemical performances, *Rare Metals.*, 30 (2011) 94–97.

[100]. H. Lin, L. Li, J. Ren, Z. Cai, Z. Yang H. Peng, Conducting polymer composite film incorporated with aligned carbon nanotubes for transparent, flexible and efficient supercapacitor, *Sci. Rep.*, 7 (2013) 1–8.

[101]. H. Wang, J. Ze, X. Shen, Conducting polymer composite film incorporated with aligned carbon nanotubes for transparent, flexible and efficient supercapacitor, *J Scl Adv Mater Dev.*, 1 (2016) 225–255.

[102]. L.-Z. Fan, J. Maier, High-performance polypyrrole electrode materials for redox supercapacitors, *Electrochem. Commun.*, 8 (2006) 937–940.

[103]. P. Asen, S. Shahrokhian, A high performance supercapacitor based on graphene/Polypyrrole/Cu_2O–$Cu(OH)_2$ ternary nanocomposite coated on nickel foam, *J. Phys. Chem.*, 121 (2017) 6508–6519.

[104]. K. Shi, I. Zhitomirsky, Influence of current collector on capacitive behavior and cycling stability of Tiron doped polypyrrole electrodes, *J. Power Sources.*, 240 (2013) 42–49.

[105]. A. Afzal, F. A. Abuilaiwi, A. Habib, M. Awais, S. B. Waje, M. A. Atieh, Polypyrrole/carbon nanotube supercapacitors: Technological advances and challenges, *J. Power Sources.*, 352 (2017) 174–186.

[106]. R. P. Mahore, D. K. Burghate, S. B. Kondawar, A nanocrystalline Co_3O_4 @polypyrrole/MWCNT hybrid nanocomposite for high performance electrochemical supercapacitors, *Adv. Mat. Lett.*, 5 (2014) 400–405.

[107]. Y. Jiang, P. Wang, X. Zang, Y. Yang, A. Kozinda, L. Lin, Uniformly embedded metal oxide nanoparticles in vertically aligned carbon nanotube forests as pseudocapacitor electrodes for enhanced energy storage, *Nano Lett.*, 13 (2013) 3524–3530.

[108]. H. T. Ham, Y. S. Choi, N. Jeong, I. Chung, Electrochemical performance of a graphene–polypyrrole nanocomposite as a supercapacitor electrode, *J of Polymer.*, 46 (2005) 1–9.

[109]. Y. Yesi, I. Shown, A. Ganguly, T. T. Ngo, L. C. Chen, K. H. Chen, Directly-grown hierarchical carbon Nanotube@Polypyrrole core–shell hybrid for high-performance flexible supercapacitors, *J. Environ. Chem.*, 9 (2016) 370–378.

[110]. M. Hughes, M. S. P. Shaffer, A. C. Renouf, C. Singh, G. Z. Chen, D. J. Fray, A. H. Windle, Electrochemical capacitance of nanocomposite films formed by coating aligned arrays of carbon nanotubes with polypyrrole, *Adv. Mater. Technol.*, 14 (2002) 382–385.

[111]. Y. W. Jua, G. R. Choia, H. R. Jung, W. J. Lee, Polypyrrole/carbon nanotube nanocomposite enhanced the electrochemical capacitance of flexible graphene film for supercapacitors, *Electrochim. Acta.*, 53 (2008) 5796–5803.

[112]. V. Khomenko, F. Frackowiak, F. Beguin, Determination of the specific capacitance of conducting polymer/nanotubes composite electrodes using different cell configurations, *Electrochim. Acta.*, 50 (2005) 2499–2506.

[113]. H. Lee, H. Kim, M. S. Cho, J. Choi, Y. Lee, Fabrication of polypyrrole (PPy)/carbon nanotube (CNT) composite electrode on ceramic fabric for supercapacitor applications, *Electrochim. Acta.*, 56 (2011) 7460–7466.

[114]. R. Xu, J. Wei, F. Guo, X. Cui, T. Zhang, H. Zhu, K. Wang, D. Wu, Highly conductive, twistable and bendable polypyrrole–carbon nanotube fiber for efficient supercapacitor electrodes, *RSC Adv.*, 5 (2015) 22015–22021.

[115]. A. A. Iurchenkova, E. O. Fedorovskaya, I. P. Asanov, V. E. Arkhipov, K. M. Popov, K. I. Baskakova, A. V. Okotrub, MWCNT buckypaper/polypyrrole nanocomposites for supercapasitor application, *Electrochim. Acta.*, 335 (2020) 135689–135700.

[116]. S. Ahmad, I. Khan, A. Husain, A. Khan, A. M. Asiri, Electrical conductivity based ammonia sensing properties of polypyrrole/MoS₂ nanocomposite, *Polymers.*, 12 (2020) 3047–3060.

[117]. D. Gopalakrishnan, D. Damien, M. M. Shaijumon, MoS₂ quantum dot-interspersed exfoliated MoS₂ nanosheets, *ACS Nano.*, 8 (2014) 5297–5303.

[118]. X. Gan, H. Zhao, X. Quan, Two-dimensional MoS₂: A promising building block for biosensors, *Biosens. Bioelectron.*, 89 (2017) 56–71.

[119]. S. V. Vattikuti, C. Byon, Synthesis and characterization of molybdenum disulfide nanoflowers and nanosheets: Nanotribology, *J. Nanomater.*, 5 (2015) 1–11.

[120]. N. A. Niaz, A. Shakoor, M. Imran, N. R. Khalid, F. Hussain H. Kanwal, M. Maqsood, S. Afzal, Enhanced electrochemical performance of MoS2/PPy nanocomposite as electrodes material for supercapacitor applications, *J. Mater. Sci. Mater. Electron.*, 31 (2020) 11336–11344.

[121]. M. Lian, X. Wu, Q. Wang, W. Zhang, Y. Wang, Hydrothermal synthesis of Polypyrrole/MoS₂ intercalation composites for supercapacitor electrodes, *Ceram. Int.*, 43 (2017) 9877–9883.

[122]. R. Mahore, D. K. Burghate, S. Kondawar, Development of nanocomposites based on polypyrrole and carbon nanotubes for supercapacitors, *Adv Mater Lett.*, 5 (2014) 400–405.

[123]. A. Moyseowicz, K. Pajak, K. Gajewska, G. Gryglewicz, Synthesis of polypyrrole/reduced graphene oxide hybrids via hydrothermal treatment for energy storage applications, *Mater.*, 13 (2020) 2273–2286.

[124]. Y. C. Eeu, H. Lim, S. P. Lim, S. A. Zakarya, Electrodeposition of polypyrrole/reduced graphene oxide/iron oxide nanocomposite as supercapacitor electrode material, *J. Nanomater.*, 9 (2013) 1–6.

[125]. M. Ates, I. Mizrak, O. Kuzgun, S. Aktas, Synthesis, characterization, and supercapacitor performances of activated and inactivated rGO/MnO₂ and rGO/MnO₂/PPy nanocomposites, *Ionics.*, 11 (2020) 1–10.

[126]. A. Davies, P. Audette, B. Farrow, F. Hassan, Z. Chen, J-Y Choi, A. Yu, Graphene-based flexible supercapacitors: Pulse-electropolymerization of polypyrrole on free-standing graphene films, *J. Phys. Chem. C.*,115 (2011) 17612–17620.

[127]. X. Lu, F. Zhang, H. Dou, C. Yuan, S. Yang, L. Hao, L. Shen, L. Zhang, X. Zhang, Ethylene glycol reduced graphene oxide/polypyrrole composite for supercapacitor, *Electrochimica Acta*, 69 (2012) 160–166.

[128]. S. Dhibar, C. K. Das, Silver nanoparticles decorated polypyrrole/graphene nanocomposite: A potential candidate for next-generation supercapacitor electrode material, *J. Appl. Polym. Sci.*, 134 (2017) 1–14.

[129]. M. Khalaj, A. Sedghi, H. N. Miankushki, S. Z. Golkhatmi, Electropolymerized polypyrrole nanocomposites with cobalt oxide coated on carbon paper for electrochemical energy storage, *Energy*, 188 (2019) 1–15.

[130]. J. Jose, S. P. Jose, T. Prasankumar, S. Shaji, S. Pillai, P. B. Sreeja, Emerging ternary nanocomposite of rGO draped palladium oxide/polypyrrole for high performance supercapacitors, *J. Alloys Compd.*, 4 (2020) 1–34.

[131]. J. Iqbal, A. Numan, M. O. Ansari, R. P. Jagadish, R. Jafer, S. Bashir, S. Mohamad, K. Ramesh, S. Ramesh, Facile synthesis of ternary nanocomposite of polypyrrole incorporated with cobalt oxide and silver nanoparticles for high performance supercapattery, *Electrochim. Acta*, 348 (2020) 1–10.

[132]. J. Li, J. Wang, Y. Wang, Flexible, free-standing TiO₂-graphene-polypyrrole composite films as electrodes for supercapacitors, *J. Chem. Eng. Data*, 52 (3) (2007) 1069–1071.

[133]. S. Kulandaivalu, Y. Sulaiman, Review of the use of transition-metal-oxide and conducting polymer-based fibres for high-performance supercapacitors, *J. Mater. Sci. Mater. Electron.*, 5 (2020) 1–12.

[134]. S. Ramesh, Y. Haldorai, H. S. Kim, J. H. Kim, A nanocrystalline Co₃O₄@ polypyrrole/MWCNT hybrid nanocomposite for high performance electrochemical supercapacitors, *RSC Adv.*, 7 (2017) 1–11.

[135]. M. Samanc, E. Das, A. B. Yurtcan, Carbon aerogel and their polypyrrole composites used as capacitive materials, *Int. J. Energy Res.*, 8 (2020) 1–19.

[136]. C. Bora, J. Sharma, S. Dolui, Polypyrrole/sulfonated graphene composite as electrode material for supercapacitor, *Am. J. Phys. Chem.*, 118 (2014) 29688–29694.

[137]. R. Oraon, A. D. Adhikari, S. K. Tiwari, T. S. Sahu, G. C. Nayak, Fabrication of nanoclay based graphene/polypyrrole nanocomposite: An efficient ternary electrode material for high performance supercapacitor, *Appl. Clay Sci.*, 118 (2015) 231–238.

[138]. A. K. Thakur, R. B. Choudhary, M. Majumder, Gupta, G, Shelke, M. V, Enhanced electrochemical performance of polypyrrole coated MoS2 nanocomposites as electrode material for supercapacitor application, *J. Electroanal. Chem.*, 782 (2016) 278–287.

[139]. X. Zhang, X. Zeng, M. Yang, Y. Qi, investigation of a branchlike moo₃/polypyrrole hybrid with enhanced electrochemical performance used as an electrode in supercapacitors, *ACS Appl. Mater. Interfaces.*, 6 (2014) 1125–1130.

[140]. W. Sun, Z. Mo, PPy/graphene nanosheets/rare earth ion6s: A new composite electrode material for supercapacitor, *Mater. Sci. Eng. C.*, 178 (2013) 527–532.

[141]. S. Paul, Y. -S. Lee, J. -A. Choi, Y. C. Kang, D. -W. Kim, A facile synthesis of polypyrrole/carbon nanotube composites with ultrathin, uniform and thickness-tunable polypyrrole shells, *Bull Korean Chem Soc B.*, 31 (2010) 1228–1232.

[142]. Y. Xie, D. Wang, D. Y. Zhou, H. Du, C. Xia, Supercapacitance of polypyrrole/titania/polyaniline coaxial nanotube hybrid, *Synth. Met.*, 198 (2014) 59–66.

[143]. X. Lu, F. Zhang, H. Dou, C. Yuan, S. Yang, L. Hao, X. Zhang, Preparation and electrochemical capacitance of hierarchical graphene/polypyrrole/carbon nanotube ternary composites, *Electrochim. Acta.*, 69 (2012) 160–166.

[144]. D. P. Dubal, S. H. Lee, J. G. Kim, W. B. Kim, C. D. Lokhande, Porous polypyrrole clusters prepared by electropolymerization for a high performance supercapacitor, *J. Mater. Chem.*, 22 (2014) 3044–3055.

[145]. K. Sun, S. Zhang, P. L. Y. Xia, X. Zhang, D. Du, F. H. Isikgor, J. O. Sun, Review on application of PEDOTs and PEDOT: PSS in energy conversion and storage devices, *J Mater Sci: Mater Electron.*, 1 (2015) 1–25.

[146]. J. Liu, J. Xu, Y. Chen, W. Sun, X. Zhou, J. Ke, Synthesis and electrochemical performance of a PEDOT: PSS@Ge composite as the anode materials for lithium-ion batteries, *Int. J. Electrochem. Sci.*, 14 (2019) 359–370.

[147]. Y. Liu, S. Liu, L. Li, C. Zhang, T. Liu, Conducting polymer composites: Material synthesis and applications in electrochemical capacitive energy storage, *Mater. Chem. Front.*, 1 (2017) 251–268.

[148]. X. Li, C. Zhou, L. Shen, W. Zhou, J. Xu, C. Luo, J. Hou, R. Tan, F. Jiang, Synthesis and electrochemical performance of a PEDOT: PSS@Ge composite as the anode materials for lithium-ion batteries *Int.* Synthesis and electrochemical performance of a PEDOT: PSS@Ge composite as the anode materials for lithium-ion batteries, *J. Electrochem. Sci.*, 14 (2019) 4632–4642.

[149]. J. Song, G. Ma, F. Qin, L. Hu, B. Luo, T. Liu, X. Yin, Z. Su, Z. Zeng, Y. Jiang, G. Wang, Z. Li, Graphene and poly (3,4-ethylenedioxythiophene) (PEDOT) based hybrid supercapacitors with ionic liquid gel electrolyte in solid state design and their electrochemical performance in storage of solar photovoltaic generated electricity, *Polymers.*, 12 (2020) 450–460.

[150]. G. J. Adekoya, R. E. Sadiku, Y. Hamam, S. S. Ray, B. W. Mwakikunga, O. Folorunso, O. C. Adekoya, O. J. Lolu, O. F. Biotidara, Pseudocapacitive material for energy storage application: PEDOT and PEDOT: PSS, *AIP Conference Proceedings*, 2289 (2020) 020073.

[151]. Z. Zhao, F. Georgia, Q. Zhu, S. Kuan, H. Chiang, *Ma*, PEDOT-based composites as electrode materials for supercapacitors, *Nanotechnology*, 27 (2016) 1–12.

[152]. D. Choa, C. Zhu, S. Fu, D. Du, M. H. Engelhard, Y. Lin, Electrochemically controlled ion-exchange property of carbon nanotubes/polypyrrole nanocomposite in various electrolyte solutions, *Electroanalysis*, 29 (2017) 929–936.

[153]. H. Zhou, H. -J. Zhai, G. Han, Superior performance of highly flexible solid-state supercapacitor based on the ternary composites of graphene oxide supported poly(3,4-ethylenedioxythiophene)-carbon nanotubes, *J. Power Sources.*, 323 (2016) 125–133.

[154]. H. Zhou, G. Han, D. Fu, Y. Chang, Y. Xiao, H.-J. Zhai, Petal-shaped poly(3,4-ethylenedioxythiophene)/sodium dodecyl sulfate-graphene oxide intercalation composites for high-performance electrochemical energy storage, *J. Power Sources.*, 272 (2014) 203–210.

[155]. P. Tang, L. Han, L. Zhang, FacileSynthesisof graphite/PEDOT/MnO₂ composites on commercial supercapacitor separator membranes as flexible and high-performance supercapacitor electrodes, *ACS Appl. Mater. Interfaces*, 6 (2014) 10506–10515.

[156]. D. Y. Fu, H. W. Li, X. M. Zhang, G. Y. Han, H. H. Zhou, Y. Z. Chang, Mater. Flexible solid-state supercapacitor fabricated by metal organic framework/graphene oxide hybrid interconnected with PEDOT, *Chem. Phys.*, 179 (2016) 166–173.

[157]. Y. Xie, H. Du, C. Xia, Porous poly(3,4-ethylenedioxythiophene) nanoarray used for flexible supercapacitor, *Micropor. Mesopor. Mat.*, 204 (2015) 163–172.

[158]. G. Qu, J. Cheng, X. Li, D. Yuan, P. Chen, X. Chen, B. Wang, H. Peng, A fiber supercapacitor with high energy density based on hollow graphene/conducting polymer fiber electrode, *Adv. Mater.*, 28 (2016) 3646–3652.

[159]. D. Fu, H. Zhou, X. -M. Zhang, G. Han, Y. Chang, H. Li, Flexible solid–state supercapacitor of metal–organic framework coated on carbon nanotube film interconnected by electrochemically-codeposited PEDOT-GO, *Chemistry Select.*, 1 (2016) 285–289.

[160]. T. S. Sonia, P. A. Mini, R. Nandhini, K. Sujith, B. Avinash, S. V. Nair, K. R. V. Subramanian, Composite supercapacitor electrodes made of activated carbon/PEDOT: PSS and activated carbon/doped PEDOT, *Bull. Mater. Sci.*, 36 (2013) 547–551.

[161]. H. J. Sim, C. Choi, D. Y. Lee, H. Kim, J. -H. Yun, J. M. Kim, T. M. Kang, R. Ovalle, R. H. Baughman, C. W. Kee, S. J. Kim, Biomolecule based fiber supercapacitor for implantable device, *Nano Energy.*, 47 (2018) 385–392.

[162]. N. Kurra, R. Wang, H. N. Alshareef, All conducting polymer electrodes for asymmetric solid-state supercapacitors, *J. Mater. Chem.*, 3 (2015) 7368–7374.

[163]. A. M. Obeidat, A. C. Rastogi, Graphene and Poly (3,4-ethylenedioxythiophene) (PEDOT) based hybrid supercapacitors with ionic liquid gel electrolyte in solid state design and their electrochemical performance in storage of solar photovoltaic generated electricity, *MRS Adv.*, 1 (2016) 3565–3571.

[164]. T. Cheng, Y. -Z. Zhang, J. -D. Zhang, W. -Y. Lai, W. Huang, High-performance free-standing PEDOT: PSS electrodes for flexible and transparent all-solid-state supercapacitors, *J. Mater. Chem.*, 4 (2016) 10493–10499.

[165]. S. Lehtimäki, M. Suominen, P. Damlin, S. Tuukkanen, C. Kvarnström, D. Lupo, preparation of supercapacitors on flexible substrates with electrodeposited PEDOT/graphene composites, *ACS Appl. Mater. Interfaces.*, 7 (2015) 22137–22147.

[166]. Y. Zhu, N. Li, T. Lv, Y. Yao, H. Peng, J. Shi, S. Cao, T. Chen, Ag-Doped PEDOT: PSS/CNT composites for thin-film all-solid-state supercapacitors with a stretchability of 480%‡, *J. Mater. Chem. A*, 6 (2018) 941–947.

[167]. M. Tahir, L. He, W. A. Haider, W. Yang, X. Hong, Y. Guo, X. Pan, H. Tang, Y. Li, L. Mai, Co-electrodeposited porous PEDOT–CNT microelectrodes for integrated micro-supercapacitors with high energy density, high rate capability, and long cycling life‡, *Nanoscale*, 11 (2019) 7761–7770.

[168]. QinYang, S.-K. Pang, K.-C. Yung, An ultra-microporous carbon material boosting integrated capacitance for cellulose-based supercapacitors, *J. Electroanal. Chem.*, 768 (2014) 1–10.

[169]. L. Manjakkal, A. Pullanchiyodan, N. Yogeswaran, E. S. Hosseini, R. Dahiya, A Wearable supercapacitor based on conductive PEDOT: PSS-coated cloth and a sweat electrolyte, *Adv. Mater.*, 32 (2020) 1–13.

[170]. X. Gao, L. Zu, X. Cai, C. Li, H. Lian, Y. Liu X. Wang, X. Cu, High performance of supercapacitor from PEDOT: PSS electrode and redox iodide ion electrolyte, *Synth. Met.*, 228 (2017) 84–90.

[171]. D. Zhao, Q. Zhang, W. C. Xin Yi, S. Liu, Q. Wang, Y. Liu, J. Li, X. Li, H. Yu, High performance of supercapacitor from PEDOT: PSS electrode and redox iodide ion electrolyte, *ACS Appl Mater Interfaces.*, 9 (2017) 13213–13222.

[172]. D. Ni, Y. Chen, H. Song, C. Liu, X. Yang, K. Cai, Free-standing and highly conductive PEDOT nanowire films for high-performance all-solid-state supercapacitors‡, *J. Mater. Chem. A.*, 7 (2019) 1323–1333.

[173]. D. Villers, D. Jobin, C. Soucy, D. Cossement, R. Chahine, L. Breau, D. Be´langer, Microwave assisted synthesis of MnO_2 on Nickel foam-graphene for electrochemical capacitor, *J. Electrochem. Soc.*, 150 (2013) 747–752.

[174]. W. Li, J. Chen, J. Zhao. C. Peng, G. A. Snook, D. J. Fray, M. S. P. Shaffer, G. Z. Chen, Carbon nanotube and conducting polymer composites for supercapacitors, *Chem. Commun.*, 5 (2006) 4629–4631.

[175]. J. Wang, Y. Xu, X. Chen, X. Du, Electrochemical supercapacitor electrode material based on poly(3,4-ethylenedioxythiophene)/polypyrrole composite, *J. Power Sour.*, 163 (2007) 1120–1125.

[176]. L.-M. Huang, T.-C. Wen, A. Gopalan, Electrochemical and spectroelectrochemical monitoring of supercapacitance and electrochromic properties of hydrous ruthenium oxide embedded poly(3,4-ethylenedioxythiophene)–poly(styrene sulfonic acid) composite, *Electrochim. Acta.*, 51 (2006) 3469–3476.

[177]. A. Vadivel Murugan, Novel organic–inorganic poly (3,4-ethylenedioxythiophene) based nanohybrid materials for rechargeable lithium batteries and supercapacitors, *J. Power Sources.*, 159 (2006) 312–318.

[178]. F. Quaranta, P. Siciliano, R. Rella, Gas sensing measurements and analysis of the optical properties of poly[3-(butylthio)thiophene] Langmuir–Blodgett films, *Sens. Actuators B*, 68 (2000) 203–209.

[179]. M. Aldissi, M. K. Ram, O. Yavuz, NO_2 gas sensing based on ordered ultrathin films of conducting polymer and its nanocomposite, *Synth. Met.*, 151 (2005) 77–84.

[180]. A. Husain, S. Ahmad, F. Mohammad, Preparation and applications of polythiophene nanocomposites, *Int. J. Eng. Comput. Sci.*, I (2020) 36–53.

[181]. M. Biswas, N. Ballav, Preparation and evaluation of a nanocomposite of polythiophene with Al_2O_3, *Polym. Int.*, 52 (2003) 179–184.

[182]. H. Vijeth, M. Niranjana, L. Yesappa, S. P. Ashokkumar, H. Devendrappa, Polythiophene nanocomposites as high performance electrode material for supercapacitor application, *AIP Conference Proceedings*, 1942 (2018) 140017.

[183]. H. Vijeth, S. P. Ashokkumar, L. Yesappa, M. Niranjana, M. Vandana, H. Devendrappa, Flexible and high energy density solid-state asymmetric supercapacitor based on polythiophene nanocomposites and charcoal, *RSC Adv.*, 8 (2018) 31414–31426.

[184]. M. Azimi, M. Abbaspour, A. Fazli, H. Setoodeh, B. Pourabbas, Investigation on electrochemical properties of polythiophene nanocomposite with graphite derivatives as supercapacitor material on breath figure-decorated PMMA electrode, *J. Electron. Mater.*, 47 (2017) 2093–2102.

[185]. K. Balakrishnan, M. Kumar, S. Angaiah, Synthesis of polythiophene and its carbonaceous nanofibers as electrode materials for asymmetric supercapacitors, *Adv Mat Res.*, 938 (2014) 151–157.

[186]. C. Fu, H. Zhou, R. Liu, Z. Huang, J. Chen, Kuang, Supercapacitor based on electropolymerized polythiophene and multi-walled carbon nanotubes composites, *Mater. Chem. Phys.*, 132 (2012) 596–600.

[187]. M. Ates, S. Caliskan, E. Ozten, *Fuller.* A ternary nanocomposite of reduced graphene oxide, Ag nanoparticle and Polythiophene used for supercapacitors, *Nanotub*, 26 (2018) 360–369.

[188]. F. Alvi, P. A. Basnayaka, M. K. Ram, H. Gomez, E. Stefanako, Y. Goswami, A. Kumar, A Review of supercapacitor energy storage using nanohybrid conducting polymers and carbon electrode materials, *J. New Mater. Electrochem. Syst.*, 15 (2012) 89–95.

[189]. P. Pascariu, D. Vernardou, M. Suchea, A. Airinei, L. Ursu, S. Bucur, E. Koudumas, synthesis of polythiophene nanoparticles by surfactant-assisted dilute polymerization method for high performance redox supercapacitors, *Mater. Des.*, 5 (2019) 108027–108036.

[190]. S. Dhibar, P. Bhattacharya, D. Ghosh, G. Hatui, C. K. Das, Graphene–single-walled carbon nanotubes–poly(3-methylthiophene) ternary nanocomposite for supercapacitor electrode materials, *Ind. Eng. Chem. Res.*, 53 (2014) 13030–13045.

[191]. J. P. Melo, E. N. Schulz, C. M. Verdejo, S. L. Horswell, M. B. Camarada, Synthesis and characterization of graphene/polythiophene (GR/PT) nanocomposites: Evaluation as high-performance supercapacitor electrodes, *Int. J. Electrochem. Sci.*, 12 (2017) 2933–2948.

[192]. A. Alabadi, S. Razzaque, Z. Dong, W. Wang, B. Tan, Graphene oxide-polythiophene derivative hybrid nanosheet for enhancing performance of supercapacitor, *J. Power Sources.*, 306 (2016) 241–247.

[193]. P. Bhattacharya, C. K. Das, Poly(3-methylthiophene)/graphene composite: *In-situ* synthesis and its electrochemical characterization, *J. Nanosci. Nanotechnol.*, 12 (2012) 7173–7180.

[194]. K. Lota, V. Khomenko, E. Frackowiak, Capacitance properties of poly(3,4-ethylenedioxythiophene)/ carbon nanotubes composites, *J. Phys. Chem. Solids*, 65 (2004) 295–301.

[195]. X. Bai, X. Hu, S. Zhou, J. Yan, C. Sun, P. Chen, L. Li, Macroporous carbon/nitrogen-doped carbon nanotubes/polyaniline nanocomposites and their application in supercapacitors, *Electrochim. Acta.*, 87 (2013) 394–400.

[196]. P. Sivaraman, A. P. Thakur, K. Shashidhara, All solid supercapacitor based on polyaniline and cross-linked sulfonated poly[ether ether ketone], *Synth. Met.*, 259 (2020) 116255–11665.

[197]. A. K. Thakur, M. Majumder, R. B. Choudhary, S. N. Pimpalkar, Supercapacitor based on electropolymerized polythiophene and multiwalled carbon nanotubes composites. *IOP Conf. Ser. Mater. Sci. Eng.*, 149 (2016) 012166–012175.

[198]. E. Bazireh, M. Sharif, Study on the structure and properties of poly(methylmethacrylate)/polypyrrole-graphene oxide nanocomposites, *Polym. Bull.*, 77 (2020) 4537–4553.

3 Graphene-Based Polymeric Supercapacitors

Soma Das

CONTENTS

3.1 INTRODUCTION

Energy storage devices can store energy in various forms, such as kinetic, chemical, electrochemical, electromagnetic, and thermal. Examples of energy storage devices include batteries, fuel cells, and various capacitors such as electrostatic, electrolytic, and electrochemical capacitors (Figure 3.1) [1–2]. These devices can be used in various research fields depending on the cost of fabrication and installation of the system, as well as the energy and power that they can store and use, respectively. Energy storage devices are used in the transportation and consumable electronic devices industries [2]. Supercapacitors are called electrochemical capacitors or ultracapacitors. Supercapacitors have exceptional characteristics, such as smaller size, good power and energy density, and longer cycle life, and because of that they have attracted more attention from researchers than the other electrochemical devices. In recent years supercapacitors have been used in several devices that can generate high power. They are also considered better devices in high-power applications. Compared

DOI: 10.1201/9781003174646-3

Batteries   **Fuel Cells**   **Supercapacitors**

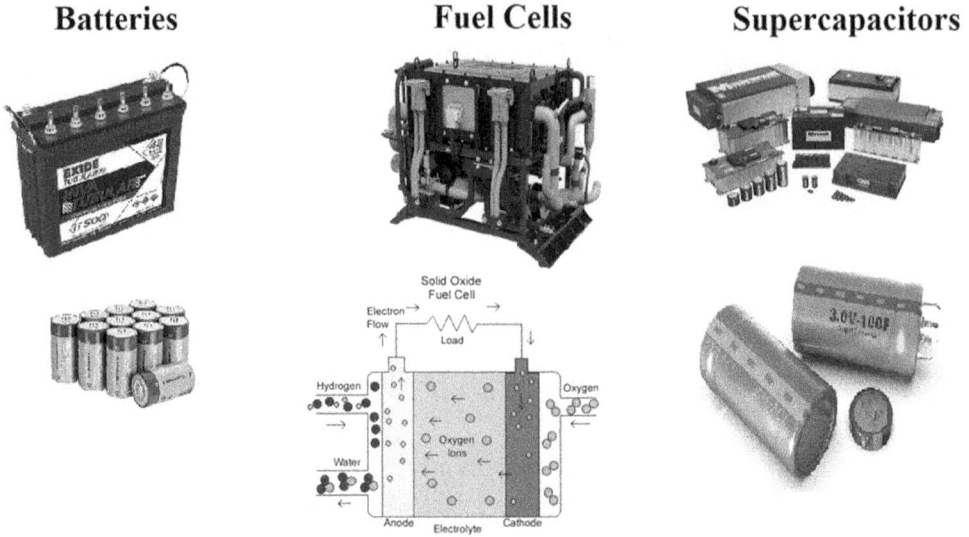

FIGURE 3.1   Energy storage devices.

to batteries, supercapacitors have few limitations in usage because they have high self-discharge, low energy density, and low operating voltage.

Supercapacitors can form connections between conventional capacitors and batteries and can improve energy and power densities; this is the principal reason why they have attracted increasing interest by researchers. In the case of supercapacitors, the electrode materials play the most crucial role to perform better electrochemical reactions [3–4]. The energy density of a supercapacitor can be calculated from the specific surface area, energy stored in per unit mass and volume of the active electrode materials, and, finally, from the operating voltage [3]. It can be increased by following two methods: (i) energy density increases with increasing capacitance and (ii) higher electrolyte voltage can improve the energy density of a supercapacitor. Power density depends on the energy conversion rate of the active electrode material. It varies with the mass, volume, and structure of the electrode materials. During the electrochemical process, ions are transferred to and from the active material surface and power density depends on the rate of transportation of those ions [5–7]. Electron density can be amplified using active electrode materials with high specific capacitance. The properties of electrode materials required for better performance of supercapacitors are as follows:

• High specific surface area
• Controlled pore size distribution
• Electrochemical stability
• Redox reactions
• Surface wettability [5–8]

There are many materials that have been studied as extensively as electrodes, such as conducting polymers, carbon and metal-based nanomaterials, and nanocomposites [1, 4]. These materials show advantages and limitations in the field of supercapacitors for various applications. Larger surface area/volume ratios, high porosity, and their crystalline nature are nanomaterials' unique properties. These properties have recently attracted attention of researchers, and in seeking to improve the performance of supercapacitors, nanomaterials have been used widely. Electrolytic materials having a wide range of operating voltages are also in demand.

Initially in the development of supercapacitors electrochemical double-layer capacitors were first used with carbon electrodes. But to improve the energy density, a considerable amount of research on the designing of supercapacitors, optimization of their operation, and performance levels has been conducted. The concept of pseudocapacitors has since been developed in which metal oxide-based nanomaterials and conducting polymers are employed as electroactive materials. These materials have excellent energy storage capacity. It has been observed that carbon-based nanomaterials like carbon nanotubes (CNTs), carbon nanorods, and graphene have brought great developments in supercapacitors. Abundant opportunities for the scientists have been provided by graphene to fabricate new electrode materials for high-performance supercapacitors [9]. CNTs, carbon nanorods and graphene-based supercapacitors have very high capacitance, power density, and mechanical stability [9–10]. Recently researchers found that graphene and conducting polymer nanocomposite-based supercapacitors have better specific capacitance than electrochemical double-layer supercapacitors. These kinds of supercapacitors also have higher stability than pseudocapacitors.

Several investigations have been made into functionalized carbon-based electrodes in supercapacitors, such as activated carbon, porous carbon, CNTs etc. Functionalized carbon-based compounds exhibit double-layer capacitance [11]. This type of supercapacitor has gained much interest because of the availability and low price of activated carbon. In recent years, mostly activated carbon synthesized from natural raw materials has been used as an electrode in supercapacitors. This type of material has high specific surface area, is economical, and is capable of being mass produced. Despite having high specific surface area, the porous carbon materials cannot be used as electrode materials very effectively because of some key limitations:

- Capacitance of functionalized carbon-based compounds is nonlinear with increasing surface area
- Functionalized carbon materials have low conductivity
- CNT-based supercapacitors have not performed up to the expected levels as they show resistance between the contacted area of current collector and electrodes

Functionalized carbon nanomaterials show low conductivity and low capacitance when applied in supercapacitors. Therefore, to improve the performance of the supercapacitors, numerous studies have been done by researchers on other morphologies of carbon materials.

Graphene, an allotropic form of carbon, is a single carbon layer with unique morphology. Graphene has a 2D structure of carbon atoms, which can be easily converted to 0D buckyballs, 1D nanotube, and 3D graphite by following wrapping, rolling, and stacking, respectively. All the processes are demonstrated in Figure 3.2. It is one of the most researched materials today and it has been found that it is a potential electrode material of supercapacitors [7, 11]. It has been reported that the specific surface area of CNTs and carbon black are lower than graphene, but activated carbon shows similarity with graphene [12].

Highest quality graphene sheets can be yielded by the chemical vapor deposition (CVD) method and those sheets have the least defects but this method is too expensive and is hardly scalable. Therefore, compare to other carbon-based nanomaterials, graphene is not a suitable candidate for supercapacitor applications. Nowadays, double-layer and multiple-layer graphene sheets are studied expansively as electrode materials in supercapacitors. Examples of such materials include graphene nanorods, nanowires, and nanoplatelets. Other graphene derivatives, like functionalized graphene, oxides of graphene, and reduced graphene oxide, are also studied extensively. The advantage of using these forms of graphene materials is that they are synthesized by low-cost methods, such as chemical, mechanical, or thermal methods, but these graphene derivatives have more surface defects and because of that they cannot be used in transistors, photodetectors, or transparent conducting electrodes [13].

Increases in the density of surface defects and an increase in electrochemical capacitance capability have made these graphene derivatives an appropriate candidate for supercapacitor applications [14]. In conclusion, graphene, the allotrope of carbon, with a larger surface area, high specific capacitance, long cycle life, and high porosity, is a better electrode material.

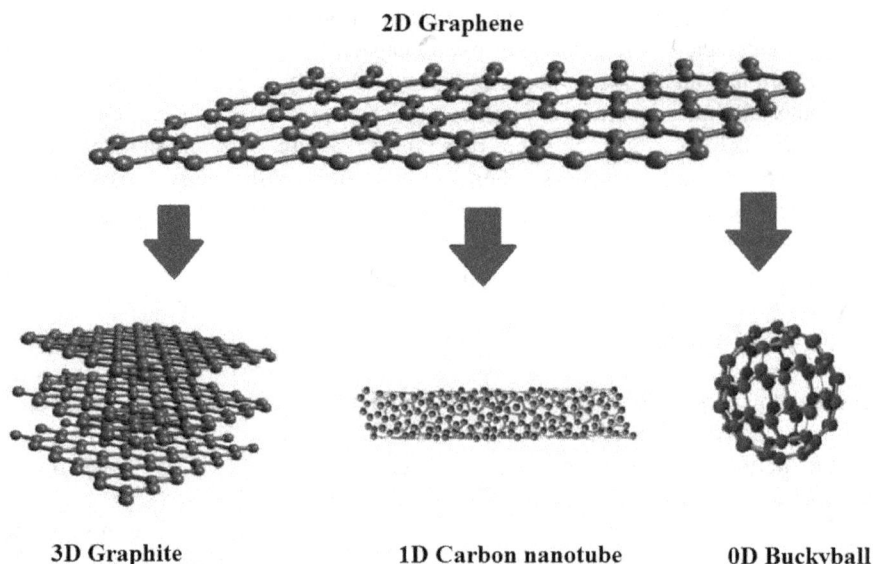

FIGURE 3.2   Conversion of 2D graphene into other forms.

Polymers have gained popularity as supercapacitor electrode materials because of their cost effectiveness, elevated capacitance, superior conductivity, and fast charge-discharge rate. They are very environmentally friendly too. The most commonly used conducting polymers as active electrodes for supercapacitors are polyaniline, polypyrrole, and the mixture of poly(3,4-ethylenedioxythiophene) and poly(styrene sulfonate) [15]. However, during charge and discharge cycles these polymers suffer from poor stability and limited cycle life because they face swelling, shrinkage, cracks, or breaking. The range of potential of the conducting polymer electrode is restricted by oxidative degradation and as a result it worsens their conducting properties. Such difficulties can be resolved by producing polymer-based nanocomposites. CNTs, and other carbon-based nanomaterials are mostly used to produce nanocomposites graphene [16]. The capacitance of the conducting polymers depends on the pH of the solution, nature of the monomers, properties of electrolytes, the substrate, and the deposition condition. Studies show that various nanostructured conducting polymers like nanotubes, nano coils, nanofibers, nanorods, and nanospheres are used as electrode materials in supercapacitors. It has also been reported that these materials can increase power density and specific capacitance effectively in supercapacitors.

Recently, graphene/polymer nanocomposites have attracted enormous attention in various fields of application, including electrochemical sensors, conductive coatings, memory devices, and catalysts. They are also used in various energy conversion and storage devices. The main advantages of using graphene/polymer nanocomposite electrodes in supercapacitors are as follows:

- Electrochemical performance of the nanocomposites increases because nanocomposite electrodes can combine both electrochemical double layer capacitance from graphene and pseudocapacitance from polymers
- Graphene can enhance mechanical, thermal, and chemical stability
- Electrical and thermal conductivity of all polymers can be enhanced by the use of graphene
- During the redox reactions in conducting polymers superior electrical conductivity of graphene assists the electron transfer
- Polymers are used as spacers. In composites, polymers separate graphene sheets from each other and ultimately improve the electrochemical properties

- Polymers act as surfactants. They can improve dispersion and process capacities of graphene layers in solvents [6]
- Pseudocapacitance effects of polymers can enhance the electrochemical performance of electrode materials

Firstly, this chapter will focus on the mechanism for the synthesis of graphene based polymeric supercapacitors. Secondly, the recent advancements in the field of graphene/polymer nanocomposites as electrodes materials to improve performance supercapacitors will be discussed. Measurement techniques to evaluate the supercapacitor performance and use of graphene/polymer nanocomposites in flexible supercapacitors will be briefly presented in subsequent sections, and finally future prospects and conclusions will be drawn.

3.2 DETAILED MECHANISM FOR THE SYNTHESIS OF GRAPHENE-BASED POLYMERIC SUPERCAPACITORS

In last two decades graphene-based supercapacitors have been studied extensively and it has been found that when graphene is functionalized with polymers it can enhance its activities as supercapacitor. Graphene as a bulk material always shows an affinity to form clusters but Stankovich et al. (2006), Dikin et al. (2006), and Wang et al. (2018) reported that performing the oxidation process and then chemical functionalization on the polymer matrix makes graphene disperse easily and also stabilizes it [17–18]. Small functional groups attached to graphene or polymer chains enhance the interactions and help them to bind firmly. Chemical functionalization also improves the solubility of graphene. To functionalize graphene, a plethora of works have been reported in the literature, like polymer wrapping, amination, esterification, salination, and isocyanate modification.

Several methods are used to prepare graphene, including arc discharge method, laser ablation method, CVD technique, reduction of graphene oxide, and electrochemical exfoliation [19]. Laser ablation techniques, High arc discharge, and CVD methods are used for the large-scale production of graphene and the produced graphene has a large surface area. But these methods require elevated temperature in synthesis procedure and the rate of production is also very high. Mechanical cleavage is another procedure that produces graphene samples with superior quality, but it cannot be considered as a suitable process to produce graphene on a large scale. Thin flakes of graphene with low defects can be produced by electrochemical exfoliation methods. However, to yield a reduced form of graphene oxide (rGO), reduction of graphene oxide (GO) is the most commonly used method among all these and this method is employed to prepare graphene/polymer nanocomposites as it is very cost effective. Park & Ruoff (2009) have reported that a soluble form of organo-modified graphene can be yielded by the reduction process of graphene oxide in the presence of stabilizers [20]. Zhu et al. (2011) reported the procedure to get modified graphene. The study found that carboxylic groups first undergo an amidation process to get a modified surface by covalent modification methods [21]. Other methods like nucleophilic substitution reaction of epoxy groups, noncovalent fictionalization method to reduce graphene oxide, and diazonium salt coupling method are also reported in the literature [18].

To synthesize graphene/polymer nanocomposites, two components are made to interact by a noncovalent or covalent approach (Scheme 3.1), and functionalization can be done in a liquid media. Functionalization of dry samples is also reported in the literature [6].

3.2.1 COVALENT APPROACH FOR THE PRODUCTION OF GRAPHENE/ POLYMER NANOCOMPOSITES FOR SUPERCAPACITORS

Graphene is functionalized with polymers by a covalent approach. Covalent functionalization involves various chemical reactions, like condensation reactions, addition reactions, and radical reactions. In both condensation and addition reactions, a surface functional group of the polymeric

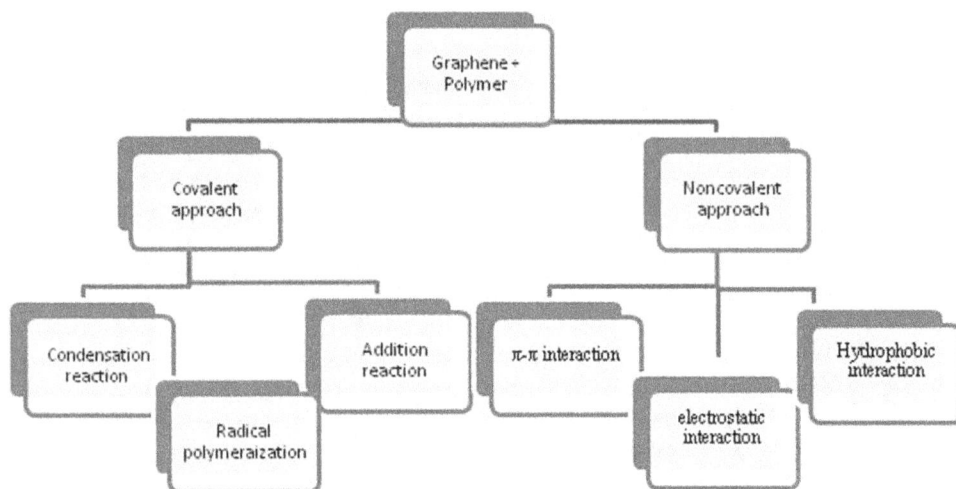

SCHEME 3.1   Covalent and noncovalent approaches for the functionalization of graphene with polymers.

matrix is embedded into the graphene wall to create covalent bonds between graphene and polymer. In radical polymerization, chemically attached initiators initiate polymerization on graphene. For covalent modification of graphene various organic compounds have been extensively used. For instance, alkyl lithium reagents, amines, imines, isocyanates, diisocyanates etc. [19]. In this approach, graphene oxides form carbamate ester and amide bonds with hydroxyl and carboxyl groups, and as a consequence of these reactions, graphene oxide can reduce its hydrophilic nature. Recently, Lonkar et al. (2015) found that through a diazonium addition reaction, the hydroxylated aryl group can be anchored to the graphene surface covalently [22]. Molecular chains of cellulose-rGO-TDI PA6 were fabricated by Xiang et al. (2019). In this reaction imino groups of PA6 and isocyanate groups of conductive cellulose-rGO-TDI undergo covalent reactions and get bonded. As a consequence, imino groups strongly attached to the cellulose backbone [23]. Compared to the noncovalent approach, covalent modification has attracted the attention of many researchers as it offers comparatively more stable and stronger composite materials [24].

3.2.2   NONCOVALENT APPROACH FOR THE PRODUCTION OF GRAPHENE/ POLYMER NANOCOMPOSITES FOR SUPERCAPACITORS

Generally, to obtain noncovalently functionalized graphene/polymer nanocomposites adsorption of polymeric molecules on the surface of graphene by π-π, hydrogen bonding, and electrostatic interactions are done. These approaches are employed to alter the properties of graphene without any kind of alteration in its chemical structure. Extended π electron systems of graphene would not be changed in a noncovalent functionalization method, and their physical properties like electrical conductivity and mechanical strength of graphene are not affected by it. These methods are also used to maintain the intrinsic properties of graphene and make it more dispersible and processible in the presence of solvents. For example, sulfonated polyaniline helps reduced graphene oxide nanosheets to be dispersed firmly in water and this is due to the presence of π-π interactions between these compounds [15]. These approaches can be completed by thoroughly mixing of all the components, or via a polymerization process by adding graphene into the solution. Other methods like vacuum filtration and layer-by-layer assembly methods are also employed to modify graphene surface. In this method agglomeration of graphene sheets can also be stopped by adding a polymeric surface functionalizing agent, for instance, poly(sodium 4-styrenesulfonate). The functionalizing agents are binded to the surface of the graphene sheets by π-π interactions, hydrogen bonding, and ionic interactions. Tensile strength, Young's modulus, and other mechanical properties of graphene/polymer nanocomposite can be improved by this technique. For example, it has been shown that when polyvinyl alcohol (PVA) is mixed with rGO to obtain PVA/rGO nanocomposites through π-π interactions and hydrogen bonding, Young's modulus value is increased by 55% and tensile strength is increased by 48% [18–19]. Moreover, restriction on the mobility of GO and rGO sheets on the top of the polymeric chains enhanced the thermal properties of the nanocomposites.

3.2.2.1   π-π Interactions

The π-π interactions have been studied vastly for the production of graphene/polymer nanocomposites for supercapacitors and mostly this method is used for noncovalent interactions. This kind of interaction occurs between the molecules that have overlapping πorbitals. In the literature it has been reported that π-π interaction can improve electromagnetic interference, tensile strength, Young's modulus value, and other mechanical properties of graphene/polymer nanocomposites for supercapacitors [25]. To study the change in mechanical properties Wang et al. (2018) synthesized graphene/polymer nanocomposites by a solution mixing method. Poly (vinyl alcohol) (PVA) and rGO, which was functionalized by a noncovalent method, were mixed thoroughly in the presence of a surface modifying agent poly(sodium 4-styrenesulfonate) (PSS). It was used to prevent the agglomeration of graphene layers. The result suggested that because of high π-π interaction and hydrogen bonding compared to PVA, PVA/rGO nanocomposites showed 48% more tensile strength and 55% more Young's modulus value [19].

3.2.2.2   Hydrogen Bonding

Researchers have found that hydrogen bonding happens in between the oxygen atom and polar groups. Graphene sheets contain oxygen atoms, and polar groups come from polymeric molecules [5]. Gupta et al. (2016) reported that the γ-radiolysis method was employed to functionalize reduced graphene oxide and poly(ethylene glycol)200 (PEG 200) was used for that purpose. The study also reported that hydrogen bonds were formed in between the oxygen atoms and hydroxyl groups. rGO was the provider of the hydroxyl groups and PEG 200 molecules contain oxygen atoms. The results also showed an enlargement into the spacing of the graphene sheets. The defect density of the entire carbon framework in the reduced graphene sheets can also be reduced by this kind of interaction [23, 26].

3.2.2.3   Electrostatic Interactions

Electrostatic interactions are another vastly used noncovalent approach to synthesize graphene/polymer nanocomposites for supercapacitor applications. It has been reported that electrostatic interactions can occur between rGO and functionalized polymers. For instance, Choi et al. (2010) showed that when amine-terminated polystyrene was used to synthesize rGO/polymer composite, the hydrophilic rGO transformed into a lipophilic composite, which was dispersed in an organic solvent. Electrostatic interactions methods have gained popularity because they can significantly increase the water resistance, anticorrosive properties, and tensile strength of graphene/polymer nanocomposites [27].

3.3   GRAPHENE/POLYMER NANOCOMPOSITES FOR SUPERCAPACITOR APPLICATION

Recently polymeric molecules with high electrical conductivity and good pseudocapacitance have received considerable attention for supercapacitors. Polymer binders are required to connect graphene sheets on the current collector and among them, polyaniline, polypyrrole, and Poly(3,4-ethylenedioxythiophene) are mostly used.

3.3.1 Graphene/Polyaniline Supercapacitors

Polyaniline (PANI) is an example of a vastly studied polymer for supercapacitor applications, and this is because of its elevated specific pseudocapacitance of 2000 Fg^{-1}, acid-base doping/de-doping properties, and unique redox properties. Usually, for the production of PANI, polymerization techniques like chemical and electrochemical have been employed. Other techniques like emulsion, seeding, and template polymerization are also used to synthesize nanostructured PANI. The sp^2 hybridized nitrogen atoms in imine, sp^3 hybridized nitrogen atoms in amine, and conjugated double bonds in backbone structure determine the conducting properties of PANI.

PANI has high conductivity, good electroactivity, high specific capacitance, and exceptional stability and thus is extensively utilized as electrode material in supercapacitors [7]. Acidic solutions or protic ionic liquids are required for PANI to be used in supercapacitors because proton helps PANI in conduction and to be properly charged and discharged during the cell reaction. Thus, polymerization reaction of aniline in the presence of graphene in acidic medium has gained more popularity for preparing graphene/PANI nanocomposites for supercapacitors. Zhang et al. (2010) prepared graphene/PANI nanocomposites by vigorous mixing of graphene oxide and PANI, and a glassy carbon electrode was used for electrochemical detection. In the first step, a graphene oxide layer-like structure was synthesized using graphite by Hummer's method and the oxidative polymerization method was employed to produce PANI nanofibers. In that method 1 M aqueous HCl acidic solution was used as reaction media and ammonium peroxydisulfate ((NH$_4$)$_2$S$_2$O) was employed as an oxidizing agent. Different concentrations of graphene oxide ingraphene/PANI nanocomposite give different composite morphologies and alter the electrochemical behaviors of the supercapacitor electrode [28].

The contribution of graphene oxide to total capacitance is very little because it is insulating in nature. Therefore, the pseudocapacitance of the PANI nanofibers is responsible for the overall capacitance of the composite. To overcome these limitation researchers started using reduced graphene oxide as it has higher specific capacitance. Graphene oxide nanoparticles undergo reduction reaction to prepare reduced graphene oxide and this method is based on various reaction conditions. Wang et al. (2010) synthesized rGO/PANI nanocomposites by a three-step method known as the "polymerization-reduction and dedoping-redoping" method [29]. The process is as follows: in the first step graphene oxide was vigorously mixed with ethylene glycol and then ultrasonicated to produce suspension of graphene oxide. The next step is drop wise addition of aniline solution with constant stirring. HCl and ammonium persulfate were added thereafter to perform polymerization

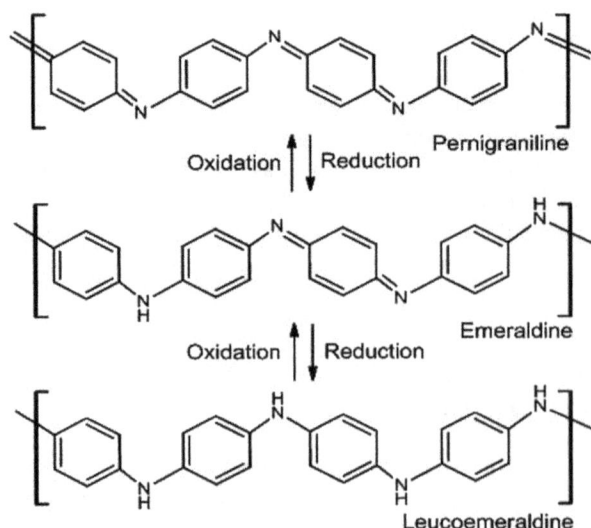

FIGURE 3.3 Chemical structure of polyaniline.

reaction and to produce GO/PANI nanocomposites. For the reduction of rGO preheated NaOH was then mixed into the GO suspension. Simultaneously PANI polymerization was de-doped to produce reduced graphene oxide(rGO) and de-doped PANI composites. Finally, HCl was added again for the re-doping process of PANI and to yield rGO-re-doped PANI composites. Thicknesses of synthesized rGO/PANI nanocomposites were 30–40 nm and size was negligible, which ultimately confirmed a larger specific area of all the nanocomposites mentioned earlier. The results also showed elevated redox peaks of rGO/PANI electrode with higher specific capacitance. The cycling retention capacity of rGO/PANI supercapacitor electrode was also progressed [7].

Following the electrochemical behavior of graphene/PANI composites Lai et al. (2012) examined the surface chemistry of graphene as electrode material of supercapacitor. GO, rGO, amine, and nitrogen doped rGO were employed as carriers. Noncovalent interactions were approached to bind these materials with PANI. Researchers reported that the surface chemistry not only controlled the development of PANI but also altered the specific capacitance. Interestingly the results showed that the amine-modified rGO has the highest specific capacitance (500 Fg^{-1}) at a very low scan rate. When nitrogen-doped rGO employed as cathodic material and amine-modified rGO/PANI composite as anodic, the supercapacitor cell gave a specific capacitance of 79 Fg^{-1}. This result also confirmed the significance of surface chemistry of graphene in the final determination of the electrochemical behavior [15, 30].

3.3.2 GRAPHENE/POLYPYRROLE-BASED SUPERCAPACITORS

Polypyrrole (PPy) is another example of a conducting polymer that attracted the attention of the researchers due to its easy production, low cost, thermal and chemical stability, high mechanical strength, flexibility, and high pseudocapacitance. Thus, it is extensively utilized as electrode material in supercapacitors. In the last few years, graphene/PPy nanocomposite electrodes have gained popularity as electrode material in supercapacitors. Solid configuration of conductive iodine doped PPy was first demonstrated by Weiss et al. in 1963 by chemical oxidative polymerization [31]. PPy shows excellent conductivity and the conductivity mechanism of the polypyrrole chain is shown in Figure 3.4. Interestingly, through doping the polymeric chain can change its conductivity from

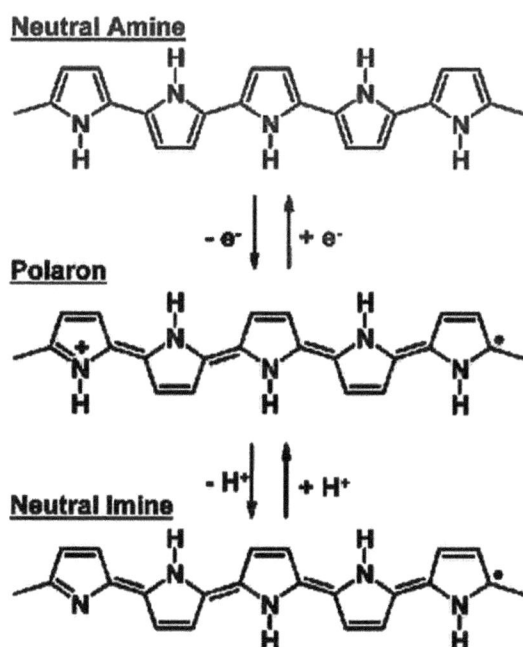

FIGURE 3.4 Conductivity of polypyrrole.

an insulating state to a metallic state. PPy is an insoluble polymer and has an inflexible polymeric backbone structure. In the last few years scientists tried to synthesize soluble PPy and many methods have been tried to produce it. According to the literature, dopant solution is prepared in alcohol or organic solvent and then mixed with organic compound with long alkyl chains, which is further subjected to polymerization reaction, to produce PPy. Example of long alkyl chain compound is, sodium bis (2-ethylhexyl) sulfosuccinate ($C_{20}H_{37}NaO_7S$).

Various graphene derivatives such as GO, rGO, amine- or imine functionalized rGO, and N-doped rGO were used to prepare nanocomposites with PPy. Graphene derivatives were merged in PPy and employed as electrode material in supercapacitors. Systematic investigations on the functionalization process on the surface and electrochemical behavior were done using this method. Literature said that, N-doping of graphene provides the best electrochemical performance than other graphene derivatives because it can enhance the efficiency to transfer electrons and also makes the surface more wettable. Experimental results showed that N-doped rGO/PPy electrode showed larger specific capacitance (394 Fg^{-1}) than others [6, 32].

Qu et al. prepared 3D noncovalent rGO/PPy foam, which can tolerate greater stress and strains without dropping elasticity and any structural deformation. These kinds of foams were used as electrodes in a highly compressible supercapacitor and showed the specific capacitance value of 360 Fg^{-1}. This work demonstrates how to fabricate supercapacitor devices with exceptional tolerance to high mechanical compression [33]. Kashani et al. synthesized Graphene/PPy composites by depositing PPy on porous nanotubular graphene electrochemically and itgave509 Fg^{-1} specific capacitance value [34]. Zhang et al. made rGO/PPy composite based flexible paper using GO and PPy. In the first step of the procedure both components were mixed thoroughly and then the vacuum filtration method applied and finally they were reduced chemically. This composite paper was employed for supercapacitor applications as binder-free electrodes and interestingly showed good capacitance of 175 $mFcm^{-2}$ [15]. Bose et al. [35] observed that compared to pure PPy film graphene nanosheet/PPy composites have much higher specific capacitance. As an electrode material, it has a better cycle life also. Literature provides a lot of evidence that the amalgamation of graphene and PPy not only enhance electrochemical properties of PPy but also strengthen it mechanically during the charge and discharge cycles by improving the structural stability of the composites.

3.3.3   GRAPHENE/POLY(3,4-ETHYLENEDIOXYTHIOPHENE)-BASED SUPERCAPACITORS

Poly(3,4-ethylenedioxythiophene) (PEDOT) was first grown in Bayer's laboratory in 1980s in Germany [5]. PEDOT was synthesized by polymerization of 3,4-ethylenedioxythiophene (EDOT). Other methods like electrochemical or chemical techniques can be employed to synthesize PEDOT. In both the methods oxidants are necessary to carry out the reactions. However, PEDOT has notably less solubility in water and therefore initially it was synthesized in a non-aqueous medium or in the presence of surfactant solutions only. But use of water-soluble polyelectrolyte could dodge this problem. Polystyrene sulfonic acid (PSS) was used for this purpose. PEDOT/PSS is one of the most widely used aqueous suspensions and has good conductivity. With a broad potential range and with better mechanical, thermal, and chemical stability than other suspensions, PEDOT has attracted the attention of scientists as a supercapacitor electrode. The cycle life of PEDOT material is quite impressive and ~85% of capacitance retain over 70,000 cycles at ambient temperature. Other advantages of PEDOT are its environment friendliness and lower band gap (1.5–1.6 eV) [5]. The chemical structures of oxidized and reduced forms of PEDOT are shown in Figure 3.5.

However, PEDOT is associated with one disadvantage. The PEDOT supercapacitor has a relatively low specific capacitance because of its high molecular weight. According to the literature, electrochemical, liquid, and vapor phase polymerization methods can be employed to yield a highly conductive PEDOT. The steps involved in liquid phase processes are thoroughly mixing the monomers and adding oxidants and inhibitors. In situ polymerization methods can yield graphene/PEDOT composites without any accumulation of graphene on the surface of the composites.

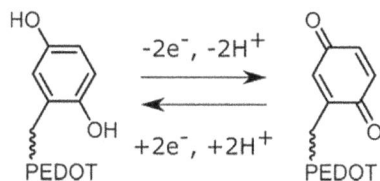

FIGURE 3.5 Oxidized and reduced forms of PEDOT.

The procedure to prepare graphene/PEDOT nanocomposites is as follows: Firstly, the monomers, PSS, and EDOT, are blended together in dilute HCl. Secondly, the degassing process is performed and then graphene or its derivatives are dispersed into the reaction mixture with continuous stirring. And finally, oxidants are added as initiators in the polymerization reaction and to form the graphene/PEDOT nanocomposites. Many oxidants like ammonium peroxydisulfate[$(NH_4)_2S_2O_8$], iron (III) chloride ($FeCl_3$), sodium persulfate ($Na_2S_2O_8$), and iron (III) sulfate[$Fe_2(SO_4)$] are used in this step. It is reported in the literature that when graphene is incorporated with PEDOT, electrical conductivity is improved more than twofold and mechanical strength is simultaneously enhanced sixfold [35–36].

Results have shown that, depending on the polymerization method, PEDOT supercapacitors possess a vast range of specific capacitance (70 to 130 Fg^{-1}) [35]. However, it has been observed that with the incorporation of graphene and its derivatives with PEDOT, both specific capacitance and cycling stability can be enhanced in graphene/PEDOT nanocomposite-based supercapacitors. For instance, Alvi et al. claimed that graphene/PEDOT supercapacitors showed better performance with respect to specific capacitance in the presence of HCl (304 Fg^{-1}) and H_2SO_4 (261 Fg^{-1}) electrolytes. The findings of Wen et al. showed higher specific capacitance of GO/PEDOT and RGO/PEDOT composites. The capacitance value 136 Fg^{-1} was shown by GO/PEDOT electrodes and 209 Fg^{-1} by RGO/PEDOT electrodes [7, 35].

The importance of graphene as an electrode material in polymer composite-based supercapacitors can be outlined as follows:

(i) PEDOT undergoes volumetric changes during the charge-discharge cycle. It swells and shrinks, and therefore it suffers from structural deformations like collapse, peeling off, and cracking. Graphene can form heterogeneous structures with PEDOT, which helps to reduce those structural deformations effectively.

(ii) Compared to PEDOT graphene or RGO are better at electric conducting. Therefore, the hybrid materials of graphene or RGO with PEDOT are more conductive in nature.

(iii) The addition of graphene gives the composite a 3D morphology that ultimately significantly improves the specific capacitance because it can provide a large surface area to the electrolyte to penetrate inside and can improvise redox reactions during the process.

3.3.4 GRAPHENE/OTHER POLYMER-BASED SUPERCAPACITORS

Along with the frequently used polymers discussed in the last section, there are other polymers that are combined with graphene to act as electrode materials for supercapacitor devices.

One interesting example is using cellulose paper in flexible energy storage devices. This has attracted the attention of scientists because of accessibility, cost effectiveness, high mechanical strength, and excellent flexibility. rGO-cellulose composite papers were made by applying vacuum filtration method where GO is dispersed through a filter paper [15]. rGO was adsorbed on a filter paper and formed a conducting framework nearby the cellulose fibers. When electrolytes are absorbed by the cellulose fibers they can act as electrolyte reservoirs that ultimately can help ion transportation. Research has reported that when these membranes are used in flexible supercapacitor devices as electrode materials, they exhibited a specific capacitance of 81 mFcm⁻¹[37].

rGO and polyselenophene-based nanocomposites were yielded by the scientists in Park's lab by polymerizing GO and selenophene monomers, and the resultant nanocomposites exhibited a larger surface area, amazing mechanical strength, and elevated electrical and thermal conductivity [15]. H_2SO_4-PVA gel was applied as a solid electrolyte in a solid-state supercapacitor and the resulting supercapacitor displayed very high capacitance values before and after bending. Conjugated polyfluoreneimidazolium ionic liquids (PILs) were combined with rGO, which is functionalized by noncovalent method. This was done to yield electrode material for high-performance graphene/polymer nanocomposite-based supercapacitors [38].

3.4 MEASUREMENT TECHNIQUES TO EVALUATE THE PERFORMANCE OF A GRAPHENE/POLYMER SUPERCAPACITOR

The efficiency of graphene/polymer supercapacitors need to be evaluated according to a few key parameters. Those performance parameters are life cycle, power density, operating voltage, internal resistance, energy density, capacitance, and time constant. Elementary electrochemical detections can be executed on the graphene/polymer supercapacitor to calculate these parameters. Electrochemical detections, namely, cyclic voltametric (CV) techniques, galvanostatic charging/discharging and, electrochemical impedance spectroscopic (EIS) measurements, are employed to evaluate the performance of supercapacitors.

3.4.1 CYCLIC VOLTAMMETRY

Cyclic voltammetry (CV) is a well-known electrochemical detection method that quantifies the current developed by a system when excess voltage is applied. CV measurements are governed by the Nernst equation. CV is performed in two steps: cycling the potential and quantifying the resulting current. To get the cyclic voltammogram working, the electrode's current is plotted against its potential.

A cyclic voltammogram of a reversible electron transfer reaction is illustrated in Figure 3.6.

CV can be used to interpret the interfacial mechanisms of electrodes and electrolytes. The behavior of both charge transfer and mass transfer methods at the boundary of electrodes and electrolytes

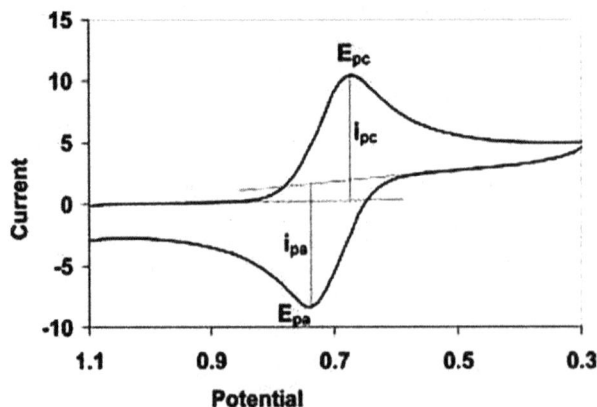

FIGURE 3.6 Cyclic voltammogram of a reversible electron transfer reaction.

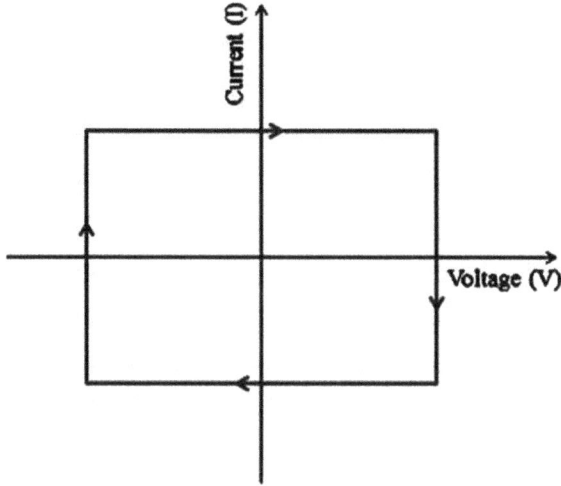

FIGURE 3.7   Cyclic voltammogram of an ideal capacitor.

can be measured qualitatively or quantitatively using CV. The equation used to measure the capacitance of a graphene/polymer supercapacitor using CV measurement is:

$$i = C\left(\frac{dv}{dt}\right)$$

where I = measured average current, C = capacitance, and dv/dt = scan rate.

For all ideal capacitors, rectangular a CV can be obtained (Figure 3.7), but most graphene/polymer supercapacitors show structural deviations. Graphene/polymer-based supercapacitors cannot show rectangular voltammograms due to the resistance offered by the electrolytic solutions and faradaic redox reactions at the electrode surfaces.

3.4.2   GALVANOSTATIC CHARGING/DISCHARGING TECHNIQUE

Specific capacitance, specific power, operating voltage, cycle life, and specific energy of the graphene/polymer supercapacitor can be measured using the galvanostatic charging/discharging technique. Galvanostatic measurement is also called *chronoamperometry*. In this technique, initially, the discharging curve shows a sudden drop, which is also known as the IR-drop, which is due to the presence of internal resistance of the cell. Linearity shows in the discharging curve for only non-faradaic electrode materials, but for pseudocapacitive electrode materials it shows deviation from linearity [5]. The Coulombic efficiency η can be quantified using the following equation:

$$\eta = \left(\frac{td}{tc}\right) \times 100$$

where, td, and tc are the charging and discharging times.

3.4.3   ELECTROCHEMICAL IMPEDANCE SPECTROSCOPIC TECHNIQUE

The electrochemical impedance spectroscopic (EIS) technique is used to understand interfacial behavior and corrosion in graphene/polymer supercapacitors. [5] Several parameters, such as interfacial charging, ohmic resistance, extracting model parameters, diffusion control, charge transfer resistance, equivalent circuit modelling, and mass transfer of graphene/polymer supercapacitors can be calculated using this technique. EIS studies depend on porous electrodes and therefore most of the supercapacitor uses porous electrodes. All the pores have resistance and double layer capacitance. Helpful data related to capacitance can be derived from EIS data using the complex models. Furthermore, EIS analysis is also determined by the frequency-dependent capacitance and the time constant.

3.5 APPLICATION OF GRAPHENE/POLYMER COMPOSITE ELECTRODES IN FLEXIBLE SUPERCAPACITORS

Flexible electronic devices have brought about a revolution in the field of electronics. Compared to conventional electronic devices flexible electronic devices possess several advantages: they are light weight, wearable, bendable, eco-friendly, economical, etc. Several attempts have been made to respond to rapid demand in the market by developing flexible supercapacitors. The electrodes used in those capacitors are mainly carbon-based materials, like carbon nanotubes, carbon nanofibers, activated carbon, and graphene. For various applications, ultrathin graphene film mixed with polyethylene terephthalate (PET) is flexible in nature and it helps to yield flexible electrode material for supercapacitors. But this graphene/PET composite is associated with few limitations such as:

(i) The PET substrate can provide only flexibility and mechanical strength towards the thin graphene sheets but cannot offer high capacitance.
(ii) Electrodes get creases and some more defects during the transfer process of graphene.
(iii) Specific surface area and capacitance of the supercapacitors decrease during the restacking process of graphene.

Other than graphene/PET there are a few more types of electrodes based on graphene materials that are used in flexible supercapacitors, for instance, graphene-based paper/foam/carbon cloth or fabric. But these materials are also associated with limitations like graphene restacking, low scale production, low power and energy density, and high cost. Hence, graphene/conducting polymer (CP) composite film has attracted more attention by researchers. Graphene/CP composite electrodes have more flexibility, are cost effective, and possess good power and energy density. Interestingly, graphene/CP film has a high tolerance for bending and twisting. Moreover, the introduction of CPs can increase the specific capacitance of the device by redox reactions [35].

3.6 CONCLUSION

Undoubtedly the application of graphene has revolutionized the research on supercapacitors because of its exceptional characteristics such as excellent electrochemical behavior, higher specific surface area, excellent electrical conductivity, high mechanical strength, and light weight [7]. When graphene mixes with binder polymers, it neutralizes many undesired properties of insulating polymers, like insulation, smaller specific surface area, and lower capacitance. Graphene provides high mechanical strength to the backbone of the conducting polymers in composites and thus significantly improves the specific capacitance and the cycling performance of the supercapacitor. In addition, flexibility of the graphene/polymer nanocomposite film can produce energy storage devices with high flexibility and wearability. In this chapter, the mechanisms of synthesis, advances in graphene/polymer nanocomposites for supercapacitors, and various measurement techniques to evaluate the performance of graphene/polymer supercapacitors have been sum-marized. However, in spite of all these novel ideas and methods in the area of graphene/polymer supercapacitor devices, there exist some problems that need to be tackled effectively. Challenges include the following:

(i) Graphene is aggregated during preparation and this is because of the strong intra layer π-π interactions. This leads to the lowering of the surface area and electrochemical performance. Finding a suitable route for mass production of low-cost graphene/polymer supercapacitor electrode without any aggregation problem is one of the major challenges.

(ii) Supercapacitor applications demand uniform dispersion of graphene within the polymeric matrix but it is still challenging. One suggested solution for this problem is to develop 3D matrix of graphene/polymer composite with selective features like large surface area, good permeability, and controlled structures.

(iii) There is still discrepancy in the use of number of electrodes into the supercapacitors. Nowadays, two and/or three electrodes are used in graphene/polymer composite-based supercapacitors.

(iv) It has been observed that although the new generation supercapacitors show good power densities, they still have poor energy densities. The reason may be the poor ionic conduc-tivities and constricted potential windows of the electrolytes and this problem might be resolved by the application of novel electrolytes as they possess excellent charge transfer properties, superior conductivities, and larger potential windows.

(v) Electrochemical properties of graphene/polymer nanocomposite-based supercapacitors rely upon the interaction between these materials and surface morphologies. To achieve larger numbers of electroactive sites the arrangement of the molecules and the boundary of the electrode materials should be optimized.

(vi) Another major challenge in the fabrication of fresh electrodes is to advance the current densities, and therefore highly conducting electrode materials are required. The conduc-tivity of any nanocomposite reduces when rGO is used as electrode material because it contains many defects. Thus, today, graphene has become a perfect alternative electrode material and often replaces rGO. The chemical vapor deposition (CVD) method is used to produce high quality graphene on a large scale, but this method is not cost effective.

(vii) In-depth knowledge of the mechanism of energy storage, the structural-electrochemical performance relationships, and the interfacial studies of electrode materials, are also highly desirable. Characterization techniques and theoretical calculations are required to explain these mechanisms.

Scientists have been making efforts to find the solution to the issues discussed in this chapter, and it is expected that in the coming decades they might be able to explain and resolve these issues. But to achieve that ultimate goal, lab researchers and industrialists should work together to solve all the challenges and produce graphene/polymer nanocomposite-based supercapacitors on a large scale. This will lead to generation of highly progressed, clean, proficient, and renewable energy storage devices headed for constructive uses in our day-to-day life.

ACKNOWLEDGEMENTS

The author is grateful to the principal of the Greater Noida Institute of Technology for allowing them to publish the book chapter. The author would also like to thank the Department of Chemistry and CoE Materials Science/Sensors & Nanoelectronics of the CMR Institute of Technology.

REFERENCES

1. Wang, G. Zhang, L. Zhang, J. (2012) 'A review of electrode materials for electrochemical supercapacitors', *Chemical Society Review* 41: 797–828. https://doi.org/10.1039/C1CS15060J
2. Conway, B. E. (1999) *Electrochemical Supercapacitors: Scientific Fundamentals and Technological Applications.* Springer, e-book, ISBN 978-1-4757-3058-6
3. Inagaki, M. Konno, H. Tanaike. (2010) 'Carbon materials for electrochemical capacitors', *Journal of Power Sources* 195: 7880–903. DOI: 10.1016/j.jpowsour.2010.06.036
4. Snook, G. A. Kao, P. Best, A. S. (2011) 'Conducting-polymer-based supercapacitor devices and electrodes', *Journal of Power Sources* 196: 1–12. https://doi.org/10.1016/j.jpowsour.2010.06.084.
5. scholarcommons.usf.edu.
6. univoak.eu.
7. nanoscalereslett.springeropen.com.
8. Basnayaka, P. A. (2013) 'Development of nanostructured graphene/conducting polymer composite materials for supercapacitor applications', Graduate Theses and Dissertations, University of South Florida. http://scholarcommons.usf.edu/etd/4864
9. Huang, Y. Liang, J. Chen, Y. (2012) 'An overview of the applications of graphene based materials in supercapacitors', *Small* 8: 1805–34. https://doi.org/10.1002/smll.201102635
10. Pan, H. Li, J. Feng, Y. (2010) 'Carbon nanotubes for supercapacitor', *Nanoscale Research Letter* 5: 654–68. https://doi.org/10.1007/s11671-009-9508-2
11. Davies, A. Yu, A. (2011) 'Material advancements in supercapacitors: From activated Arbon to Arbon nanotube and graphene', *The Canadian Journal of Chemical Engineering* 89: 1342–57. https://doi.org/10.1002/cjce.20586
12. Bonaccorso, F. Colombo, L. Yu, G. Stoller, M. Tozzini, V. Ferrari, A. C. (2015) '2D materials. Graphene, related two-dimensional crystals, and hybrid systems for energy conversion and storage', *Science* 347: 1–9. DOI: 10.1126/science.1246501
13. Hibino, T. Kobayashi, K. Nagao, M. Kawasaki, S. (2015) 'High-temperature supercapacitor with a proton-conducting metal pyrophosphate electrolyte', *Scientific Reports* 5: 1–7. DOI: 10.1038/srep07903
14. Quinlan, R. A. Cai, M. Outlaw, R. A. Butler, S. M. Miller, J. R. Mansour, A. N. (2013) 'Investigation of defects generated in vertically oriented graphene', *Carbon* 64: 92–100. DOI: 10.1016/j.carbon.2013.07.040
15. Xiaoyan, Z. Paolo, S. (2017) 'Graphene/Polymer Nanocomposites for Supercapacitors' Chemistry of Nanomaterials for energy', *Biology and More* 3: 362–72. https://doi.org/10.1186/s11671-017-2150-5
16. Zhang, J. Zhao, X. S. (2012) 'Conducting polymers directly coated on reduced graphene oxide sheets as high-performance supercapacitor electrodes', *Journal of Physical Chemistry C* 116: 5420–26. https://doi.org/10.1021/jp211474e
17. Stankovich, S. Dikin, D. A. Dommett, G. H. B. Kohlhaas, K. M. Zimney, E. J. Stach, E. A. Piner, R. D. Nguyen, S. T. Ruoff, R. S. (2006) 'Graphene-based composite materials', *Nature* 442: 282–87. https://doi.org/10.1038/nature04969
18. Wang, X. Liu, X. Yuan, H. Liu, H. Liu, C. Li, T. Yan, C. Yan, X. Shen, C. Guo, Z. (2018) 'Non-covalently functionalized graphene strengthened poly(vinyl alcohol)', *Materials and Design* 139: 372–79. DOI: 10.1016/j.matdes.2017.11.023
19. Abdulazeez, T. L. (2020) 'Recent progress in graphene based polymer nanocomposites', *Cogent Chemistry* 6: 1–50. DOI: 10.1080/23312009.2020.1833476
20. Park, S. Ruoff, R. S. (2009) 'Chemical methods for the production of graphenes', *Nature Nanotechnology* 4: 217–22. https://doi.org/10.1038/nnano.2009.58
21. Zhu, J. Li, Y. Chen, Y. Wang, J. Zhang, B. Zhang, J. Blau, W. J. (2011) 'Graphene oxide covalently functionalized with zinc phthalocyanine for broadband optical limiting', *Carbon* 49: 1900–5. https://doi.org/10.1016/j.carbon.2011.01.014
22. Lonkar, S. P. Yogesh, S. D. Ahmed, A. A. (2015) 'Recent advances in chemical modifications of graphene', *Nano Research* 8(4): 1039–74. https://doi.org/10.1007/s12274-014-0622-9
23. Xiang, M. Yang, R. Yang, J. Zhou, S. Zhou, J. Dong, S. (2019) 'Fabrication of polyamide 6/reduced graphene oxide nano-composites by conductive cellulose skeleton structure and its conductive behavior', *Composites Part B: Engineering* 167: 533–43. https://doi.org/10.1016/j.compositesb.2019.03.033
24. Yu, D. Yang, Y. Durstock, M. Baek, J. B. Dai, L. (2010) 'Soluble P3HT-grafted graphene for efficient bilayer- heterojunction photovoltaic devices', *ACS Nano* 4: 5633–40. https://doi.org/10.1021/nn101671t
25. Wei, L. Zhang, W. Ma, J. Bai, S. L. Ren, Y. Liu, C. Simion, D. Qin, J. (2019) 'π- π stacking interface design for improving the strength and electromagnetic interference shielding of ultrathin and flexible water-borne polymer/sulfonated graphene composites', *Carbon* 149: 679–92. https://doi.org/10.1016/j.carbon.2019.04.058

26. Gupta, B. Niranjan, K. Kalpataru, P. Ambrose, A. M. Shailesh, J. Sitaram, D. Ashok, K. T. (2016) 'Effective noncovalent functionalization of poly(ethylene glycol) to reduced graphene oxide nanosheets through γ-radiolysis for enhanced lubrication', *The Journal of Physical Chemistry C* 120: 2139–48. https://doi.org/10.1021/acs.jpcc.5b08762

27. Choi, E. Y. Han, T. H. Hong, J. Kim, J. E. Lee, S. H. Kim, H. W. Kim, S. O. (2010) 'Noncovalent functionalization of graphene with end-functional polymers', *Journal of Materials Chemistry* 20: 1907–11. DOI: 10.1039/B919074K

28. Zhang, K. Zhang, L. L. Zhao, X. Wu, J. (2010) 'Graphene/polyaniline nanofiber composites as supercapacitor electrodes', *Chemistry of Materials* 22: 1392–1401. https://doi.org/10.1021/cm902876u

29. Wang, H. Hao, Q. Yang, X. Lu, L. Wang, X. (2010) 'A nanostructured graphene/polyaniline hybrid material for supercapacitors', *Nanoscale* 2: 2164–70. DOI: 10.1039/c0nr00224k

30. Lai, L. Yang, H. Wang, L. I, B. K. Zhong, J. Chou, H. Chen, L. Chen, W. Shen, Z. Ruoff, R. S. Lin, J. (2012) 'Preparation of supercapacitor electrodes through selection of graphene surface functionalities', *ACS Nano* 6: 5941–5 1. https://doi.org/10.1021/nn3008096

31. Bolto, B. A. McNeill, R. Weiss, D. E. (1963) 'Electronic conduction in polymers. III. Electronic properties of polypyrrole', *Australian Journal of Chemistry* 16: 1090–103. DOI: 10.1071/CH9631090

32. Lai, L. Wang, L. Yang, H. Sahoo, N. G. Tam, Q. X. Liu, J. Poh, C. K. Lim, S. H. Shen, Z. Lin, J. (2012) 'Conducting polymer composites: Material synthesis and applications in electrochemical capacitive energy storage', *Nano Energy* 1: 723–3 1. https://doi.org/10.1039/C6QM00150E

33. Zhao, Y. Liu, J. Hu, Y. Cheng, H. Hu, C. Jiang, C. Jiang, L. Cao, A. Qu, L. (2013) 'Highly compression-tolerant supercapacitor based on polypyrrole-mediated graphene foam electrodes', *Advanced Materials* 25: 591–9 5. https://doi.org/10.1002/adma.201203578

34. Kashani, H. Chen, L. Ito, Y. Han, J. Hirata, A. Chen, M. (2016) 'Bicontinuous nanotubular graphene-polypyrrole hybrid for high performance flexible supercapacitors', *Nano Energy* 19: 391–400. DOI: 10.1016/j.nanoen.2015.11.029

35. Gao, Y. (2017) 'Graphene and polymer composites for supercapacitor applications: A review', *Nanoscale Research Letters* 12: 387–40 3. https://doi.org/10.1186/s11671-017-2150-5

36. Yoo, D. Kim, J. Kim, J. H. (2014) 'Direct synthesis of highly conductive poly(3,4ethylenedioxythiophene): Poly(4-styrenesulfonate)(PEDOT:PSS)/composites and their applications in energy harvesting systems', *Nano Research* 7: 717–3 0. https://doi.org/10.1007/s12274-014-0433-z

37. Zhang, X. Y. Ciesielski, A. Richard, F. Chen, P. Prasetyanto, E. A. De Cola, L. Samori, P. (2016) 'Modular graphene-based 3D covalent networks: Functional architectures for energy applications', *Small* 12: 1044–5 2. https://doi.org/10.1002/smll.201503677

38. Park, J. W. Park, S. J. Kwon, O. S. Lee, C. Jang, J. Jang, J. Park, J. W. Park, S. J. Lee, C. Kwon, O. S. (2014) 'In situ synthesis of graphene/polyselenophene nanohybrid materials as highly flexible energy storage electrodes', *Chemistry of Materials* 26: 2354–6 0. https://doi.org/10.1021/cm500577v

4 Carbon Nanotube (CNT)-Based Polymeric Supercapacitors

Sobhi Daniel

CONTENTS

4.1   INTRODUCTION

Energy storage devices have received more attention in recent years owing to the precipitous expansion of international monetary, shortage of fossil fuels, and ever-expanding environmental pollution. This leads to the rapid and urgent growth of extremely efficient energy storage devices especially electrochemical storage devices alleged as supercapacitors (SC), having superior power density, incredible reversibility, prolonged cycle life, fast charge-discharge, and quick mode of operation [1–3]. Supercapacitors are expected to be one of the promising and emergent materials having immense privilege in compact electronics, electric vehicles, dense equipment, systems aerial sites, end-user electronics, medical electronics, electrical efficiencies, conveyance, and military defense, and are expected to be used in satellite systems [4–5]. Besides, SCs facilitate the enrichment of the life of existing batteries. The development of energy storage devices has been subverted by the concept of composite materials in the form of composite supercapacitors and hybrid battery-supercapacitor devices. The efficiency of energy storage devices can be tremendously enhanced by the integrated coupling of diverse materials having inherent energy density, specific surface area, pseudocapacitance, electrical conductivity, etc., with carbon nanotubes (CNTs). Thus, the emergence of composite material CNTs will generate a novel class of promising materials having a high power density, in terms of greater capacitance, and specific features reducing the contact resistance between the electrode and current collector needed for supercapacitor applications.

One of the key aspects in the fabrication of SCs is the appropriate selection of electrode materials, which depicts an influential involvement in enhancing the execution of SCs. The foremost allotropic structures of carbon like activated carbon, graphene, CNTs, and nanocarbon dots, have been broadly explored for concocting SC electrodes [6–8]. The inherent properties of carbon and its allotropic forms are perfect candidates for the design of electrode materials and supercapacitor applications due to their excellent conductivity, adequate corrosion endurance, minimal density, exceptional stability, and low cost [9–11]. Also, the porosity and physical structure of carbon-based materials can be easily assembled by employing oxidizing agents at temperate processes, referred

DOI: 10.1201/9781003174646-4

to as stimulation. Supercapacitors can be categorized into three major types based on the energy storage mechanism and nature of electrode material employed during the fabrication process. These are recognized as electric double-layer capacitors (EDLCs), pseudocapacitors, and hybrid capacitors [12].

The charging technique in an EDLC is non-faradaic, and preferably, electron transference occurs by no means. The electrostatic adsorption/desorption in EDLC is a substantial progression and is incredibly arising quickly in high power density as well as extended life cycles, unlike batteries. In the case of a pseudocapacitor, an electrical charge is generated through electron transfer and will engender a change in chemical state or oxidization number of the electroactive species according to Faraday's laws. Thus, the origin of energy storage in a pseudocapacitor is faradaic charge transferal [13]. In pseudocapacitors, the redox reactions taking place at the electrode materials were responsible for the energy storage mechanism. Owing to the battery-like performance of pseudocapacitors, they were also symbolized as redox supercapacitors. In hybrid supercapacitors, materials such as activated carbon, CNTs, conducting polymers, various transitional metal oxides, etc., were incorporated along with the electrode materials and can exhibit both electrostatic responses and reversible faradaic-type charge transfer process.

The performance of different energy storage devices can be compared with the assistance of the Ragone plot, in which energy density (Wh/kg) is plotted against the power density (W/kg). Generally, the horizontal and vertical axes are plotted in logarithmic scale, and it exemplifies an imprint of the performances of the supercapacitors. The advancement of supercapacitors bridges the gap between batteries and capacitors in terms of both power and energy densities. The energy density of CNT polymer composites was found to be higher compared to other carbon-based materials. In the case of CNT polymer-based supercapacitors, they exploit both faradaic and non-faradaic processes to store the charge and their power densities were found to be superior to EDLCs without loss in cyclic permanence and stability. The CNT polymer-based materials accelerate a capacitive double layer of charge and offer a high surface area backbone that enhances the interaction between the deposited pseudocapacitive materials and electrolytes. The pseudocapacitive materials such as polymers were proficient enough to increase the capacitance of the composite electrodes through faradaic reactions. Figure 4.1 represents the Ragone plot for various energy storage devices.

Allotropic forms of carbon are frequently employed as electrode raw material in the case of EDLCs, and the mechanism of charge storage occurs electrostatically at the electrode-electrolyte interface. The electrode constituents typically explored are transition metal-oxides and conducting polymers in the case of pseudocapacitors, and capacitance is due to fast-faradaic responses arising at the exterior of electrode and electrolyte ions and integration of EDLC and pseudocapacitors

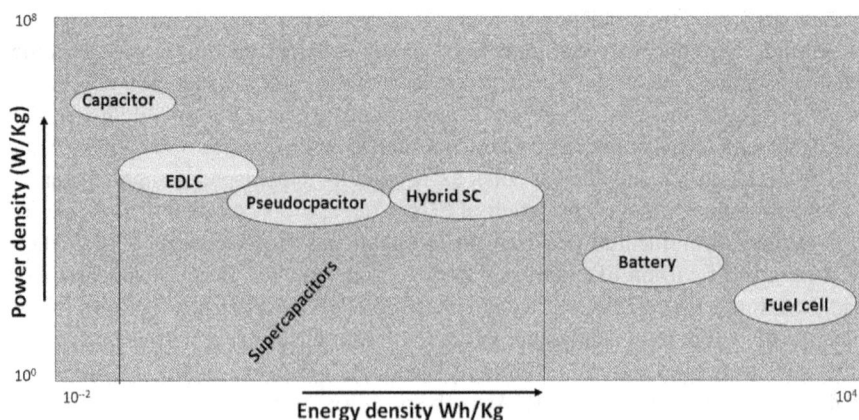

FIGURE 4.1 Ragone plot for various energy storage devices.

transpires in the case of hybrid supercapacitors [14]. CNTs are extensively employed for the fabrication of electroactive raw material in supercapacitors owing to their reduced specific capacitances compared to activated carbon. Properties like exceptional conductivity, improved specific energy, and increased power density can be achieved by adopting the CNTs possessing mesoporous configuration. In supercapacitor applications, blending of CNTs with conducting polymers or metallic oxides establishes a sharper specific capacitance value compared to CNT materials alone [15]. This chapter highlights the significance of CNTs and their nanocomposites in conjunction with polymers in supercapacitors.

4.1.1 CARBON NANOTUBES

CNTs, the one dimensional allotropic form of carbon, have fascinated researchers ever since their invention due to their exceptional material characteristics. CNTs were deemed as a new-fangled form of fullerenes and were discovered by Sumio Iijima in 1991. Iijima discovered CNTs while he was analyzing the new carbon formations on a cathode shallow in an electric arc discharge technique. CNTs can be deliberated as molecules comprised of 60 atoms of carbon assembled in stifled bundles and can be imagined as a plagiaristic coalition of carbon fibers and fullerene. The electronic band assembly of a nanotube can be explained by contemplating the bonding of carbon atoms organized in a hexagonal lattice. Each carbon atom $(Z = 6)$ is covalently bonded to three adjacent carbon atoms through sp2 hybridized molecular orbitals. The fourth valence electron, present in the Pz orbital, hybridizes with all the other Pz orbitals to form a delocalized band. The unit cell of graphene has two carbon atoms with an even number of electrons that are contained in the basic nanotube structure, which subsequently can be either metallic or semiconducting. CNTs exhibit different properties depending on how the nanotubes are rolled to yield the cylindrical shape and structurally, CNT can be articulated as a one-atom-thick sheet of graphite rolled in the tubular form with a diameter of one nanometer [16–17]. According to the number of graphene layers present in the CNTs, they are mainly categorized into two types, single-walled carbon nanotubes (SWCNTs) and multiwall carbon nanotubes (MWCNTs). A schematic representation of the structure of SWCNTs and MWCNTs is shown in Figure 4.2. The diameter of SWCNTs and MWCNTs differs from 0.4 to 2.5 nm and the length can vary from a few nanometers to 100 nm respectively. The layers in MWCNTs are held together by dint of van der Waals forces of attraction and diversity of arrangements of two-dimensional crystal structures with different mechanical, electrical, and optical properties, establishing the multi-layered CNTs to deliver diverse physical spectacles and device functionality.

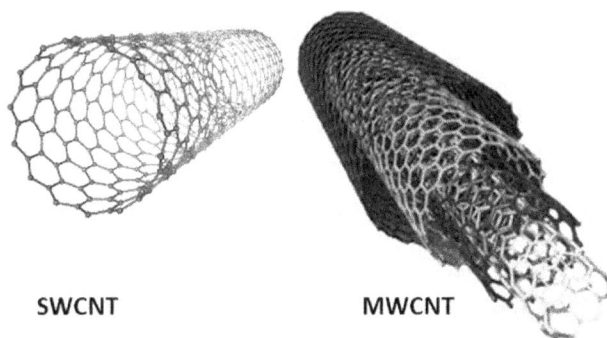

SWCNT   MWCNT

FIGURE 4.2 Structure of carbon nanotubes.

The actual composition of SWCNT can be imagined to be rendered up of hexangular benzene-sort rings of carbon atoms and rolled up in the form of a cylindrical shell of a graphene sheet. The graphene sheets can be portrayed as harmonious cylinders originating from a honeycomb matrix, demonstrating a single atomic stratum of crystalline graphite. An MWCNT is a bundle of graphene sheets rolled up into concentrical cylinders. Every single nanotube is a distinct molecule comprised of zillions of atoms and the magnitude of this molecule can be tens of micrometers long with diameters as small as 0.7 nm. SWCNTs naturally include only 10 atoms around the boundary and the width of the tube is only one atom thick. Nanotubes frequently have a significant length-to-diameter ratio (aspect ratio) of about 1000, so they can be measured as nearly single-dimensional structures [18–19]. MWCNTs are bigger and encompassed with several single-walled tubes stuffed one inside the other. MWCNT is confined to a nano configuration with an outward diameter of less than 15 nm and dominates a very high degree of order having three-dimensional crystallinity in comparison to SWCNTs. Every Single cylinder, or shell, of MWCNT nests effortlessly in the structure with spacing similar to the interplanar distance in crystalline graphite. Augment to the dual unique primary structures, three distinct possible types of carbon nanotubes can exist. They are categorized as armchair, zigzag, and chiral carbon nanotubes. This discrepancy in carbon nanotubes arises due to the nature of the rolling up of graphene sheets during its construction procedure. Wrapping of graphene sheets is controlled through diverse preferences of the chiral vector prominent to distinct CNT geometries. Chiral vector is exemplified with indices pair (n, m), and these two integers are parallel to the number of unit vectors along with the two directions in honeycomb crystal trellis of graphene. SWCNT is called an armchair when the chiral indices are comparable in magnitude (n = m), and when the chiral angle is 30°. SWCNT can be named zigzag, if one of the chiral indices is zero (n, 0) or (0, m) and a nanotube can be designated as chiral, when its chiral angle is $0° < \theta < 30°$ and (n ≠ m).

4.1.2 PROPERTIES OF CNTS

The excellent physical properties of CNTS endowed them with a unique category of novel nanomaterial and expedited them in innumerable applications. The tensile strength of CNTs was found to be higher than steel and Kevlar and this strength can be attributed to the sp2 hybridized bonds present in the individual carbon atoms of CNTs. In supplement to the mechanical potency, CNTS are also elastic. The nanotubes can be stretched and bend and will regain their original shape by the removal of the applied force. CNTs possess excellent electrical properties and can be either conducting or semiconducting. A single graphite sheet can be a semimetal, which exhibits properties in between that of a semiconductor, and can behave either as metal or non-metal. The conducting nature of carbon nanotubes can also be ascribed to the sp2 hybridized bonds present between the carbons atoms in CNTs [20–23] The underlying conducting estates of a graphene tubule are contingent upon the type of wrap up (chirality) and diameter of the CNTs. SWCNTs can direct electrical waves at a speed up to 10 GHz when they are used as interlocks on semi-conducting gadgets. Their electronic assets can be tuned by the application of peripheral magnetic fields, mechanical force, etc. CNTs can withstand elevated temperatures and can act as promising thermal conductors. The carbon nanotubes are shown to convey over 15 times the magnitude of a watt per minute as compared to copper wires. Moreover, the macroscopic quantum tunneling effect of the one-dimensional CNTs can restrict the transport of electrons from scattering during conduction and is known as ballistic transport [24].

Since absolute CNTs are composed of graphitic tubular fences of carbon atoms, they are found to be non-polar. The facades of CNTs (particularly MWCNTs) are extremely hydrophobic and is having a terrific attraction in the direction of non-polar objects such as hydrocarbons, organic solvents, paraffin, or oils. Also, these materials possess inherent hydrophobicity and exhibit comparatively exalted specific surface areas while fabricating in three-dimensional architecture. Non-polar nature and hydrophobicity restrict the applications of these wonder materials in many device applications especially in the synthesis and fabrications of composites of CNTs [25]. This limitation can be overcome by adopting a suitable strategy for the functionalization of the nano-tubes. Functionalization can be made possible by the linkage of selective functional groups on the sides or edges of the CNTs by making use of suitable chemical reactions and this will help to over-come the barriers and will become a fascinating material in the world of nanoscience.

4.2 CHEMICAL MODIFICATIONS OF CNTS

With the intrinsic inertness and non-compatibility of CNTs with nearly all solvents, a boundless effort has been taken to alter its surface characteristics by either covalently or noncovalently graft-ing different functional groups or biomolecules on its surface. Strong van der Waals forces of inter-actions that tightly bind the graphene layers together, forming bundles are accountable for the aura of insoluble nature and hydrophobicity of the CNTs. An ideal approach towards the integration of CNTs with other materials is the chemical modification/functionalization of the surface of the CNTs [26–29]. The chemical functionalization will improve the solubility of the CNTs in a vari-ety of solvents and will generate innovative hybrid materials which are hypothetically suitable for diverse applications. To enhance dispersity in aqueous media and diminish toxicity, CNTs are usu-ally functionalized along with different active functional groups (e.g., OH and COOH) by cova-lent as well as noncovalent approaches. Functionalization necessitates an ideal interfacial contact between nanomaterial and reactants, to achieve a good chemical interaction and better reproducible yield of the chemical reaction products.

In current years, four different methodologies evolved for the functionalization of carbon nano-tubes such as covalent sidewall functionalization, noncovalent functionalization, defect functional-ization, and endohedral functionalization. CNTs functionalized using a covalent approach permits attachment of functional groups at the terminates or sidewalls of CNTs. The spots of greatest chem-ical reactivity in CNT edifice are the caps, which have a partial fullerene resembling arrangement. The chemical functionalization of CNT tips has been implemented primarily based on oxidative treatments. The oxidation reactions of CNT generate opened tubes with oxygen-comprising func-tional groups (primarily carboxylic acid) at both the sidewall and tube wind-ups. These carboxyl groups can effectively be utilized as chemical anchors for additional derivatization processes. Even though CNTs are having poor solubility in organic solvents, they can be suspended in suitable sol-vents to facilitate the functionalization process. Commonly used solvents for suspending CNTs are shown in Figure 4.3.

Covalent functionalization of CNTs will take the lead to the creation of stable chemical connec-tions, which in turn may disrupt the graphitic core and electronic properties of CNTs. While in the litigation of noncovalent functionalization, core structure and electronic properties will be retained, and functionalization reactions occur through the wrapping of functional moieties through the van der Waals forces of attraction [30]. The existence of sidewall flaws such as vacancies or pentagon-heptagon pairs present in the vicinity increases the chemical acuteness of the graphitic nanostruc-tures. The covalent sidewall functionalization of CNTs creates sp^3 carbon positions and will disrupt the band-to-band transitions of π electrons, which will diminish novel properties of CNTs like con-ductivity and incredible mechanical assets. With an ever-increasing degree of functionalization, the CNTs can ultimately be converted into an insulating material. In this context, one must be conscious that the electrical and mechanical estates of the CNT are irreversibly unaffected after the chemical modification process.

In the noncovalent approach, functionalization occurs through physical adsorption or wrap-up of polynuclear molecules/biomolecules/polymers, etc. through van der Waals interactions. The key benefit of noncovalent functionalization of CNTs, as contrasted to covalent ones, is that the chemical functionalities can be launched to CNTs without modifying the structure and electronic arrangement of tubes. Noncovalent functionalization of CNTs can be accomplished through pow-erful molecules possessing aromatic groups [31]. The functionalization of CNTs was managed by ascertaining precise and directional π-π stacking alliances between aromatic molecules and the

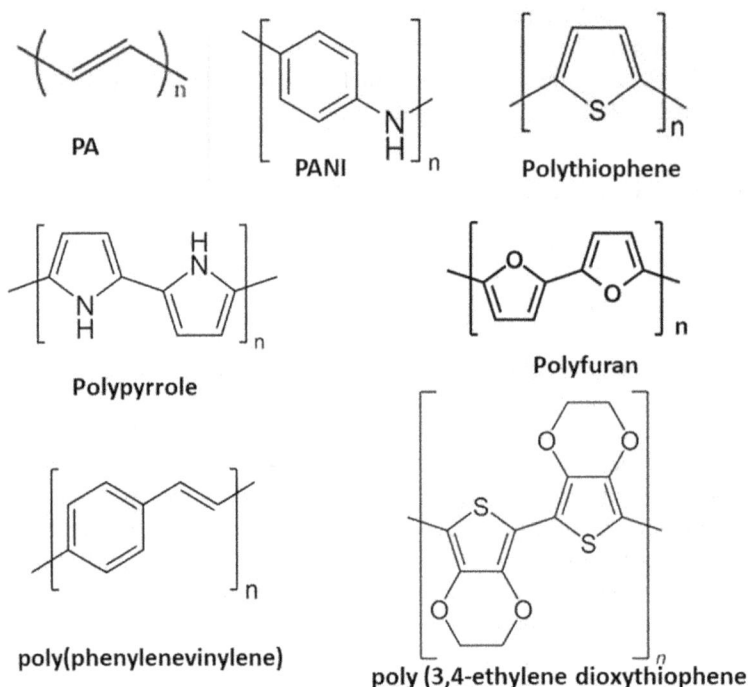

FIGURE 4.3 Commonly used solvents for suspending CNTs [27].

graphitic surface of carbon nanotubes. Noncovalent derivatization of CNTs. can be furnished by providing π-π interactions (using aromatic compounds or polymers), electrostatic interactions, and CH-π interactions between CNTs and the attaching molecules.

The mechanism of energy storage in CNT-based nanomaterials is due to the accumulation of electrostatic charge at the electrode/electrolyte interface, and their performance is powerfully related to the effective surface area. The electrode materials made up of CNTs possess promisingly higher specific surface area, pore size as well as specific capacitance. The wetting competence of the electrode materials also significantly improves the supercapacitance behavior of CNTs.

4.3 IMPORTANCE OF CNT-POLYMER COMPOSITES IN THE FABRICATION OF SCS

CNTs can be effortlessly integrated into polymeric matrices deprived of altering structural morphologies of CNTs distribute great interest among researchers and have been expansively explored in recent years [32–34]. The brilliant mechanical properties of CNTs, such as tensile strength and strain to fracture, high elastic modulus, competence to tolerate cross-sectional and twisting falsifications, and compression without fracture, are chiefly voyaged to obtain structural materials with lightweight, great elastic modulus, high tensile and compressive strength, and stiffness, etc in the case of CNT-polymer structural composites. Fascinating properties of CNTs, such as high electrical and thermal conductivity can be explored to establish functional materials with energy storage performances. SWCNTs and MWCNTs can be integrated with polymeric matrices and were widely explored in the fabrication of supercapacitors. CNTs were found to be attractive in the field of supercapacitors owing to their high specific surface area, low electrical resistance, low mass density, and high cyclic stability.

Composites encircling CNTs and an electroactive phase such as conducting polymers exhibit the pseudocapacitive properties and exemplify a convincing revolt in the fabrication of a novel generation of supercapacitors. Compared to other forms of carbonaceous materials, the percolation of the electroactive particles is more efficient in CNTs. The open mesoporous network formed by the entanglement of nanotubes permits the ions to diffuse effortlessly to the active surface of the composite components. Also, the CNT-polymer composite electrodes can easily acclimate to the volumetric changes during the charging/discharging process and will significantly improve the cycling performance of the SCs. Thus, the composites of CNT integrate the large pseudocapacitance of conducting polymers with the fast charging/discharging double-layer capacitance and outstanding mechanical properties of CNTs [35].

The integrated application of nanotubes in composite fabrication hinges on the capability to disband CNTs homogeneously throughout the polymeric matrix, its compatibility with matrix, and these components are crucial in the operation of fillers in polymer composites of CNTs [36]. Interaction mechanisms between CNTs and polymeric matrices are mainly grouped into three types. They are micro-mechanical interlocking, chemical bonding amongst the nanotubes and matrix, and the weak van der Waals bonding between CNT and the polymeric matrix. The addition of nanotubes to polymeric backbone outcomes the high level of the external surface to volume proportion and substantially enhances macroscopic properties of the polymer including enhanced mechanical properties and flexibility [37]. Also, the incorporation of carbon nanotubes into polymeric matrices will significantly enhance their electrical, thermal, and optical properties. CNTs having exceptional pore construction, virtuous mechanical strength, thermal steadiness, greater electrical assets, and consistent mesopores structure permit a constant charge circulation in comparison to other allotropic forms of carbon. More excellent conductivity and improved charge transmission cylinders of CNTs deliver the greatest encouraging element toward energy-saving purposes. In modern generations, Thus, by compositing carbon nanomaterials with other materials having pseudocapacitances such as conducting polymers, the energy density can be largely improved, but their rate capability and cyclic stability may decrease to 60–90% after 1000 cycle.

4.4 APPLICATIONS OF CNT-POLYMER COMPOSITE IN SUPERCAPACITORS

The supercapacitor is an alternative and smart option for energy storage applications in portable and distant devices contrasted to batteries and traditional capacitors [38]. The supercapacitor has achieved substantial relevance due to its rapid charging/discharging speed, superior power density, and prolonged cycling stability in comparison to antiquated batteries [39–42].

Carbon, both in conducting and dispersed form, is extensively exploited for fabricating saleable electrode material for supercapacitors. CNTs, conducting and porous allotropic forms of carbon were investigated widely in the fabrication of new generation supercapacitors owing to their exceptionally high mechanical strength, excellent electrical property, extraordinary specific area, and elevated dimensional ratios. The exceptional estates of CNTs and particularly their high surface area make them magnificent electrode materials for energy storage rationale. Also, CNTs have limited dissemination of pores due to vertical orientation and most of the surface area is due to mesopores which facilitate the electrolyte to access all existing surface areas. Mesopores also furnish the freedom to use electrolytes with distinct molecular sizes [43–46]. Other allotropic forms of carbon such as activated carbon, fullerenes, and carbon dots are promising candidates for electrical double layer supercapacitor electrodes owing to a larger surface area [47]. Existing literature reports indicate that EDLCs stranded on pristine CNTs showed superior-rate proficiencies and cyclic stabilities, simultaneously with rectangular cyclic voltammograms and proportionate triangular galvanostatic charge-discharge contours, implying high performance for charge storage [48]. Carbon nanotubes with unique structures can be combined with polymeric materials to display synergetic impacts for their electrochemical properties and have surfaced as hybrid supercapacitors with exceptional characteristics. In hybrid supercapacitors, blending of the faradaic embolism on the cathode and non-faradaic surface reaction on an anode creates an occasion to accomplish both spiraling energy and power densities even without negotiating the cycling permanence and affordability. The integration of CNTs with electrically conducting polymers creates composites that amalgamate superior pseudocapacitance of polymers with unresolved mechanical and structural properties of the nanotubes and are thus exceedingly proficient in novel supercapacitors with excellent and power density. The exceptional nanostructure and redox charge storing competence of conducting polymers with incredible surface area and nano porosity of CNTs broaden its application in the arena of supercapacitors. The alteration of CNTs with conducting polymers is one way to upsurge the capacitance of the composite resulting from the redox contribution of the conducting polymers. In the case of supercapacitors fabricated through CNT/conducting polymer composite, CNTs will behave as electron acceptors and the conducting polymer serves as electron donors. CNTs were widely explored as electrodes for supercapacitors and batteries follow the EDLC mechanism of charge storage at the electrode/solution boundary. The pictorial representation of a CNT-polymer-based supercapacitor is shown in Figure 4.4. Conducting polymers can also be electrochemically deposited on the surface of CNTs and can effectively perform the supercapacitance behavior and these composites can be used as an electrode for supercapacitors. The extraordinary properties of CNTs, especially their high surface area make them efficient electrode materials for energy storage purposes. CNTs possess a narrow distribution of pores due to the vertical alignment and most of the surface area is due to the mesopores which help the electrolyte to access all available surface areas. Mesopores also provides the freedom to use electrolytes independent of molecular size. CNTs integrated with conducting polymers, such as polypyrrole and polyaniline, were found to be promising electrode materials because the entangled mesoporous network of nanotubes in the composite can adapt to the volume changes. Thus more stable capacitance values with cycling stability were obtained by the integration of CNTs together with conducting polymers.

Conducting polymers (CP) can be effectively exploited for the fabrication of CP- CNT composites and were found to have immense applications in the field of electrochemical devices like photovoltaic cells, solar batteries, and energy storage devices [49]. The stability of the nanostructured objects was found to score an upsurge by the inclusion of CNTs with CPs. Also, the excellent elasticity

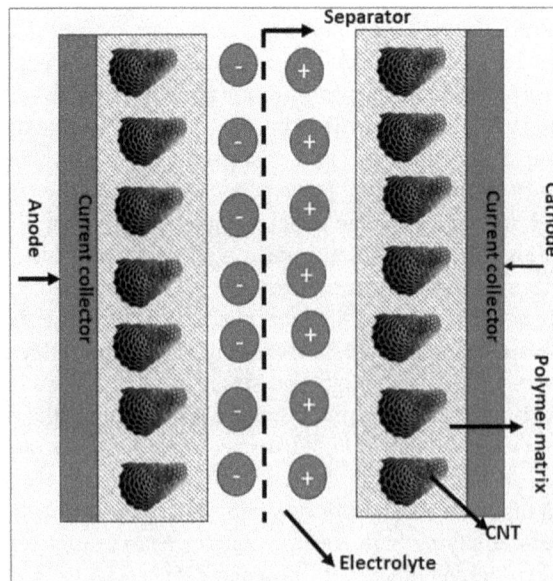

FIGURE 4.4   CNT-based polymeric supercapacitor.

of CP accelerates the flexibility of the nanocomposites. The composite electrodes merged with CNTs and conductive polymeric materials blend both chemical and physical charge storage mechanisms promptly in a particular electrode. Also, CNT materials will advance capacitive double-layer of charge and will deliver a high-surface-area backbone that boosts the interaction between the electrolyte and deposited pseudocapacitive materials. Pseudocapacitive materials can further intensify the capacitance of composite electrode materials through faradaic reactions. Several literature reports have established that composite electrode material can achieve higher capacitances than either a pristine CNTs or pristine conducting polymer-based electrode. The superior performance of composite materials is accredited to the availability of the entangled material structure that permits a consistent coating of conducting polymers and a three-dimensional distribution of charge. Additionally, structural harmony of intertwined material has been exposed limit towards mechanical stress caused by integration and elimination of ions in deposited CPs. Thus, in comparison to CPs, CNT- polymer composites are proficient enough to accomplish cycling stability

Cyclic voltammetry (CV) appraises both quantitative and qualitative data concerning the electrochemical processes taking place at surfaces of the electrode materials. The principle of CV involves the application of a potential to the working electrode, concerning the reference electrode's fixed potential, which linearly sweeps back and forth between the two predefined potentials. In addition, CV is a universal method for determining the electrochemical performance of a supercapacitor. The rectangular-shaped cyclic voltammograms are peculiar for EDLCs and that of the pseudocapacitors exhibit a broad redox peak with small peak-to-peak separation instead of a rectangular-like shape. Thus CV measurements were found to be an ideal approach to distinguish the behaviors of distinct types of SCs. The existing literature reports reveal that CNT-based polymer composites exhibit both EDLC and pseudocapacitor-like voltammograms. The enhancement in the electrochemical properties of the CNT composites can be attributed to the presence of micro-and nanometer pores in the CNT polymer composites, which offer corridors for the transport of ions and solvent molecules within the composite films. Also, the CVs of the composites displayed capacitive features with almost straight and vertical current variations at the end potentials, which suggest a fast charge-discharge switching resulting from high electronic and ionic conductivity. Thus, the CNT-based polymeric composites exhibit enhanced electrical conductivity and other benefits for EDLCs or pseudocapacitance electrodes. Due to the synergetic effects of EDLC and pseudocapacitance behavior on electrochemical performance, CNT-polymer nanocomposites play substantial roles on supercapacitor devices that integrate its electrochemical and mechanical characteristics.

Identical to CV, galvanostatic charge-discharge (GCD) is an alternative technique to assess the capacitance of the electrochemical material. The GCD measurements were carried out at constant current density and responsive potential concerning time will be recorded. Normally, in GCD studies, the working electrode is charged to a pre-set potential and the discharge process is then monitored to assess the capacitance. In parallel to CV curves, both EDLC and pseudocapacitance materials express divergent responses. In the case of EDLC materials, the charge and discharge process occurs linearly, while in pseudocapacitive materials it occurs through a nonlinear pathway and can be attributed to the redox reactions. Most of the reported EDLCs based on pure CNTs showed high-rate capabilities and cyclic stabilities together with rectangular CV and symmetric triangular GCD profiles, indicating high performance for charge storage. But in the case of CNT -polymer composite-based supercapacitors, the GCD curves exhibit a triangular shape with a small deviation from linearity, indicating mixed electrical double layer and pseudocapacitive contributions.

The literature reports [50–52] demonstrate that the addition of CNTs to the conductive polymers significantly increases the conductivity of the composite and its stability, reducing one of the serious disadvantages of the conductive polymer. CNTs were also employed as a matrix for many composites: with poly-pyrrole (PPy), poly (3,4-ethylene-1,4-dioxythiophene)(PEDOT), poly (3-octylthiophene), polyphenylvinylinylene (PPV) [53–55], polyacrylonitrile (PAN), and also polyaniline (PANI) [53–54]. The commonly used CPs in the fabrication of CNT composites supercapacitors includes polypyrrole,(PPy), poly (3,4-ethylene dioxythiophene) (PEDOT), polyacetylene

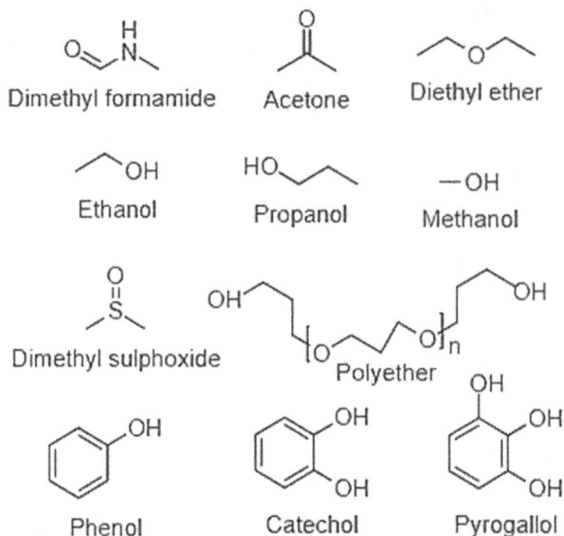

FIGURE 4.5 Commonly used polymers explored in the fabrication of CNT-composite SCs.

(PA), polyaniline (PANI),), polyfuran (PF), polythiophene (PTH), poly(phenylenevinylene) (PPV), etc., and the chemical structures of few of the conducting polymers extensively explored for the fabrication of CNT-polymer composites is shown in Figure 4.5.

The supercapacitance of SWCNT-polypyrrole composites were recently reported by Matei et al. [55]. Covalent functionalization of CNTs and pyrrole units occurs through esterification and coupling of acyl chloride functionalized SWCNTs with N-(6-hydroxyhexyl) pyrrole. Electro polymerizable pyrrole group chemically attached to the SWCNT mainstay through a hydroxy hexyl chain and assisted as a flexible spacer for enabling chemical link flanked by pyrrole radical cations during the polymerization process. The composite films fabricated exhibited improved electrochemical responses and capacitance per symmetrical electrode surface area (F cm^{-2}) with a capacitance value of 0.226 F cm^{-2}.

Muhammad Rakibul et al. [56] synthesized SWCNT bolstered PVA surfactant-free nanocomposites (PVA/SF-SWNT) by solution-cast method and composite material exhibited enhanced electrochemical performance. Composite of PVA/SF-SWNT flashed a specific capacitance value of 26.4 F g^{-1}), which is fourfold greater than that of PVA (6.1 F g^{-1}) while applying a current density of 0.5 mA g^{-1}). Hoe-Seung Kim et al. effectively synthesized Polypyrrole/graphene nanosheet/MWCNT composites by a facile method of in situ polymerization [57]. The specific capacitances of the PPy/GNS/MWCNT composites were astonishingly improved, compared to individual PPy and GNS. Enhancement in specific capacitance of the composite can be ascribed to the cooperative effect amongst the carbon-based materials and PPy. The synthetic procedure adopted for the fabrication of PPy/GNS/MWCNT composites is shown in Figure 4.6.

Foivos Markoulidiss et al. [58] explored composite electrodes with excellent energy storage from activated carbon and MWCNT adorned with silver nanoparticles. Polyacrylonitrile PAN)/Cu (OAc)$_2$-CNTs composite nanofiber concocted via electrospinning technique using DMF by Dawei Gao et al. [59]. PANI-graphene nanoribbon (GNR)-carbon nanotube (CNT) composite, PANI-GNR-CNT having a three dimensional (3D) structure was produced by onsite polymerization of aniline monomer on the surface of GNR-CNT hybrid by Mingkai Liu et al. and electrochemical investigations revealed the hierarchical PANI-GNR-CNT composite based on the two-electrode cell embraces an advanced specific capacitance (890 F g^{-1}) than GNR-CNT hybrid (195 F g^{-1}) and pure PANI (283 F g^{-1}) at a discharge current density of 0.5 A g^{-1} [60].

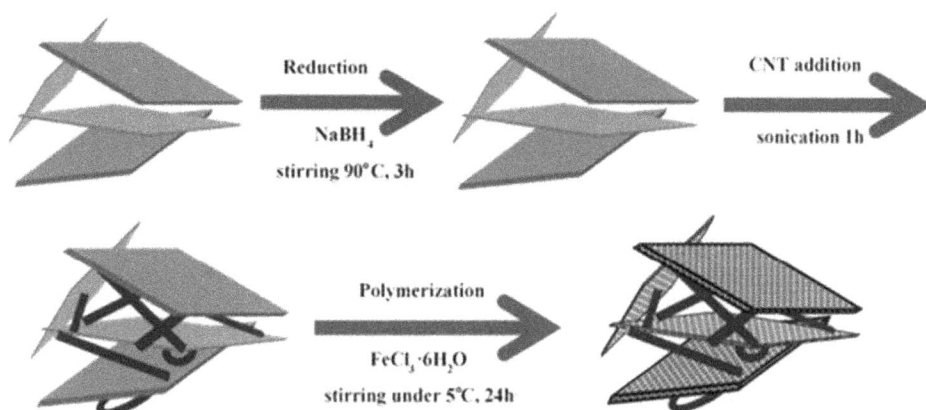

FIGURE 4.6 Synthesis route adopted for the fabrication of PPy/GNS/MWCNT composites [57].

Zhihong Ai et al. reported novel composite electrode material, $CoNi_2S_4$/CNT synthesized through a two-step route, incorporating deposition of Co-Ni precursor on CNT and conversion of Co-Ni precursor via anion exchange process [61]. The obtained $CoNi_2S_4$/carbon nanotubes composites exhibited ultrahigh specific capacitance of 2094 F g^{-1} at 1 A g $^{-1}$ and good rate capability (72% capacity retention at 10 A g^{-1}). Outstanding mechanical properties, high electrical conductivity, large surface area, and functionality entrusted CNTs as promising candidates for the fabrication of flexible supercapacitor electrodes. This category of flexible SCs can be explored for the fabrication of wearable devices. CNTs can gather a variety of macroscopic materials with different proportions. Flexible CNT assemblies including different dimensional structures, aerogels, and sponges on various design approaches and construction techniques were recently reviewed by Sheng Zhu et al. [62]. Ferrocene functionalized multi-walled carbon nanotubes (Fc-MWCNTs) were magnificently blended in two steps at low-temperature by Gomaa A. M. Ali et al. [63]. The Fc-MWCNTs electrode demonstrated brilliant retention capacity (90.8% over 5000 cycles) and a specific capacitance of 50 F g^{-1} at 0.25 A g^{-1} contrasted to the $MWCNTs-NH_2$.

Nitin Muralidharan et al. [64] demonstrated the strengthening of an epoxy matrix having ion-conducting properties with CNTs and steel mesh electrodes covered with insulating Kevlar materials. CNTs sprouted on the conductive structural template were utilized as an instinctively encouraging interface and multifunctional energy storage device. The findings indicated that the elastic modulus of nanocomposite materials was greater than 5 GPa, with a specific energy of 3 mWh/kg. Also, results supported the astonishing potential of CNTs as two-fold fortifying and energy storage materials towards the next prototype of composite structures. CNT reinforced structural supercapacitor material that can be adapted in a vehicle framework is shown in Figure 4.7. Evgeny Senokos et al. [65] fabricated a novel structural composite supercapacitor by assimilating thin sandwich structures of CNT fiber veils and an ionic liquid-based polymer electrolyte between carbon fiber strands, followed by infusion and curing of an epoxy resin. The obtained structure performed simultaneously as an electric double-layer capacitor and a structural composite, with a modulus of 60 GPa and a strength of 153 MPa, combined with 88 mF/g of specific capacitance having the highest power (30 W/kg) and energy (37.5 mWh/kg). A brief overview of CNT-Ni-Co-O-centered composite material and their benefits over single-phase CNT, Ni-Co-O as energy material towards supercapacitor applications were recently reviewed by Soumya Mukherjee [66]. Wen Lu et al. [67] reported a brand-new category of nanocomposite electrodes towards the progress of efficient supercapacitors with ionic liquid green electrolytes. The inherent surface area of activated carbon, carbon nanotubes, and ionic liquids as cohesive primitive components substantially improve charge storage, delivery facilities and fabricated composite possess a superior capacitance value of 188 F/g).

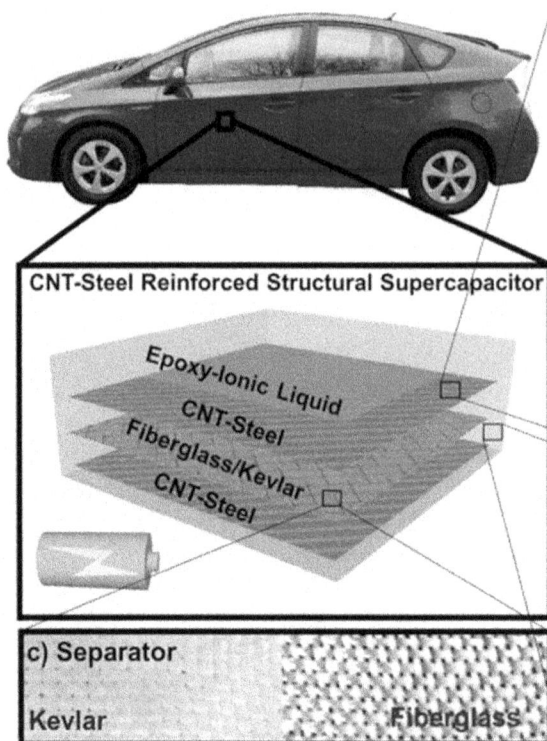

FIGURE 4.7 Design of a CNT reinforced structural supercapacitor with a vision of a reinforced composite material in a vehicle chassis [64].

Lota et al. [68] reported a novel composite material prepared from a homogenous mixture of polymer poly(3,4-ethylene dioxythiophene; PEDOT) and CNTs by chemical or electrochemical polymerization of EDOT directly on CNTs. An et al. [69] demonstrated that the SWCNT/PPy (1/1 in weight) nanocomposite electrode with a higher specific capacitance value compared to pure PPy and pristine SWCNT electrode. Recently, Yanfang Xu et al. [70] fabricated a structural supercapacitor (SSC) based on aligned discontinuous carbon fiber and a solid polymer electrolyte. The obtained SSCs showed a maximum specific capacitance of 0.128 mF/cm2 (11.62 mF/g) and a power density of $1.19 \times 10-2$ W/cm2 which were comparable to the performance of SSCs based on unaltered woven carbon fiber electrodes.

A high-performance solid-state supercapacitor (SSSC) constructed on an amphiphilic comb polymer) solid electrolyte and an electrode containing porous one-dimensional (1D) hierarchical CNTs were reported by Lee et al. [71]. The SSSC fabricated with the comb polymer electrolyte exhibited a high specific capacitance of 239.3 F g $^{-1}$, Yaping Zhu et al. [72] established all-solid-state supercapacitors by using aligned CNT/conducting polymer (Ag-doped poly(3,4-ethylene dioxythiophene)-poly(styrene sulfonate)) composites as electrodes and polyvinyl alcohol-based electrolytes. The obtained all-solid-state supercapacitors displayed a high specific capacitance of 64 mF cm^{-2} and maintained 98% of their original capacitance even with a tensile strain as high as 480%.

A novel stretchable, wearable coiled CNT/MnO$_2$/polymer fiber solid-state supercapacitors were created by Changsoon Cho et al. [73]. Parayangattil Jyothibasu et al. [74] fabricated a composite film based on PPy and CNTs through a green synthetic route from plant-derived material curcumin as a template. The synthesized composite film was used as a free-standing electrode for supercapacitors. Flexible supercapacitors based on a PANI/CNT/EVA composite was synthesized by the in situ growth of a high-performance all-solid-state electrode and was reported by Xipeng Guan et al. [75]. The electrode was comprised of polyaniline deposited on a CNT and a poly (ethylene-co-vinyl acetate) film. And these hybrid electrodes exhibited excellent mechanical and electrochemical performance.

4.5 CONCLUSIONS

The design and development of novel green electrode materials is an emerging field of research targeting energy storage devices and supercapacitors. Supercapacitor electrodes engineered from carbon-based composite materials can transmit a better specific surface area, superior electrical conductivity, low mass density, and prolonged cyclic stability. The strategies to breed non-toxic, safe, and green CNTs and their polymeric composite remain a momentous challenge in the device fabrication of energy storage devices. Integration of distinct types of CNTs such as SWCNTs and MWCNTs in polymeric matrix displayed a diversity of structural and physiochemical reinforcement characteristics and enhancement of material's strength, conductivity, flexibility, and biocompatibility. With advancements in synergetic impacts upon electrochemical performance, CNT-polymer nanocomposites compete for substantial objectives on supercapacitors that incorporate interfacial adsorption, redox reaction, mechanical strength, hierarchical microstructure, conductivity, and flexibility.

By incorporating carbon nanotubes with other materials having pseudocapacitance properties such as conducting polymers, metal oxides, or hydroxyls, energy density can be primarily enhanced. The emergence of blended supercapacitors can satisfy the breach between a supercapacitor and a battery by enhancing both energy and power density in a single electrochemical device. Integrating CNT with suitable conducting polymers has been exhibited to be an operative methodology towards the development of SCs with brilliant flexibility and strain resistance while preserving their electrochemical performance and conductivity. The interfacial bonding of carbon nanotubes to polymeric matrices accelerates stress transfer from matrix and CNTs. Integration of conducting polymer-CNT composites blend pseudocapacitance property of polymers with mechanical and structural properties of the nanotubes and have emerged as a promising novel supercapacitor material with revolutionary capacitance and power density values. The hybrid supercapacitors comprising CNT and polymers will surely emerge as a novel platform for the fabrication of stretchable and flexible supercapacitors. Also, towards the promotion of green chemistry and renewable assets, the assimilation of CNT with polymers and green synthesized quantum-sized carbon dots must be explored. The interfacial interaction and bonding between the nano-sized materials and the polymeric matrices were also considered and it still needs more understanding at the microscale dimensions. Thus, the evolution of multiple redox nanostructures with synergic effects and novel electrochemical performances will open new horizons in the development of efficient supercapacitors in near future.

REFERENCES

1. Zhenhui, L., Ke, X., Yusheng, P., 2019. Recent development of supercapacitor electrode based on carbon materials. *Nanotechnology Reviews* 8: 35–49.
2. Meng, Q., Cai, K., Chen, Y., Chen, L., 2017. Research progress on conducting polymer-based supercapacitor electrode materials. *Nano Energy* 36: 268–285.
3. Sun, J., Wu, C., Sun, X., Hu, H., Zhi, C., Hou, L., et al., 2017. Recent progresses in high-energy-density all pseudocapacitive-electrode materials-based asymmetric supercapacitors. *Journal of Materials Chemistry A* 5: 9443–9464.
4. Sharma, K., Arora, A., Tripathi, S. K., 2019. Review of supercapacitors: Materials and devices. *Journal of Energy Storage* 21: 801–825.
5. Amin, M., Saleem, Vincent, D., Peter, E., 2016. Performance enhancement of carbon nanomaterials for supercapacitors. *Journal of Nanomaterials* Article ID 1537269, 17 pages.
6. Yu, J. Lu, W., Pei, S., Gong, K., Wang, L., Meng, L., et al. 2016. Omnidirectionally stretchable high-performance supercapacitor based on isotropic buckled carbon nanotube films. *ACS Nano* 10: 5204–5211.

7. Dubal, D., Ayyad, O., Ruiz, V., Gómez-Romero, P., 2015. Hybrid energy storage: The merging of battery and supercapacitor chemistries. *Chemical Society Reviews* 44: 1777–1790.
8. Ayesha, K., 2020. Nanocomposite material for supercapacitor application. *American Journal of Applied Physics* 4: 1–8.
9. Jeong, H. K., Ji-Young, H., Ha, R. H., Han, S. K., Joong, H. L., Jae-Won, S., et al., 2018. Simple and cost-effective method of highly conductive and elastic carbon nanotube/polydimethylsiloxane composite for wearable electronics. *Scientific Reports* 8: 1375. DOI:10.1038/s41598-017-18209w
10. Byrne, M. T., Gun'ko, Y. K., 2010. Recent advances in research on carbon nanotube-polymer composites. *Advanced Materials* 22: 1672–1688.
11. Sultan, A., Rafat, M., Ahsan, A., 2018. Nitrogen doped activated carbon derived from orange peel for supercapacitor application. *Advances in Natural Sciences: Nanoscience and Nanotechnology* 9: 035008 (8pp).
12. Chuang, P., Shengwen, Z., Daniel, J., George, Z. C., 2008. Carbon nanotube and conducting polymer composites for supercapacitors, *Progress in Natural Science* 18: 777–788.
13. Chunsheng, D., Ning, P., 2007. Carbon nanotube-based supercapacitors. *Nanotechnology Law and Business* 4: 569–577.
14. Murat, A., Aysegul, A. E., Bulent, E., 2017. Carbon nanotube-based nanocomposites and their applications. *Journal of Adhesion Science and Technology* 31: 1977–1997.
15. Enkeleda, D., Zhongrui, L., Yang, X., Viney, S., Alexandru, R. B., et al., 2009. Carbon nanotubes: Synthesis, properties, and applications. *Particulate Science and Technology* 27: 107–125.
16. Kapil, D., Patel, Rajendra, K. S., Hae-Won, K., 2019. Carbon-based nanomaterials as an emerging platform for theranostics. *Materials Horizons*. DOI: 10.1039/c8mh00966
17. Saifuddin, N., Raziah, A. Z., Junizah, A. R., 2013. Carbon nanotubes: A review on structure and their interaction with proteins. *Journal of Chemistry* Article ID 676815, 18 pages.
18. Qiu, H., Yang, J., 2017. Structure and properties of carbon nanotubes. *Industrial Applications of Carbon Nanotubes*: 47–69. doi:10.1016/b978-0-323-41481-4.00002-2
19. Huang, L., Cao, D. P., 2012. Mechanical properties of polygonal carbon nanotubes. *Nanoscale* 4: 5420–5424.
20. Neto, H. C., Guinea, F., Peres, N. M., Novoselov, K. S., Geim. A. K., 2009. The electronic properties of graphene. *Reviews of Modern Physics* 81: 109–162.
21. Ren, F., Yu, H., Wang, L., Saleem, M., Tian, Z., Ren, P., 2014. Current progress on the modification of carbon nanotubes and their application in electromagnetic wave absorption. *RSC Advances* 4: 14419. DOI: 10.1039/c3ra46989
22. Mittal, G., Dhand, V., Rhee, K. Y., Park, S. J., Lee, W. R., 2015. A review on carbon nanotubes and graphene as filers in reinforced polymer nanocomposites. *Journal of Industrial and Engineering Chemistry* 21: 11–25.
23. Nikolaos, K., Nikos, T., 2010. Current progress on the chemical modification of carbon nanotubes. *Chemical Reviews* 110: 5366–5397.
24. Valery, N., K., Merlyn, X. P., 2006. Chemical modification of carbon nanotubes. *Mendeleev Communications* 16: 61–66.
25. Bin, Z., Hui, H., Aiping, Y., Daniel, P., Robert, C. H., 2005. Synthesis and characterization of water soluble single-walled carbon nanotube graft copolymers. *Journal of the American Chemical Society* 127: 8197–8203.
26. Sergio, M., Jean-Christophe, P. G., 2019. Methods for dispersing carbon nanotubes for nanotechnology applications: Liquid nanocrystals, suspensions, polyelectrolytes, colloids and organization control. *International Nano Letters* 9: 31–49.
27. Sarat K. S., Itishree, J., 2010. Polymer/carbon nanotube nanocomposites: A novel material. *Asian Journal of Chemistry* 22: 1 1–15.
28. Du, J. H., Bai, J., Cheng, H. M., 2007. The present status and key problems of carbon nanotube-based polymer composites. *Express Polymer Letters* 1: 253–273.
29. Satish, G., Prasad, V. V. S., Koona, R., 2017. *Manufacturing and characterization of CNT based polymer composites*. JVE international Ltd. ISSN Print 2351–5279.
30. Wei, W., Susan, L., Ming, L., Qian, Z., Yuhe, Z., 2014. Polymer composites reinforced by nanotubes as scaffolds for tissue engineering. *International Journal of Polymer Science*. Article ID 805634, 14 pages.
31. Li, Q., Yongkang, C., Yongzhen, Y., Lihua, X., Xuguang, L., 2013. A study of surface modifications of carbon nanotubes on the properties of polyamide 66/multiwalled carbon nanotube composites. *Journal of Nanomaterials* Article ID 252417, 8 pages.

32. Hyeong, T. H., Chong, M. K., Sang, O. K., Yeong, S. C., In, J. C., 2004. Chemical modification of carbon nanotubes and preparation of polystyrene/carbon nanotubes composites. *Macromolecular Research* 12: 384–390.

33. Shaffer, M. S. P., Sandler, J. K. W., 2006. Carbon nanotube/nanofibre polymer composites. In *Processing and properties of nanocomposites*, ed. S. G. Advani. World Scientific, New York, 1–60.

34. Junjie, C., Longfei, Y., Wenya, S., Deguang, X., 2018. Interfacial characteristics of carbon nanotube-polymer composites: A review. *Composites Part A: Applied Science and Manufacturing* 114: 149–169.

35. Emilia, G., Krzysztof, W., 2018. Recent progress on electrochemical capacitors based on carbon nanotubes. http://dx.doi.org/10.5772/intechopen.71687.

36. Jon, S. B., 2017. A technical overview on the interface between carbon nanotubes and a polymer matrix. *Intelligent Concrete.*

37. Junjie, C., Baofang, L., Xuhui, G., Deguang, X., 2018. A review of the interfacial characteristics of polymer nanocomposites containing carbon nanotubes. *RSC Advances* 8: 28048–28085.

38. Zhenhui, L., Ke, Xu., Yusheng, P., 2019. Recent development of supercapacitor electrode based on carbon materials. *Nanotechnology Reviews* 8: 35–49.

39. Holdren, J., 2007. Energy and sustainability. *Science* 315: 737.

40. Wang, H., Dai, H., 2013. Strongly coupled inorganic—nano-carbon hybrid materials for energy storage. *Chemical Society Reviews* 42: 3088–3113.

41. Yu, G., Xie, X., Pan, L., Bao, Z., Cui, Y., 2013. Hybrid nanostructured materials for high-performance electrochemical capacitors. *Nano Energy* 2: 213–234.

42. Fitri, A. P., Muhammad, A. I., Satria, Z. B., Ferry, I., 2021. Carbon-based quantum dots for supercapacitors: Recent advances and future challenges. *Nanomaterials* 11: 91–126.

43. Brian, K. K., Serubbable, S., Aiping, Y., Jinjun, Z., 2015. Electrochemical supercapacitors for energy storage and conversion. In *Handbook of clean energy systems*, pp. 1–25.

44. Amin, M. S., Vincent, D., Peter, E., 2016. Performance enhancement of carbon nanomaterials for supercapacitors. *Journal of Nanomaterials* Article ID 1537269, 17 pages.

45. Chuang, P., Shengwen, Z., Daniel, J., George, Z. C., 2008. Carbon nanotube and conducting polymer composites for supercapacitors. *Progress in Natural Science* 18: 777–788.

46. Chunsheng, D., Ning, P., 2007. Carbon nanotube-based supercapacitors. *Nanotechnology Law and Business* 4: 569–577.

47. Xuli, C., Rajib, P., Liming, D., 2017. Carbon-based supercapacitors for efficient energy storage. *National Science Review* 4: 453–489.

48. Lee, J., Kim, J., Hyeon, T., 2006. Recent progress in the synthesis of porous carbon materials. *Advanced Materials* 18: 2073–2094.

49. Wang, H., Dai, H., 2013. Strongly coupled inorganic—nano-carbon for energy storage. *Chemical Society Reviews* 42: 3088–3113.

50. Ajayan, P. M., Stephan, O., Colliex, C., Trauth, D., 1994. Aligned carbon nanotube arrays formed by cutting a polymer resin—nanotube composite. *Science* 265: 1212–1214.

51. Schadler, L. S., Giannaris, S. C., Ajayan, P. M., 1998. Load transfer in carbon nanotube epoxy composites. *Applied Physics Letters* 73: 3842–3844.

52. Wagner, H. D., Lourie, O., Feldman, Y., Tenne, R., 1998. Stress-induced fragmentation of multiwall carbon nanotubes in a polymermatrix. *Applied Physics Letters* 72: 188–190.

53. Ago, H., Petritsch, K., Shaffer, M. S. P., Windle, A. H., Friend, R. H., 1999. Composites of carbon nanotubes and conjugated polymers forphotovoltaic devices. *Advanced Materials* 11: 1281–1285.

54. Kymakis, E., Amaratunga, G. A., 2002. Single-wall carbon nanotube/conjugated polymer photovoltaic devices. *Journal of Applied Physics Letters* 80: 112–114.

55. Matei, R., Alina, P., Luisa, P., 2013. Supercapacitance of single-walled carbon nanotubes-polypyrrole composites. *Journal of Chemistry.* Article ID 367473, 7pages.

56. Muhammad, R. I., Nazmus, S. P. S. M., Rabeya, B. A., Saiful, I. K., 2020. Enhanced electrochemical performance of solution-processed single-wall carbon nanotube reinforced polyvinyl alcohol nanocomposite synthesized via solution-cast method. *Nano Express* 1: 030013.

57. Hoe-Seung, K., Yongju, J., Seok, K., 2017. Capacitance behaviors of conducting polymer-coated graphene nanosheets composite electrodes containing multi-walled carbon nanotubes as additives. *Carbon Letters* 23: 63–68.

58. Foivos, M., Nadia, T., Rossana, G., Constantina, L., Christos, T., 2019. Composite electrodes of activated carbon and multiwall carbon nanotubes decorated with silver nanoparticles for high Power Energy Storage. *Journal of Composites Science* 3: 97–110.

59. Dawei, G., Lili, W., Chunxia, W., Yuping, C., Pibo, M. X., 2017. Preparation and characterization of carbon-based composite nanofibers for supercapacitor. *Autex Research Journal* 17: 129–134.
60. Mingkai, L., Yue, E. M., Chao, Z., Weng, W. T., Zhibin, Y., Huisheng, P., Tianxi, L., 2013. Hierarchical composites of polyaniline—graphene nanoribbons—carbon nanotubes as electrode materials in all-solid-state supercapacitors. *Nanoscale* 5: 7312–7320.
61. Zhihong, A., Zhonghua, H., Yafei, L., Mengxuan, F., Peipei, L., 2016. Novel 3D flower-like CoNi2S4/carbon nanotubes composite as high-performance electrode materials for supercapacitor. *New Journal of Chemistry* 40: 340–347.
62. Sheng, Z., Jiangfeng, N., Yan, L., 2020. Carbon nanotube-based electrodes for flexible supercapacitors. *Nano Research* 13: 1825–1841.
63. Gomaa, A. M. A., Elżbieta, M., Piotr, C., Mohammad, R. T., Jan, R., Algarni, H., Kwok, F, C., 2020. Ferrocene functionalized multi-walled carbon nanotubes as supercapacitor electrodes. *Journal of Molecular Liquids* 318: 114064.
64. Nitin, M., Eti, T., Andrew, S. W., Deanna, S., AnatItzhak, M. M., Gilbert, D. N., Cary, L. P., 2018. Carbon nanotube reinforced structural composite supercapacitor. *Scientific Reports* 8: 17662.
65. Evgeny, S., Yunfu, O., Juan, J. T., Federico, S., Carlos, G., Rebeca, M., Juan, J. V., 2018. Energy storage in structural composites by introducing CNT fber/polymer electrolyte interleaves. *Scientific Reports* 8: 3407.
66. Soumya, M., 2020. CNT-Ni-Co-O based composite for supercapacitor applications by cyclic voltametry analysis: A short quick glimpse. *Material Science Research India* 17: 16–24.
67. Wen, L., Rachel, H., Liangti, Q., Liming, D., 2011. Nanocomposite electrodes for high-performance supercapacitors. *Journal of Physical Chemistry Letters* 2: 655–660.
68. Lota, K., Khomenko, V., Frackowiak, E. J., 2004. Capacitance properties of poly(3,4-ethylenedioxythiophene)/carbon nanotubes composites. *Journal of Physical Chemistry Solids* 65: 295–301.
69. An, K. H., Jeon, K. K., Heo, J. K., Lim, S. C., Bae, D. J., Lee, Y. H. J., 2002. High-capacitance supercapacitor using a nanocomposite electrode of single-walled carbon nanotube and polypyrrole. *Electrochemical Society* 149: A1058.
70. Xu, Y., Pei, S., Yan, Y., Wang, L., Xu, G., Yarlagadda, S., Chou, T. W., 2021. High-performance structural supercapacitors based on aligned discontinuous carbon fiber electrodes and solid polymer electrolytes. *ACS Applied Materials & Interfaces* 13(10): 11774–11782.
71. Lee, C. S., Ahn, S. H., Kim, D. J., Lee, J. H., Manthiram, A., Kim, J. H., 2020. Flexible, all-solid-state 1.4 V symmetric supercapacitors with high energy density based on comb polymer electrolyte and 1D hierarchical carbon nanotube electrode. *Journal of Power Sources* 474: 228477.
72. Zhu, Y., Li, N., Lv, T., Yao, Y., Peng, H., Shi, J., Chen, T., 2018. Ag-doped PEDOT:PSS/CNT composites for thin-film all-solid-state supercapacitors with a stretchability of 480%. *Journal of Materials Chemistry A* 6: 941–947.
73. Changsoon, C., Shi, H. K., Hyeon, J. S., Jae, A. L., Young, C., Youn, T. K., Xavier, L., Geoffrey, M., 2015. Stretchable, weavable coiled Carbon Nanotube/MnO2/Polymer fiber solid-state supercapacitors. *Scientific Reports* 5: 9387–9392.
74. Parayangattil, J., J., Chen, M. Z., Lee, R. H., 2020. Polypyrrole/Carbon Nanotube freestanding electrode with excellent electrochemical properties for high-performance all-solid-state supercapacitors. *ACS Omega* 5: 6441–6451.
75. Xipeng, G., Debin, K., Qin, H., Lin, C., Peng, Z., Huaijun, L., Zhidan, L., Hong, Y., 2019. In situ growth of a high-performance all-solid-state electrode for flexible supercapacitors based on a PANI/CNT/EVA composite. *Polymers* 11: 178–190.

5 Flexible and Stretchable Supercapacitors

Praveena Malliyil Gopi, Kala Moolepparambil Sukumaran, and Essack Mohammed Mohammed

CONTENTS

5.1 INTRODUCTION

Of the numerous energy storage systems, supercapacitors are the most successful candidates for a smart wearable. Supercapacitors can store much energy because of their speedier charge-discharge time, a maximum power density of 10 kW/kg, simpler structure, and prolonged cycle life. So they are ideal for new micro storage devices. Furthermore, flexible and stretchable supercapacitors can achieve high electrochemical performance, and their applications can be extended by adding novel functionalities. Flexible and stretchable energy storage systems are increasingly necessary to power integrated active devices as personal wearable electronics become smaller. However, realizing devices with high storage capacity, adequate mechanical stability and stretchability is a critical problem. The smaller scale, high performance, biocompatibility, lightweight, and versatility are the criterias for these energy storage devices [1–5].

Material substances with greater surface area, including nanocarbon-based materials and structured foam materials, are being used to shrink weight and size, thus maximizing the efficiency and quantity of energy stored on the electrode surface. Supercapacitors possess the unique property of ease of production in flexible/stretchable composition and are equipped with beneficial capabilities because of the flexible and stretchable nature of the electrode materials. However, because they have significant drawbacks such as low specific energy density and operating voltage, considerable work has been put into improving their electrochemical efficiencies, such as enhanced capacitance, electrode potential window, and specific energy density [6].

DOI: 10.1201/9781003174646-5

5.1.1 THE MECHANISMS OF ENERGY STORAGE IN SUPERCAPACITORS

The energy contained in a supercapacitor is typically saved at the interface between electrode and electrolyte through the use of a chemical pseudocapacitive faradaic reaction and the physical capacitive ion adsorption-desorption, and they have high energy densities but lower power densities, depending on the configuration of electrolyte, anode, cathode and a separator. The electrical-double-layer capacitance (EDLC) energy is stored by accumulating charges at the interface between electrode and electrolyte even with no chemical process. Pseudocapacitance, but on the other hand, energy is stored via a reversible redox reaction on the electrode's surface. The redox reaction changes the oxidation state of the electrode; however, there is no phase change as a consequence of the electrode. The electrode surface area in electrical-double-layer capacitance is critical to the capacitor's efficiency, as higher power densities, faster charge-discharge operations, and extraordinary cycling stabilities are all possible, but energy densities are limited.

Graphene, CNTs, activated carbon, graphene, carbon nanofibers, carbide-derived carbon, and mesoporous carbon are examples of carbon materials with impressive conductivity and large surface areas that have been commonly used in electrical-double-layer capacitance superconductors. In Pseudocapacitive superconductors, composite materials made up of electrically conductive polymers and carbon nanomaterials are widely used. Polyethylene dioxythiophene (PEDOT) [3–4], polyaniline (PANI), and metal oxides like RuO, MnO_2, and NiO are some of the examples.

Furthermore, conductive polymer materials have a short cycle life because of their unique inherent instabilities in the structure. Metal oxides have a reduced power density and lack flexibility due to their decreased electrical conductivity and inherent rigidity. Stretchable supercapacitors may benefit from hybrid capacitors to achieve high electrochemical efficiency and mechanical stability. It is essential to improve the performance of supercapacitors by carefully selecting suitable electrode materials. Additionally, obtaining flexibility and stretchability and the ability to deform the shape of portable devices in response to the external stress triggered by muscles or joint motion movements is also important [7–8].

5.1.2 STRETCHABLE AND FLEXIBLE SUPERCAPACITORS

Flexible and stretchable electronics have received much interest and demand in the last ten years for a variety of applications, including thin, flexible, foldable-portable electronics and lightweight, electronic tattoo sensors, skin sensors, compatible surgical tools, and wearable electronics, as well as other electro-mechanical devices with exceptional durability, foldability, flexibility, and stretchability. Currently, researchers working on flexible and stretchable energy storage devices like supercapacitors are focusing on three primary goals: (i) fabricating electrodes and designing them, (ii) achieving steady electrochemical properties, and (iii) increasing power and energy densities. The aforesaid barriers motivate researchers to look for new techniques for flexible and stretchable energy storage system advancements [9]. Stretchable logic devices, field-effect transistors, photodetectors, organic, inorganic light-emitting diodes, and other applications could be made possible by converting solid rigid materials into softer or elastic materials. Supercapacitors that can withstand enormous mechanical strains with no deterioration are known as stretchable supercapacitors. Flexible supercapacitors are among the most attractive power sources in stretchable electrical appliances because of their characteristic properties such as lower energy density, higher power density, and the ability for fast charging-discharging. They also have a simple device structure that is safe, robust, and relatively easy to design, with no hazardous or flammable materials. We tried to describe the most current indicators of advancement on flexible or stretchable supercapacitors, with the primary goal of sustaining good electrochemical performance in conjunction with a significant trend toward wearable and portable electronics, which necessitates that they be lightweight, narrow, and flexible.

The rapid advancement of technology for the fabrication of supercapacitors has resulted in significant changes in the design of devices that uses it. Both one dimension and two-dimensional, wearable micro-supercapacitor systems consisting of all-solid-state gel electrolytes, electrode materials, and flexible/stretchable substrates has been built to replace the typical sandwich-type design of collector, electrode and separator. The foundation of wearable devices is a flexible or extremely stretchable substrate. Due to their excellent physicochemical stability and mechanical flexibility, hydrogel, paper, metal film, polymer plastic, silicone, and carbon clothes/fiber are commonly used as traditional substrates.

Elastomeric polymers (polydimethylsiloxane or rubber) have been used as the principal material in stretchable supercapacitors and lithium-ion batteries. The application of elastomer materials is required to manufacture flexible and stretchable supercapacitors utilizing traditional electrode materials. This kind of supercapacitors may be made by printing or depositing the electrode materials onto elastomer substrates without the need of pre-straining procedures. By using poly(3,4-ethylenedioxythiophene) and polystyrene sulfonate (PEDOT:PSS) and doping it with silver (Ag), Zhu et al. created an electrode based on aligned carbon nanotubes deposited onto a polydimethylsiloxane layer. The developed supercapacitor electrode could be stretched up to 480 percent [10]. A combination of elastomers with conductive materials could be used to make flexible and stretchable electrodes. The preparation steps may be greatly shortened since composite-based supercapacitors do not need a transfer step to achieve its stretchability. The prepared composite electrode material would really be stretchable in several directions without compromising its performance. Nanowires with a high aspect ratio are more favorable among the many nanomaterials because they quickly construct a conductive channel along the wire that surpasses the penetration threshold. Acrylamide, PDMS, and different copolymers are by far the most often utilized polymers for stretchy and flexible composite electrodes [11].

However, this material have many disadvantages: the device's volume and weight increase by using elastomeric polymers, rendering it unsuitable to be used in handheld devices and wearable electronics; the device's specificity is limited. The other consequences are low capacitance and energy density, poor mechanical properties, and low temperature of operation. When conductive materials are combined with elastomers, the conductivity of the material is necessarily diminished in return for the flexibility gained. In addition, elastomeric materials may obstruct ionic transport. With their high ionic and electronic conductivity, large surface area, and mechanical strength, hydrogel-based composites are used as both electrolytes and electrodes in stretchable supercapacitors. Hydrogels with intrinsic capabilities such as stimulus reactivity or self-healing characteristics have recently been described. There are two types of all-solid-state supercapacitors, namely, the sandwich model and the twisted fiber model. The gel electrolytes are placed in between the two electrodes in a sandwich model. Stretchability is achieved in the fiber type model by covering the electrode material in the hydrogel electrolyte [12–13].

Polyethylene terephthalate (PET) is a popular choice among the frequently utilized bendable substrate materials due to its high clarity, versatility, stability in weak acids and alkalis, and low cost [14]. Even though thermal evaporation and photolithography are applicable for PET substrate, the extreme limit for PET processing is that its highest breakdown temperature is about 150 °C. Thus the dispersion of electrode or sensing material over the PET substrate is very hard using the chemical synthesis methods other than a spin coating, which will lead to lower physical interaction between materials and substrates.

Polyimide (PI) easily breaks the constraints due to its high heat resistance and does not influence flexibility or stretchability. This polymer allows for the usage of various synthesis techniques, including chemical vapor deposition on a PI substrate, hydrothermal synthesis, and electro-spinning, all of which leads to the formation of a better chemical reaction between active materials and the substrate, resulting in improved electrochemical performance. Material growth and device development can also be achieved by using mica plate and ultrathin Si film as heat stabilized bendable substrate. Materials like graphite paper and carbon paper have been attracting much attention for printed substrate because they possess novel features such as bendable, foldable, and rollable, which can help minimize electronic waste for future generations of wearable electronics. Stainless-steel mesh has recently been demonstrated to be stretchable, and it usually experiences low strain. Polyurethane

(PU), dimethylpolysiloxane (PDMS), thermoplastic polyurethanes (TPU)/thermoplastic elastomer (TPE), and elastic threads are a few of the major substrates that are stretchable. This material has high transparency, is stable in poor alkali and acids, and is highly stretchable [14–15].

5.2 ELECTRODE MATERIALS FOR STRETCHABLE AND FLEXIBLE SUPERCAPACITORS

The electrode materials for supercapacitors include hybrid composites, conducting polymers, metal oxides and carbon materials. The easiest and efficient technique to improve the electrochemical performance of supercapacitors is to adopt novel electrode materials. Considerable effort has been made to develop new electrode materials, or materials can be improved using various physical and chemical techniques. In the case of an electrical double-layer capacitor, the electrode/electrolyte interface is responsible for the charge storage, and this is influenced by the specific surface area of electrode material and its mesoporous quantity. As a result, carbon materials with a particular porous structure, strong electrical conductivity, and high specific surface area have been extensively studied. Activated carbon, carbon aerogels, graphene, graphene hydrogel and carbon nanotubes (CNTs) are the most commonly used carbon materials as electrode materials [16–17].

Metal/metal oxides, 2-dimensional nanomaterials like graphene, MoS_2, carbon, conducting polymers, and hybrid composite materials have already been for use as suitable electrode materials in the fabrication of flexible/stretchable supercapacitors thus far. The general method for making stretchable electrodes is to coat the stretchable substrates with electroactive materials, such as polymer films, carbon nanotube, graphene, polymer hydrogels, and composites. Metal/metal oxide electrode fabrication is a promising alternative for foldable and bending electronics because of its low cost, strong electrical conductivity, and great stability. Graphene has gained a lot of interest in flexible/stretchable supercapacitors because it is chemically inert and possesses a high surface area of 2,600 m^2/g, significant electrical conductivity, and has high mechanical flexibility. Likewise, carbon-based nanomaterials like CNTs have huge surface areas with outstanding electrical conductivities and possess excellent flexibility, which motivates the fabrication of flexible/stretchable supercapacitors. Conducting polymers including PANI, PEDOT, and PPy are popular electrode material choices in supercapacitors because they have a reasonably high specific capacitance. Still, they have low cyclic stability because of the structural flaws in between the charge-discharge cycles, which limits their use in energy storage devices. These materials were combined to make composite materials for achieving high performance and to resolve the cycling stability problems.

Traditional elastic or stretchable substrates primarily focus upon the materials' mechanical characteristics and deformation capacity to achieve their flexibility and stretchability. Furthermore, most of these substrates have limited conductance or capacitance, making higher energy and power densities challenging to attain. Unlike conventional substrates with flexibility and stretchability, an optimal design could be realized by selecting appropriate active elastomer materials and adjusting the material's flexibility and stretchability through structure construction. As a result, it would offer a practical way to expand the range of materials applications and overcome their mechanical property limitations [18, 12–13].

The highest specific capacitance of NiO makes it a suitable electrode material for fabrication of flexible and stretchable supercapacitors. But the mono metal/metallic oxide electrode materials exhibit lower electrical conductivity than bimetallic oxide electrodes. The bimetallic electrodes are highly expected to solve the low electrical conductivity issues in supercapacitors, and they can achieve significant energy density values. For example, it is reported that, when compared to the monometallic oxides NiO and Co_3O_4, the spinel bimetallic oxide $NiCo_2O_4$ has a double- or triple-fold electrical conductivity value. Several bimetallic oxide electrode materials have begun to appear as prospective flexible supercapacitor electrodes, including $CuCo_2O_4$ $CoMoO_4$, $NiCo_2S_4$, and $NiFe_2O_4$ [19–20].

Ruthenium oxide (RuO_2) is a widely researched electrode material for flexible and stretchable supercapacitors. This metal oxide possesses high thermal stability, good specific capacitance value, strong electrical conductivity, and good cycle stability. Furthermore, RuO_2's remarkable corrosion protection ability with acidic and basic chemicals enhances the performance of the supercapacitor. However, practical applications of this metal oxide are limited because of the agglomeration during the charging-discharging process. Numerous efforts have been made according to the literature to achieve structurally strong, flexible, and stretchable supercapacitors with high energy density [21]. A cathodic electrodeposition approach to make a RuO_2/Graphene/Cu ternary electrode for flexible-bendable super-capacitor applications was used by Hyungsang et al. [22]. Electrochemical performance of the supercapacitor under the bent condition was improved by the addition of graphene. This electrode exhibits an enhanced specific capacitance value of 1,561 F/g with a power density value of 21 kW/kg.

MnO_2 possess several advantages among metal oxide-based electrodes, and it includes high theoretical specific capacitance, a broad potential range, and is environmentally friendly. As an alternative, electrodes made up of conducting polymers are often selected. These conducting polymers have theoretical capacities of 100 to 140 mAh/g, adjustable redox activity, wide voltage window, environmental compatibility, and remarkable storage capacity/reversibility; conducting polymer-based electrode materials are often chosen. Conducting polymers are helpful as innovative materials because they typically change color in response to different factors such as ligand interaction, ions, temperature, solvent, and pH value. The fabrication of a series of polymer SCs devices has been completed. Peng and co-workers, for example, have given detailed descriptions of polyaniline (PANI) electrochromism [23–24].

Carbon materials have greater charge transport capacity together with electrolyte accessibility because of their large surface area and porous structure, which makes them suitable for flexible/stretchable supercapacitor devices. Yani Teresa et al. developed a flexible 3D electrode employing carbonized flax fabric with in situ generated CNTs. CNTs were grown over the 3-dimensional flexible carbonized flax fabric surfaces, which could provide a broad surface area and also in situ nucleation, resulting in a tightly connected interface. After 5,000 cycles, the prepared electrode has a surface area of 580 m^2/g and it could achieve a specific capacitance value of 191 F/g at 0.1 A/g [25].

5.2.1 Structural Configurations of Flexible and Stretchable Electrodes

Researchers have been trying to improve and stabilize the electro-mechanical characteristics of flexible and stretchy electrodes for many years, starting with creating unique electrode structures and combining them with appropriate conductive materials and substrates. This section discusses some basic structures of flexible and stretchable electrodes.

The possible structures of stretchable electrodes are wavy, nano-network, helical, serpentine-like, and textile. The wavy texture structure for a stretchable electrode can be obtained by coating a pre-stretched elastic substrate with conductive materials and releasing it. The stretching and releasing of the substrate generates wrinkles, which is the reason for wave formation. Another explanation is that conductive materials develop on wavy substrates before being moved to elastic substrates. Nanomaterials distributed on an elastic substrate form network structures classified into two disordered and ordered network structures. The one-dimensional linear metal nanowires are commonly used to make disordered network structured tensile electrodes.

The helical structure is preferable for one-dimensional conductive materials. As the stretchable electrode with helical structure is extended, the distance between conductive materials progressively widens, resulting in a loss of mutual control. As a result the stretchable electrodes' internal resistance does not show any significant improvement. The helical structure can tolerate more tensile deformation. The electro-spinning method was used to prepare the yarn for textile structure, then weave yarn into stretchable, breathable, wearable fabrics. The serpentine-like structure, in which conductive materials are distributed on an elastic substrate in a specific ordered configuration, is perhaps the most typical and frequently used in stretchable electrodes [26, 57]. Various processes such as screen printing, inkjet printing, and spray coating were used to create stretchable electrodes on elastic substrates. The development of stable dispersions is an important concern that these fabrication procedures will aid with [27–28].

The current flexible electrodes are classified into three categories according to the microstructures and macroscopic patterns: fiber-like, paper-like, and three-dimensional (3D) porous flexible electrodes. Fiber-like structure is made up of two fiber-like flexible electrodes twisted around each other, while an inner electrode with a coaxial structure and an outer electrode is made up of a fiber-like flexible electrode and a paper-like flexible electrode (otherwise known as "film electrode" and "membrane electrode"). In most of the supercapacitors, the space between two electrodes contains electrolyte and a separator. For a liquid electrolyte, a separator is often needed, but in the case of solid gel electrolyte, it can act as a separator and resist short circuits within supercapacitors. Liquid gel electrolytes also have the disadvantage of electrolyte leakage. The fiber-like flexible electrode should be highly flexible and have excellent electrochemical properties. Electrochemically active powder materials are coated on a metal foil prior to the electrochemical performance tests. This coated metal foil can be considered a paper-like flexible electrode. The paper-like flexible electrodes are split into free-standing, flexible electrodes, and the flexible substrate supports a flexible electrode [29–30].

A novel stretchable electrode based on the composite film was developed by Wang et al. [31], consisting of acrylate rubber/multiwall carbon nanotubes (ACM/MWCNTs) backed by poly (1, 5-diaminoanthraquinone) or polyaniline. Acrylate rubber/multiwall carbon nanotubes with poly (1, 5-diaminoanthraquinone) are denoted as ACM/MWCNTs@PDAA, and these were used as the anode. Acrylate rubber/multiwall carbon nanotubes with polyaniline is noted as ACM/MWCNTs@ PANI, and this was used as the cathode. The ACM/MWCNTs film seems to have a high electrical conductivity (9.6 S/cm) and elastic resilience because it contains 35 percent multiwall carbon nanotubes. The anode has the basic volumetric capacitance of 20.2 F/cm^3, and that of ACM/MWCNTs@ PANI cathode is 17.02 F/cm^3. The electrochemical performance of the prepared electrode (stretchable) was analyzed by assembling an organic asymmetric stretchable supercapacitor using ACM/ Et4NBF4-AN as the semi-solid-state electrolyte. The prepared asymmetric stretchable supercapacitor possesses superior energy density with a value of 2.14 mW/cm^3, and it outperforms several other supercapacitors in terms of cycle stability, even when subjected to static and 50 percent strain.

Wang et al. [23] created an elastomeric solid-state supercapacitor composed of polyaniline hydrogel electrodes rather than the conventional solid electrode materials. The combined conducting polymer hydrogel changes its electrical conductivity leads to inherent conducting porous frameworks that facilitate the charges, ions, and molecules transportation. In these two electrodes configured prototype flexible solid-state supercapacitor, the capacitance value for the polyaniline hydrogel electrode is 430 F/g, which is a pretty remarkable value. In addition, the bendability, rate capability, and cyclic stability of this supercapacitor are all excellent. Furthermore, this supercapacitor can effectively move a glow armlet, demonstrating that the system has much potential for real-time applications.

The surface-modified nanocellulose fiber (NCF) substrates for making supercapacitor electrodes was prepared by Wang et al. [24]. The prepared electrodes could achieve the highest full electrode-normalized gravimetric capacitance of 127 F/g and volumetric capacitances of 122 F/cm^3 at high current densities of 300 mA/cm^2 33 A/g with mass loading of up to 9 mg/cm^2. When unaltered or carboxylic groups are functionalized, NCFs are employed as polymerization substrates, the macropore volume of PPy-NCF composites can be lowered, although the micro and mesopores can be preserved at the same level. Before that polypyrrole (PPy) polymerization process, the surface of NCFs was modified by adding quaternary amine groups. Device-particular volumetric energy value and power density of 3.1 mWh/cm^3 and 3 W/cm^3 have been documented for electrode material made up from conducting polymer used in an aqueous electrolyte. The effectiveness of the instruments was evaluated using a red LED (light-emitting diode).

To overcome the drawbacks of each individual material, hybrid composite materials combine several nanomaterials in/on different polymer backbones. For example, introducing metal oxides into a graphene or conductive polymer hybrid composite can help hybrid electrodes have greater mechanical, electrical, and electrochemical capabilities. Hybrid composite materials combining carbon compounds with metallic oxides or polymers (conducting) possess the characteristic properties of both its components. Metal oxides or polymers have a higher capacitance value, while carbon compounds provide a large specific surface area and strong conductivity. As a result, high-performance supercapacitors can use electrodes made from hybrid materials. Choi et al. created yarn supercapacitors out of a MWNT/MnO_2 composite with a very high specific capacitance of 25.4 F/cm^3 [30, 32].

Soft materials such as nickel foam, carbon cloth and metal mixed cloths are indeed the best for achieving good comfortability, flexibility as well as stretchability. Even though supercapacitors have come a long way, attaining stable performance even when subjected to stretching, twisting, and bending is still a major challenge in the development of flexible/stretchable supercapacitors.

5.2.2 THE FABRICATION TECHNIQUES OF ELECTRODE MATERIALS

Pseudocapacitance is an interfacial property that is closely related to prepared electrode materials morphology, and the importance of morphology in boosting electrochemical efficacy is widely acknowledged. As a result, different synthesis conditions, including the growth temperature, reactant concentration, and growth time, must be controlled to tailor the dimension and morphology of optimal electrode materials selectively. Furthermore, researchers have shown that several synthesis methods, such as chemical precipitation, electrodeposition, mechanochemical, chemical bath deposition, hydrothermal and sol-gel and all of this have been widely used to obtain intended characteristics in electrode materials for supercapacitors [33].

The porous electrode materials are mostly produced with the aid of chemical vapor deposition technique. This method is employed for the synthesis of defect-free graphene structure. In the typical process, a template is required in order to form a new graphene layer and the whole process takes place in the presence of vapor. The vapor formed graphene was heated to a high temperature (800–1,000 °C) in the presence of a target substrate and then uniformly deposited over the substrate [34].

The electrodeposition technique is mainly utilized for non-toxic compounds and requires just modest processing conditions. PANI, polypyrene, and poly (3,4-ethylenedioxythiophene) are among the conductive polymers synthesized using this method. The low cyclability issue of conductive polymers can be resolved by combining with EDLC materials including graphene, CNT, and activated carbon. Recently, graphene has been employed to improve device performance by combining a one-layer-thick carbon sheet with conducting polymers to produce high conductivity and a large surface area [35].

The preparation of hybrid graphene electrodes is done using the hydrothermal technique. An autoclave and a temperature controller are used in this approach to regulate the crystallization process of the material. Graphene and vanadium-based hybrid supercapacitor electrode was prepared by Lee et al. by employing the Hummers method. while the hydrothermal approach is used to make VO_2 [36].

Direct coating is another simpler and quick process most often utilized approach for fabricating supercapacitors electrode materials. The active material is deposited onto the substrate material. The adherence of the substrate to the active material is crucial in this technique. The adhesion can be improved by adding binders like polyvinyl fluoride with the active material. It is also possible to retain the conductivity of the electrode by mixing activated carbon or carbon black with the material prior to the coating onto the substrate [37].

5.3 ELECTROLYTES USED FOR STRETCHABLE AND FLEXIBLE SUPERCAPACITORS

Electrolytes are generally known as electrically conducting solutions, consisting of solute molecules dissolved in any polar solvents such as water, and they can separate cations and anions following dissolution. Choosing the proper electrolytes is also critical for achieving high electrochemical efficiency. The electrolytes possess the following functionalities as an ion source, an electrode particle adhesive, and an electric charge conductor, higher electrochemical stability with a wide range of voltage windows for enhancing the energy density, and it prevents depletion issues by high ionic concentrations, high electrochemical stability, nontoxicity, low volatility, cost-effective, and not viscous. The selection of mixed solvents leads to the optimization of electrolyte conductivity. Interactions between electrolytes and electrodes, as well as those between ions and solvents, have the ability to determine cycle's lifespan and the self-discharge of supercapacitor in a mixed solvent. Furthermore, aqueous electrolytes and solid/quasi-solid-state electrolytes are the two primary forms of electrolytes [38–39].

Ultimately wearable energy storage devices, especially supercapacitors, rely heavily on all-solid-state gel electrolytes with excellent stretchable properties. The chance of liquid leakage can be reduced by the gel electrolyte with the characteristics like reliability, simplicity, and dependability, making supercapacitor fabrication easier by omitting additional substrate and separator. Gel polymer electrolytes are more preferred than dry solid-polymer electrolytes because of their higher ionic conductivity in atmospheric conditions. Modern gel electrolytes include organic ions electrolytes and solvent-based electrolytes. They exhibit excellent air stability and improved electrochemical performance [40].

There are two types of electrolytes: polyelectrolytes based on hydrogels and organic or solvent-based ions. The polyelectrolytes degrade rapidly after one day in the air but organic solvent-based gel electrolytes have more than two weeks of air stability. One of the examples for solvent-based gel electrolytes is reported by Ha et al. [12]. They created a solvent that is not aqueous, and it is a poly (methyl methacrylate)-propylene carbonate-lithium-perchlorate gel electrolyte. A polymeric matrix, an aqueous dispersing media, and conducting ions from an electrolytic salt/acid/alkali, make up hydrogel electrolytes. The majority of hydrogel polymer electrolytes, such as PVA-acid/alkali/salt systems, are used in flexible superconductors, leading to poor mechanical strength and low structural integrity.

Bu et al. [40] studied a polymerized zwitterionic molecule and demonstrated the in situ synthesis of a collection of hybrid cross-linked zwitterion-containing copolymer deep eutectic solvent (DES) gels within DES. They have shown that this could be possible by the free radical copolymerization of poly (ethylene glycol) diacrylate (PEGDA), acrylic acid (AA), and sulfobetaine vinyl imidazole (VIPS) monomers induced UV. The molar ratio of AA: VIPS and contents of the copolymer in the copolymer network are the two parameters that can change the ionic conductivity as well as the mechanical properties of the poly (AA-co-VIPS) DES gels. The value of tensile strength (28–176 kPa) and fracture strain has increased in P (AA-coVIPS) DES gels (720–1370 percent). The copolymer DES gels possess a very high ionic conductivity value (2.7–4.1 mS/cm) as the copolymer content in the gel increased from 25 percent to 45 wt percent. This particular gel offers excellent capacitive output across a wide range of temperatures, and also it enables the supercapacitor to produce a capacitance value of 71.52 F/g with maintaining 97 percent of its capacitance value even after 2000 cycles.

To obtain a suitable conductive polymer, two main conditions should be considered. The first is that the polymer must have conjugated double bonds, which are double and single bonds that alternate. Step two is the doping method in which the polymer is disrupted by injecting electrons or extracting them into it. The three conductive polymers that have been used as stretchable conductors are polypyrrole (PPy), poly (3, 4-ethylene dioxythiophene):poly (styrene sulfonate) (PEDOT:PSS), and polyaniline (PANI).

5.3.1 PANI

PANI possess excellent electrical conductivity and pseudocapacitive nature that is useful for enhancing the supercapacitor's electrochemical efficiency. PANI and graphene are being used as electrode materials in a stretchy all-solid-state supercapacitor by Yang Chai et al. The entire supercapacitor exhibits high durability and stretchability due to the wavy structure and the use of H_3PO_4-polyvinylalcohol (H_3PO_4-PVA) as an electrolyte. Furthermore, the device's electrochemical performance can be relatively constant under high bending and tensile strain of 30 percent. Using a borated PANI-poly(vinyl alcohol) hydrogel conductive polymer and a freeze-thaw cycling technique, Mingming et al. created a flexible supercapacitor. After five freeze-thaw cycles, the manufactured electrode has increased stretchable strength and also the elongation break. With a capacitance of 420 mF/cm^2 and an energy density of 18.7 Wh/kg, electrode outperformed the others. The quality of freeze-thaw cycles for improving the performance of functional hydrogel electrodes was disclosed by this flexible supercapacitor [41].

Xie et al. prepared a graphene and polyaniline based stretchy electrode for supercapacitor application. In a PVA/H_3PO_4 electrolyte, they tested electrochemical performance of the supercapacitor. The resulting electrode has a high specific surface area and porosity. The electrode measures around 100 micrometers in thickness. At a current density of 0.38 Ag^{-1}, the specific capacitance was 261.24 Fg^{-1}[42].

5.3.2 POLYPYRROLE

Polypyrrole (PPy) is a popular conductive polymer along with its simplicity of fabrication, lack of toxicity, strong adhesion to a variety of substrates, and high conductivity. The Wuhan Textile University's Dong Wang group fabricated a human breath detection strain sensor prepared by depositing PPy upon a Polyurethane (PU) elastomer. It leads to the development of a stretchy capacitor with excellent sensitivity but also reproducibility. This stretchable conductor could exceed the value of electrical resistivity to 8.364 Ωcm. PPI is also a good candidate for gas sensors (CO_2, N_2, CH_4, H_2S, NH_3, etc.), and it is also using the electronic nose. The in situ polymerization process was used to make a flexible and stretchable electrode using PPy and knitted cotton fabric for wearable electrical device application. At 1 mA/cm^2, this dual functionable electrode could achieve a areal capacitance value of 1,433 mF/cm^2. This flexible electrode also has excellent strain capacitances, indicating that it could be used in wearable electrical devices in the future [43].

5.3.3 POLY(3,4-ETHYLENE DIOXYTHIOPHENE) POLYSTYRENE SULFONATE

Poly (3, 4-ethylene dioxythiophene) polystyrene sulfonate (PEDOT:PSS) is one of the attractive electrode materials that have many appealing properties, including visible spectral clarity, high air and thermal stability, and tunable conductivity in the range of 10^{-4} to 10^{-3} S/cm. The stretchable transparent electrodes in the form of a thin film using PEDOT:PSS mixed with Zonyl fluorosurfactant on polydimethylsiloxane (PDMS) substrates were studied Zhenan Bao et al. The synthesized film can maintain conductivity and reversible stretchability even under the very high value of strain. PEDOT:PSS has also been employed in organic light-emitting diodes (OLEDs) and stretchable sensors.

Elastic polymers with inherent stretchability, such as polyurethane (PU), PDMS, acrylate rubber, and silicone rubber (Ecoflex®), are the most commonly used stretchable substrates. After structural adjustment, less stretchable or non-stretchable textiles and cellulose may also be used as substrates. Metals, notably gold, silver, and copper, are used as the current collector materials. The high conductivity possessed by these metallic materials makes them ideal for rising applications. On the other hand, bulk metal materials have an intrinsic stiffness that causes conductivity to decrease at low pressure, restricting their usage in stretchable items. Some wavy shapes have been designed, and nanostructured metals have been developed to improve the stretchy feature. Carbon-based products, such as since the introduction of carbon nanotubes (CNTs) and graphene due to their existing collectors, into stretchable supercapacitor flexibility and high conductivity. The more commonly used electrolyte for stretchable supercapacitors is polyvinyl alcohol (PVA), which has a mild deformation capability. Other electrolytes include polyacrylamide hydrogels that were created to accommodate considerable strain. The electrode materials for stretchable supercapacitors include transition metal oxide, conductive polymers, and carbon compounds [44, 13].

The substrate materials also play a vital role in the fabrication of flexible and stretchable supercapacitors. The mechanical strength and flexibility of electrode materials are provided by the substrate material. Carbon cloth, nickel foam, graphite sheets, and aluminum foils are just a few of the substrate materials discussed in previous studies. These substrate materials possess high conductivity, flexibility with porosity. Substrates serve as a current collector in flexible supercapacitors and hence can provide a conductive framework for the adhesion of electrodes. Carbon nanofibers prepared from electrospinning methods have recently been adopted as a substrate material due to their superior mechanical flexibility, high conductivity. The substrate ought to be soft enough to enable out-of-plane deformation of conductive materials in general [45–46].

5.4   RECENT STUDIES ON THE APPLICATION OF FLEXIBLE AND STRETCHABLE SUPERCAPACITORS

The quest for renewable energy sources is one of the most relevant and fascinating difficulties confronting technology and science in the 21st century. The practical, eco-friendly and renewable energy generation and use have generated a lot of interest worldwide. Devices like batteries, fuel cells, and electrochemical supercapacitors are critical for supplying renewable energy for portable, stationary, and transportation applications. Electrochemical supercapacitors are a new type of energy storage system among the many available. They have a higher energy storage capacity than traditional capacitors and deliver more power than batteries [47–48]. Some of the recent studies involving stretchable and flexible supercapacitors, their electrochemical performance and applications are described here.

Supercapacitor-based flexible integrated systems have gotten much attention in recent years because they can provide noninvasive real-time monitoring through detectors and sensors. To date, several integrated systems are aimed at portable and wearable electronics, such as supercapacitors, sensor devices and self-driven all-in-one devices (along with a nanogenerator, a photovoltaic cell or wireless charging units). The kit of flexible/stretchable supercapacitors is discussed in this section [13].

The textile supercapacitors based on MnO_2 nanoparticles in CNT provide superior power density and energy density, but MnO_2 nanoparticles can delaminate at the charge-discharge process, which will cause significant capacitance loss. Yan et al. [49] tried to avoid MnO_2 nanoparticle delamination by covering polypyrrole on CNT textile supercapacitor with MnO_2 nanoparticles deposited on the top. Figure 5.1 demonstrates the prepared supercapacitor's electrochemical performance. There was a 38-percent increase in electrochemical energy capacity (461 F/g) and a boost in cyclic efficiency.

For the investigated system, the energy density of 31.1 Wh/kg and power density of 22.1 k W/kg was obtained. An in situ electrochemical and mechanical analysis was reported. The prepared textile supercapacitor was subjected to 21-percent tensile strain, and it was found that it could preserve 98.5 percent of all its original energy capacity. When it was subjected to 13-percent bending, there was no measurable change in the value of energy storage capacity. The applicability of these supercapacitors by operating a green LED is shown in Figure 5.2. This study showed that when tensile strain and bending strains are performed in situ, the prepared composite textile will perform even at maximum power and energy densities.

The development of substrate-free solid-state supercapacitors has ignited interest in conductive polymer-based hydrogels (CPHs), but most polymer-based hydrogels are not used because they exhibit poor electric conductivity and low shelf stability. Wang et al. [50] developed CPH with stretchability, which can be used to produce any polymer solid-state supercapacitors. Moreover, the author claimed that this type of supercapacitor could give high specific capacitance, better electric conductivity, and extended storage stability. Polyaniline and soft poly (vinyl alcohol) are interactively cross-linked with boronate bonds to create this CPH. The flexible network architecture of

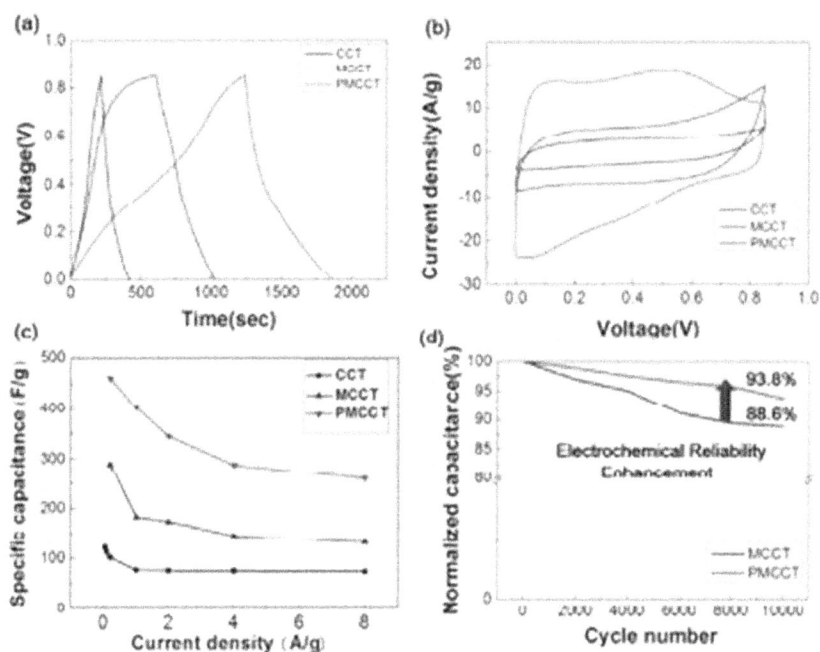

FIGURE 5.1 (a) tTe charge-discharge test, (b) cyclic voltammetry test, (c) specific capacitance normalized by mass, and (d) 10,000 cycle electrochemical reliability testing for the polypyrrole, MnO_2 coated CNT-textile (PMCCT), MnO_2 coated CNT-textile (MCCT) and CNT-cotton (CCT) electrodes. Reproduced with permission from ACS.

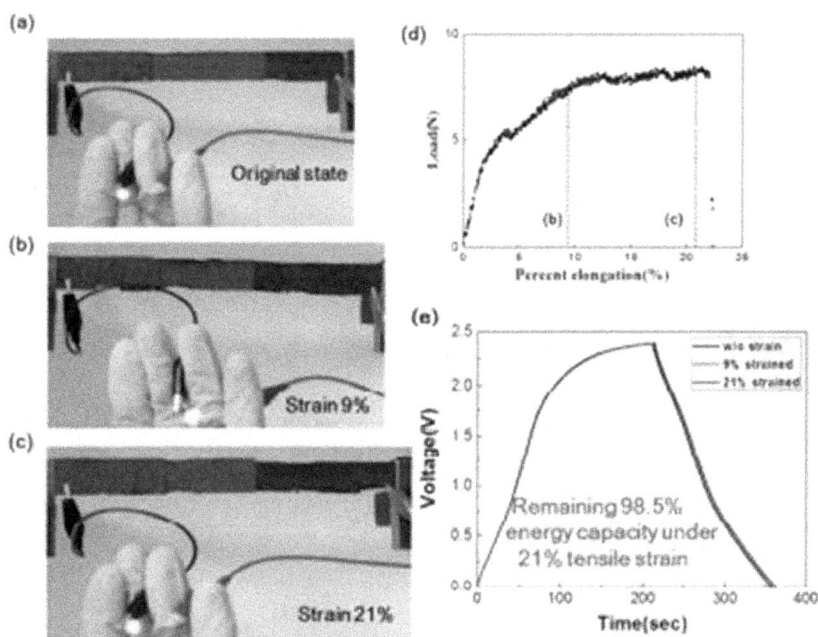

FIGURE 5.2 (a) Strain-free operation of a green LED utilizing prepared textile supercapacitor. (b) 9-percent and (c) 10-percent tensile strains, as well as the related (d) normalized load and percent elongation graphs. (e) In situ changing charge-discharge time measurements under tensile strains of 0 percent, 9 percent, and 21 percent. Reproduced with permission from ACS.

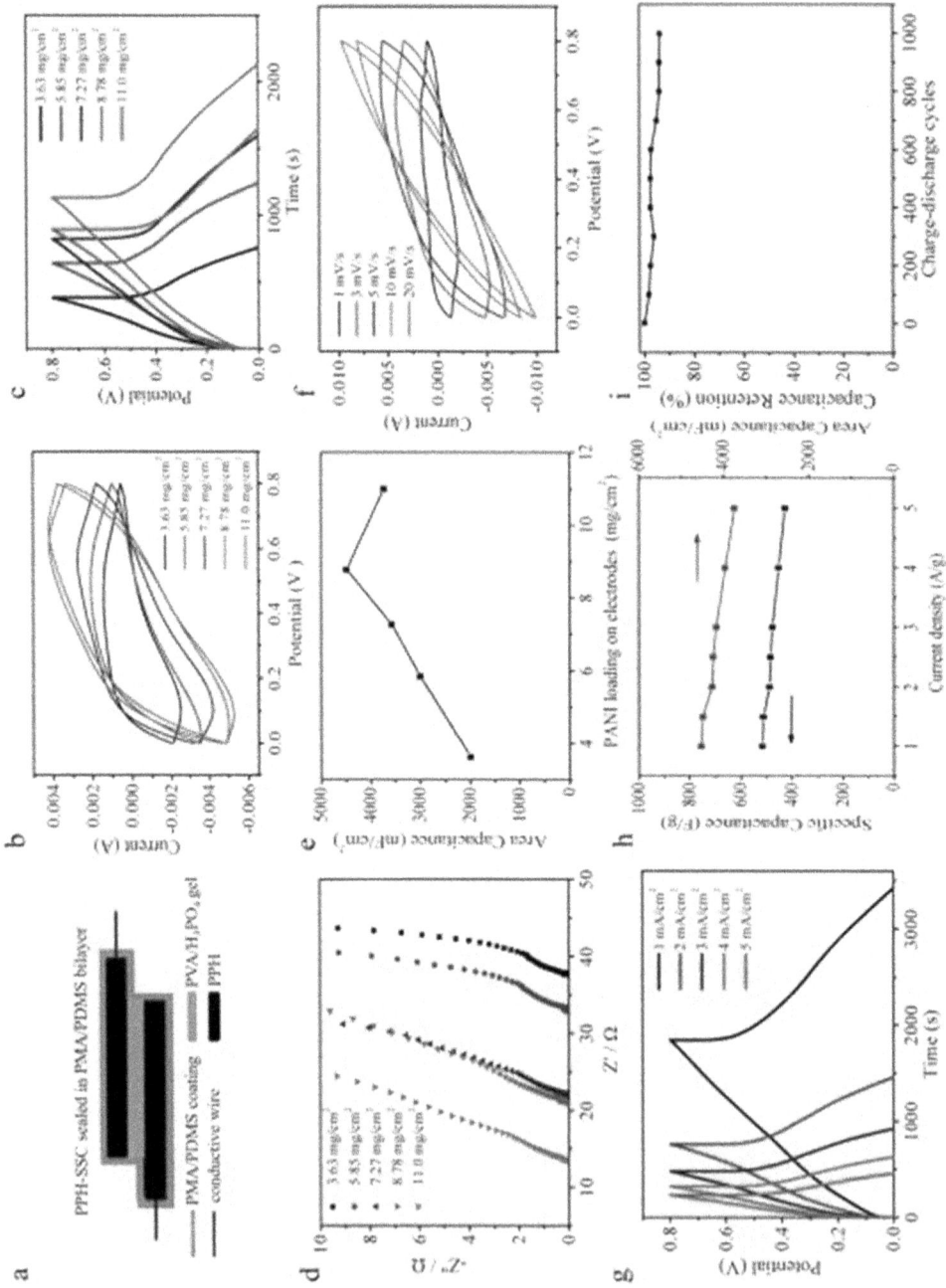

FIGURE 5.3 (a) Electrochemical characterization of the PPH-SSC (b, c) CV, GCD, and (d) impedance graphs with varied PANI loadings on areal capacitance. (f–i) With 8.78 mgcm⁻² PANI loading, the CV, GCD, specific capacitance, areal capacitance, and capacitance retention graphs are shown. (e) The effect of varied PANI loading on areal capacitance. Reproduced with permission from ACS.

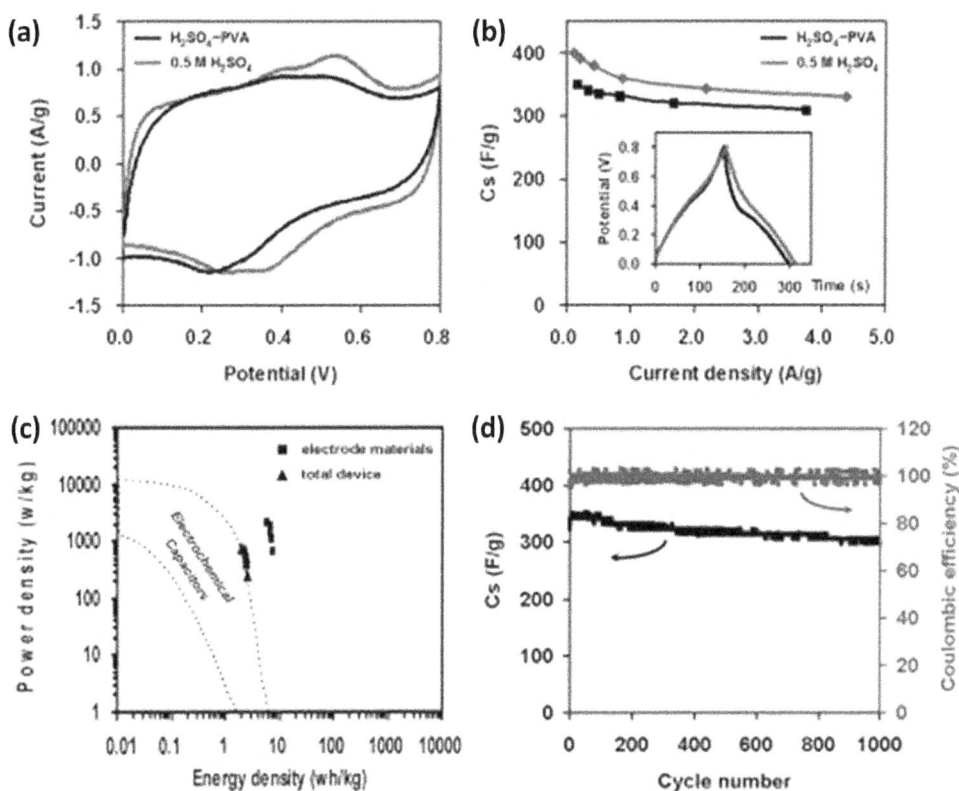

FIGURE 5.4 The electrochemical performances of the devices are presented. A and B represent the evaluation of discharge abilities and voltammetry at 5 mV/s. One cycle of galvanostatic charge-discharge graphs at 1 A/g is shown in the inset in (B). Ragone plots for electrode materials and the whole all-solid-state device (C) and (D) The all-solid-state device's cycle stability at 1 A/g. Reproduced with permission from ACS.

this CPH can dissipate destructive energy during physical or electrochemical distortion, making it resilient and powerful; this even aids in polyaniline use, allowing it to have high electric conductivity and specific capacitance values. The electrochemical performance of the supercapacitors is illustrated in Figure 5.3.

Meng et al. [51] demonstrate that an ultrathin all-solid-state supercapacitor can be made with an exceedingly easy procedure employing two slightly spaced polyaniline-based electrodes thoroughly solidified in the gel electrolyte based on H_2SO_4-polyvinyl alcohol. The entire device is around the same thickness as a piece of A4 print paper (commercially available). In its twisting-flexible state, the integrated device displays a specific capacitance value of 350 F/g for the electrodes, good cycle stability over thousand cycles, and a very low leakage current value of 17.2 μA. The total specific capacitance of the system is 31.4 F/g. Figures 5.4 and 5.5 highlight the electrochemical performance and application of the constructed flexible paper-like device.

Saborio et al. [52] designed a flexible electrode using a two-way technique including poly(3,4-ethylene dioxythiophene) (PEDOT) microparticles and a poly-glutamic acid (-PGA) hydrogel matrix. PEDOT microparticles are initially added into the -PGA matrix when interacting between the biopolymer chains and the cystamine cross-linker. PEDOT particles are used again for amperometric (Chrono) preparation of poly(hydroxymethyl-3,4-ethylene dioxythiophene) polymerization nuclei in the aqueous medium. The cyclic voltammetry analysis for the synthesized supercapacitor is displayed in Figure 5.6. The capacitance properties of the electrode composites depend

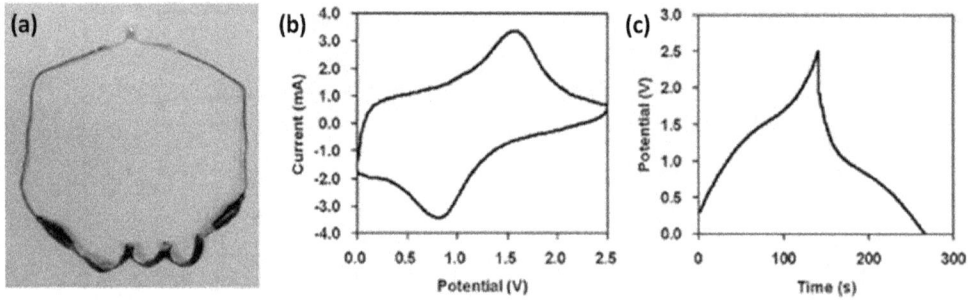

FIGURE 5.5 The very flexible paper-like technology is demonstrated in action. A digital image of three highly flexible devices connected in series to illuminate a red LED well (A). CV plot at 5 mV/s (B), and galvanostatic charge-discharge plot at 5 mA (C). Reproduced with permission from ACS.

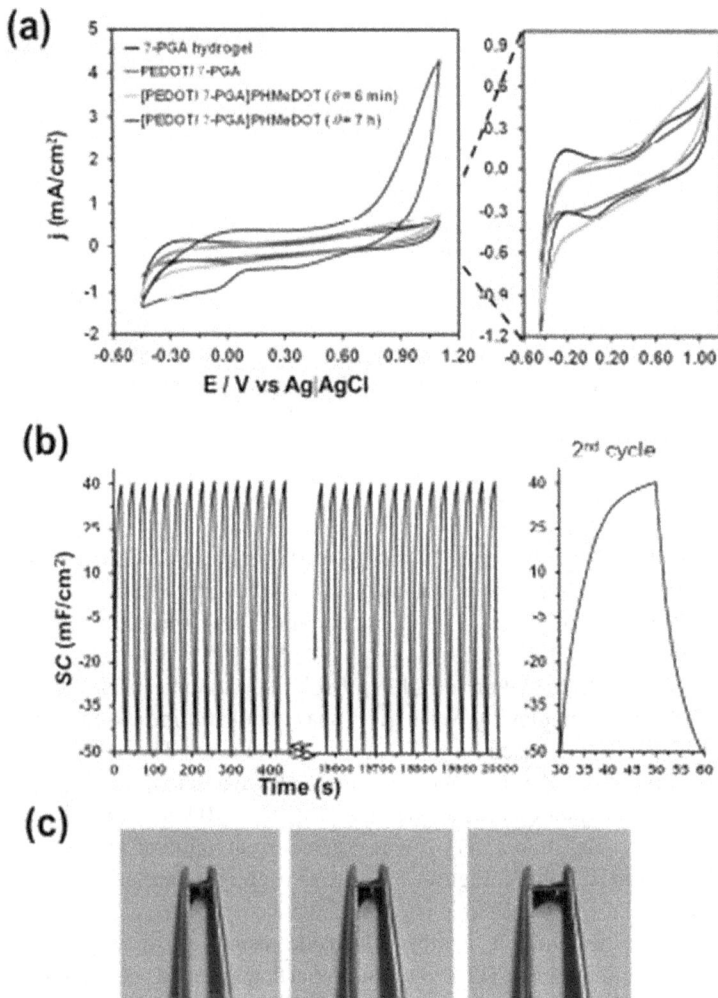

FIGURE 5.6 (a) Control voltammograms (2nd cycle) obtained for γ-PGA, PEDOT/γ-PGA, [PEDOT/γ-PGA] PHMeDOT(θ = 6 min) and [PEDOT/γ-PGA]PHMeDOT(θ = 7 h) with a scan rate of 100 mV/s. (b) Galvanostatic charge-discharge plots for [PEDOT/γ-PGA]PHMeDOT (θ = 7 h) recorded at 0.1, and (c) photographs illustrating the mechanical strength and compression behavior of [PEDOT/-PGA]PHMeDOT(= 7). Reproduced with permission from ACS.

on the time of polymerization utilized to prepare poly(hydroxymethyl-3,4-ethylene dioxythiophene) within the preloaded -PGA matrix, according to electrochemical experiments. The produced flexible electrodes have a polymerization time of seven hours and a specific capacitance value of 45.40 mF/cm^2 based on voltammetric and charge-discharge long term stability studies. The author demonstrated the ability of these electrodes to power an LED bulb in a lighter and portable energy-harvesting system appropriate for energy-independent, low-power, reusable electrical gadgets. Figure 5.7 demonstrates the fluctuation in electrical conductivity and the use of a supercapacitor to power an LED.

Electronic textiles have gotten a lot of interest recently as cutting-edge technology for smart wearable of the future. Current power sources are incompatible with wearable devices because of their restricted versatility, high price, and lack of environmental friendliness. Sundriyal et al. [53] showed in their work that how bamboo fabric can be used to create supercapacitor devices that could easily be incorporated into smart wearables. This work uses a variety of metal oxide inks to print straight on fabric substrates made of bamboo, revealing a repeatable printing method. In order to construct a hybrid battery-supercapacitor unit, MnO_2-$nico_2o_4$ serves the positive electrode, the negative electrode is rGO, while the solid-state electrolyte is LiCl/PVA gel on bamboo fabrics. The MnO_2-$nico_2o_4$/rGO supercapacitor has obtained enhanced electrochemical properties such as electrochemical efficiency (1,766 F/g) at 2 mA/cm^2, areal capacitance value obtained is 2.12 F/cm^2, the energy density value is 37.8 mW/cm^3, and long cycle life with power density value achieved is 2,678.4 mW/cm^3. It could retain its electrochemical properties throughout the mechanical deformation conditions, showing the high mechanical strength together with flexibility. The proposed strategy would make it easier to create long-lasting electronic textiles for portable devices.

Huang et al. [54] used a simple fabrication technique to create a PANI pseudocapacitor capable of high rates, good flexibility, and good ability to stretch. This supercapacitor made from PANI and PVA/H$_3$PO$_4$ gel polymer electrolyte shows superior pseudocapacitance activity within the potential range of 0 to 1.4 V. In its initial relaxed state, the supercapacitor can reach a capacitance of 369 F/g, which is maintained in various deformation states, and it was explained as the cumulative effect originated from the conducting substrate and also the elastic gel electrolyte. Electrodeposition technique with PVA/H$_3$PO$_4$ gel electrolyte was used for the development of the supercapacitor. The reported supercapacitor exhibits significant mechanical properties and rate capability. The supercapacitor has a specific capacitance obtained is 282 F/g for 2.5 A/g applied currents. The supercapacitor has shown pseudocapacitance activity for scan rate value reaches 50 mV/s, and scanning rates of up to 5000 mV/s are available with capacitive characteristics.

Wang et al. [55] showed that the nanocellulose fibers with surface modifications are used as substrates for developing electrodes in supercapacitors. This can provide a full electrode-normalized gravimetric (127 F/g) and volumetric (122 F/cm^3) capacitances for a higher current density in the range of 300 mA/cm^2 ~~ 33 A/g. Prior to polypyrrole (PPy) polymerization processing, quaternary amine groups were added to the surface of nanocellulose fibers. The volume of macropore of the prepared nanocomposites can be decreased. While the micro and mesopores volume can be preserved to a level, as polymerization substrate, carboxylate groups functionalized nanocellulose fibers are employed. Device-specific volumetric energy and power densities of 3.1 mW/cm^3 and 3 W/cm^3 have been documented for conductive polymer electrodes used in aqueous electrolytes. When the supercapacitor system is indifferent to mechanically demanding states, the usability of the systems is checked by powering a red light-emitting diode.

The synthesis of Au nanograin designed aligned multiwall carbon nanotube (CNT) sheets, followed by the addition of polyaniline, was described by Xu et al. [26] (PANI). The Au nanograins in the linear electrodes facilitate rapid radial ion diffusion while also improving axial electron transport. The supercapacitor made by twisting two PANI@Au@CNT yarns has obtained a volumetric capacitance value of 6 F/cm^3 for a 10 V/s scan rate and has superior electrochemical efficiency. Buckled linear electrodes made by wrapping PANI@Au@CNT sheets on elastic rubber fibers are also used to make highly stretchable supercapacitors with significant rate efficiency, cycling and

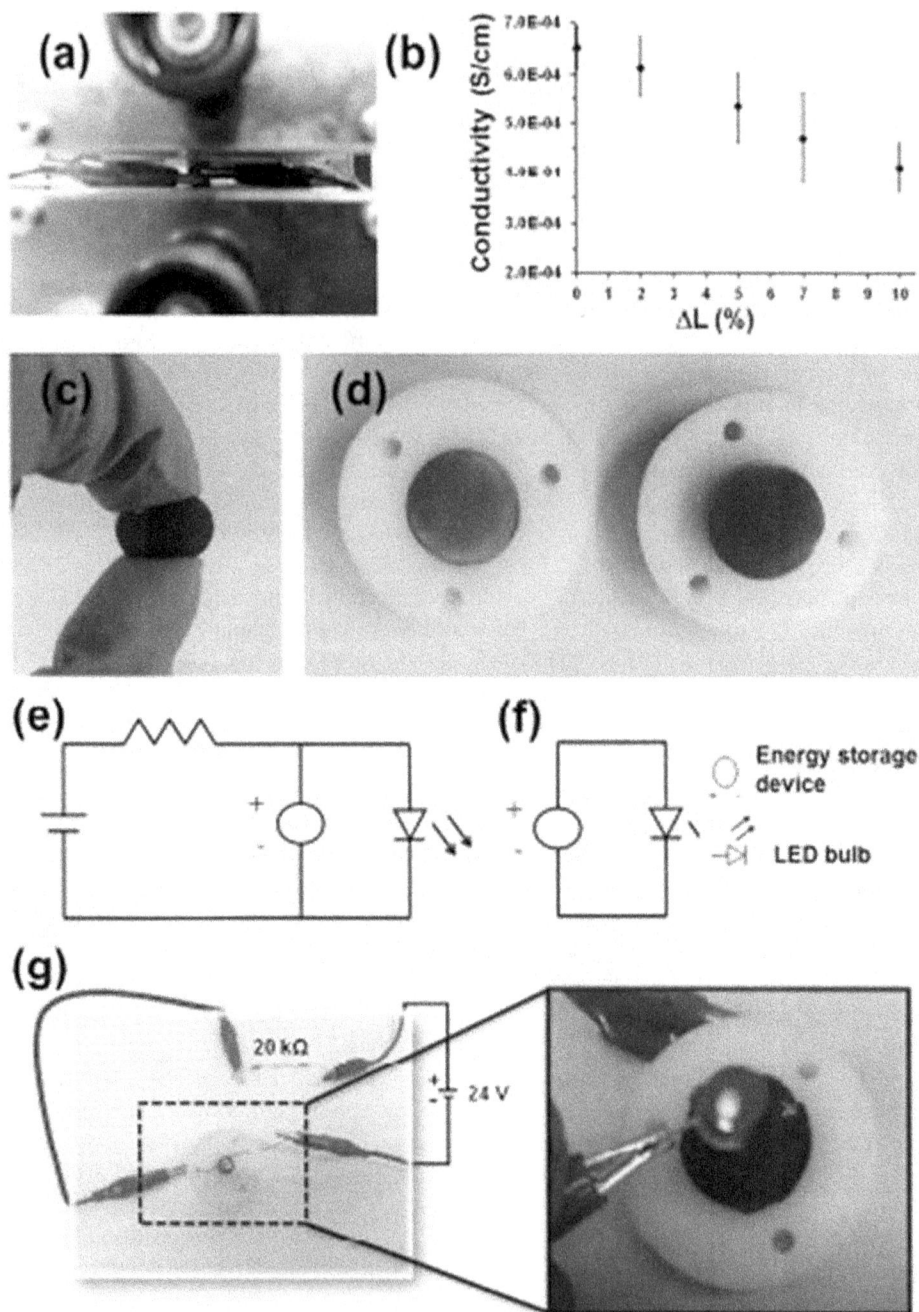

FIGURE 5.7 Photograph and fluctuation of electrical conductivity with strain for [PEDOT/-PGA] PHMeDOT(= 7 h) electrode are shown in (a) and (b), respectively. (c) The electrode powers the LED bulb. (d) The [PEDOT/-PGA]PHMeDOT(= 7 h) electrode was used to develop an energy harvesting system. The circuit utilized to (e) charge and (f) power the LED is shown schematically. g) Images of the apparatus that powered the LED bulb. Reprinted with permission from ACS.

stability. The supercapacitor has a robust total volumetric capacitance of 0.2 F/cm^3 and exceptional capacitance retention of about 95 percent over 1000 stretch/release cycles at a scan rate of 1 V/s and 400-percent pressure.

The design of strain-sensors with highly sensitive using graphene aerogel (GA) and polydimethylsiloxane (PDMS) nanocomposites was proposed by Wu et al. [48], with the primary goal of tuning the sensitivity of the sensors through control of the manufacturing processes by adjusting the cellular microstructure. Nanocomposite sensors that result have a high level of sensitivity and can achieve a gauge factor value of 61.3. The strain sensors' sensitivity can easily be improved by modifying the concentration of the dispersion of graphene oxide and GA procession freezing temperature. The findings show that cell size and cell-wall thickness of the resulting GA are the two parameters, which may be connected to the strain sensors' sensitivity variations. The concentration of graphene oxide has an inverse dependence on the sensitivity of the resulting nanocomposite strain sensor. When the freezing temperature is raised from -196 °C to -20 °C, the sensitivity rises to a maximum of 61.3 at -50 °C, then falls when the freezing temperature is raised further to -20 °C. To fulfil the needs for wearable electronics, It is vital to use strain sensors with a high elastic limit and sensitivity.

Kang et al. [56], used ordinary printing paper as an electrode in a high-performance flexible and foldable electrochemical supercapacitor, with water-dispersible conductive polymer polyaniline-poly(2-acrylamide-2-methyl-1-propane sulfonic acid) (PANI-PAAMPSA) and poly(vinyl alcohol) (PVA) serving as the conducting agent and polymer matrix, respectively. The conversion of insulating paper to a conductive substrate was done with the aid of PANI-PAAMPSA, whereas PVA offers electrolyte ion channels and mechanical durability for the paper substrate. Supercapacitors made of paper have a high capacity for electrochemical energy storage. The proposed supercapacitors can attain a maximum mass and area-specific capacitances of 41 F/g and 45 mF/cm^{-2}, respectively, for 20 mV/s. Also, bending tests revealed good mechanical toughness as well as flexibility. As paper-based supercapacitors are bent gradually from 0° to 100°, their real capacitance changes by up to 16 percent compared to the original value. The high water dispersibility and conductivity achieved by PANI- PAAMPSA are responsible for the paper-based supercapacitors' excellent electrochemical stability. PVA is used as a rigid polymer matrix that enables electrolyte ion pathways to ensure high mechanical durability. This research could pave the way for the development of future paper-based electronics and energy sources.

5.5 CONCLUSION

Because of their extended cycle life and higher power densities, supercapacitor devices are becoming more popular as energy storage devices. However, the rapid advancement of wearable electronics necessitates the use of flexible/stretchable supercapacitors that are foldable, stretchy, and twistable as the energy source. We looked at the latest developments in the field of flexible/stretchable electrodes, electrolytes used for supercapacitors and their practical application by studying the electrochemical performance. The use of stretchable and flexible substrates or electrode materials for structural stability and stretchability and the high-power density of supercapacitors makes them appealing from an application point. Graphene and carbon nanotubes are the head up materials in flexible/stretchable supercapacitors. Power receivers made of polymeric materials and metal oxides have also proved effective. The literature highlighted hybrid supercapacitors electrodes as emerging new possibilities for flexible/stretchable electrical systems because of the improvement in electronic, mechanical, and electrochemical properties. Enhanced mechanical strength, energy, and power density under foldable and bendable situations should be the next priority. In the ultimate performance of foldable/bendable supercapacitors, the electrolyte is critical. Currently, the supercapacitor electrode's potential window and flexibility/stretchy nature are limited due to its extensive usage of the PVA gel as the electrolyte. As a result, only a few solid polyelectrolytes with exceptional flexibility/stretchability but a narrow working potential window evolved. Despite recent research innovations in the engineering and development of flexible/stretchable supercapacitors, many unresolved problems remain to be investigated.

ACKNOWLEDGMENTS

The authors gratefully acknowledge the Kerala State Council for Science, Technology and Environment (KSCSTE), Kerala, India, for the financial support. We thank the Head of the Department of Physics and the Principal, Maharaja's College, Ernakulam, and the Head of the Department of Physics and the Principal, St.Teresa's College, Ernakulam.

REFERENCES

1. Yuan, H., Wang, G., Zhao, Y., Liu, Y., Wu, Y. and Zhang, Y., 2020. A stretchable, asymmetric, coaxial fiber-shaped supercapacitor for wearable electronics. *Nano Research*, *13*, pp. 1686–1692.
2. Yang, D., 2012. Application of nanocomposites for supercapacitors: Characteristics and properties. *Nanocomposites-New Trends Developments*, pp. 299–328.
3. Wang, Z., Zhu, M., Pei, Z., Xue, Q., Li, H., Huang, Y. and Zhi, C., 2020. Polymers for supercapacitors: Boosting the development of the flexible and wearable energy storage. *Materials Science and Engineering: R: Reports*, *139*, p. 100520.
4. Guo, T., Zhou, D., Liu, W. and Su, J., 2021. Recent advances in all-in-one flexible supercapacitors. *Science China Materials*, *64*(1), pp. 27–45.
5. Zhang, Y., Bai, W., Cheng, X., Ren, J., Weng, W., Chen, P., Fang, X., Zhang, Z. and Peng, H., 2014. Flexible and stretchable lithium-ion batteries and supercapacitors based on electrically conducting carbon nanotube fiber springs. *Angewandte Chemie International Edition*, *53*(52), pp. 14564–14568.
6. Zhao, S., Li, J., Cao, D., Zhang, G., Li, J., Li, K., Yang, Y., Wang, W., Jin, Y., Sun, R. and Wong, C. P., 2017. Recent advancements in flexible and stretchable electrodes for electromechanical sensors: Strategies, materials, and features. *ACS Applied Materials & Interfaces*, *9*(14), pp. 12147–12164.
7. Kim, B. K., Sy, S., Yu, A. and Zhang, J., 2015. Electrochemical supercapacitors for energy storage and conversion. *Handbook of Clean Energy Systems*, pp. 1–25.
8. Li, L., Lou, Z., Chen, D., Jiang, K., Han, W. and Shen, G., 2018. Recent advances in flexible/stretchable supercapacitors for wearable electronics. *Small*, *14*(43), p. 1702829.
9. Chee, W. K., Lim, H. N., Zainal, Z., Huang, N. M., Harrison, I. and Andou, Y., 2016. Flexible graphene-based supercapacitors: A review. *The Journal of Physical Chemistry C*, *120*(8), pp. 4153–4172.
10. Zhu, Y., Li, N., Lv, T., Yao, Y., Peng, H., Shi, J., Cao, S. and Chen, T., 2018. Ag-Doped PEDOT:PSS/CNT composites for thin-film all-solid-state supercapacitors with a stretchability of 480%. *Journal of Materials Chemistry A*, *6*(3), pp. 941–947.
11. Keum, K., Kim, J. W., Hong, S. Y., Son, J. G., Lee, S. S. and Ha, J. S., 2020. Flexible/Stretchable supercapacitors with novel functionality for wearable electronics. *Advanced Materials*, *32*(51), p. 2002180.
12. Hao, G. P., Hippauf, F., Oschatz, M., Wisser, F. M., Leifert, A., Nickel, W., Mohamed-Noriega, N., Zheng, Z. and Kaskel, S., 2014. Stretchable and semitransparent conductive hybrid hydrogels for flexible supercapacitors. *ACS Nano*, *8*(7), pp. 7138–7146.
13. Jeerapan, I. and Poorahong, S., 2020. Flexible and stretchable electrochemical sensing systems: Materials, energy sources, and integrations. *Journal of The Electrochemical Society*, *167*(3), p. 037573.
14. Kandula, S., Kim, N. H., Lee, J. H. and Lee, J. H., 2019. Polymer-based flexible electrodes for supercapacitor applications. *Nanomaterials for Electrochemical Energy Storage Devices*, pp. 573–624.
15. Wang, Y., Ding, Y., Guo, X. and Yu, G., 2019. Conductive polymers for stretchable supercapacitors. *Nano Research*, pp. 1–10.
16. Yu, A., Chabot, V. and Zhang, J., 2013. *Electrochemical supercapacitors for energy storage and delivery: Fundamentals and applications* (p. 383). Boca Raton: Taylor & Francis.
17. Keum, K., Kim, J. W., Hong, S. Y., Son, J. G., Lee, S. S. and Ha, J. S., 2020. Flexible/Stretchable supercapacitors with novel functionality for wearable electronics. *Advanced Materials*, *32*(51), p. 2002180.
18. Zhao, S., Li, J., Cao, D., Zhang, G., Li, J., Li, K., Yang, Y., Wang, W., Jin, Y., Sun, R. and Wong, C. P., 2017. Recent advancements in flexible and stretchable electrodes for electromechanical sensors: Strategies, materials, and features. *ACS Applied Materials & Interfaces*, *9*(14), pp. 12147–12164.
19. An, C., Zhang, Y., Guo, H. and Wang, Y., 2019. Metal oxide-based supercapacitors: Progress and prospectives. *Nanoscale Advances*, *1*(12), pp. 4644–4658.
20. Yu, Z., Tetard, L., Zhai, L. and Thomas, J., 2015. Supercapacitor electrode materials: Nanostructures from 0 to 3 dimensions. *Energy & Environmental Science*, *8*(3), pp. 702–730.
21. Lee, H., Cho, M. S., Kim, I. H., Do Nam, J. and Lee, Y., 2010. RuO_x/polypyrrole nanocomposite electrode for electrochemical capacitors. *Synthetic Metals*, *160*(9–10), pp. 1055–1059.

22. Cho, S., Kim, J., Jo, Y., Ahmed, A. T. A., Chavan, H. S., Woo, H., Inamdar, A. I., Gunjakar, J. L., Pawar, S. M., Park, Y. and Kim, H., 2017. Bendable RuO$_2$/graphene thin film for fully flexible supercapacitor electrodes with superior stability. *Journal of Alloys and Compounds*, *725*, pp. 108–114.
23. Wang, K., Zhang, X., Li, C., Zhang, H., Sun, X., Xu, N. and Ma, Y., 2014. Flexible solid-state supercapacitors based on a conducting polymer hydrogel with enhanced electrochemical performance. *Journal of Materials Chemistry A*, *2*(46), pp. 19726–19732.
24. Wang, Z., Carlsson, D. O., Tammela, P., Hua, K., Zhang, P., Nyholm, L. and Strømme, M., 2015. Surface modified nanocellulose fibers yield conducting polymer-based flexible supercapacitors with enhanced capacitances. *ACS Nano*, *9*(7), pp. 7563–7571.
25. Zhang, Y., Mao, T., Wu, H., Cheng, L. and Zheng, L., 2017. Carbon nanotubes grown on flax fabric as hierarchical all-carbon flexible electrodes for supercapacitors. *Advanced Materials Interfaces*, *4*(9), p. 1601123.
26. Xu, J., Ding, J., Zhou, X., Zhang, Y., Zhu, W., Liu, Z., Ge, S., Yuan, N., Fang, S. and Baughman, R. H., 2017. Enhanced rate performance of flexible and stretchable linear supercapacitors based on polyaniline@ Au@ carbon nanotube with ultrafast axial electron transport. *Journal of Power Sources*, *340*, pp. 302–308.
27. Jeong, H. T., 2015. *Fabrication of stretchable and flexible supercapacitor using nanocarbon based materials.*
28. Zhang, X., Zhang, H., Lin, Z., Yu, M., Lu, X. and Tong, Y., 2016. Recent advances and challenges of stretchable supercapacitors based on carbon materials. *Science China Materials*, *59*(6), pp. 475–494.
29. Palchoudhury, S., Ramasamy, K., Gupta, R. K. and Gupta, A., 2019. Flexible supercapacitors: A materials perspective. *Frontiers in Materials*, *5*, p. 83.
30. Dong, L., Xu, C., Li, Y., Huang, Z. H., Kang, F., Yang, Q. H. and Zhao, X., 2016. Flexible electrodes and supercapacitors for wearable energy storage: A review by category. *Journal of Materials Chemistry A*, *4*(13), pp. 4659–4685.
31. Wang, X., Yang, C., Jin, J., Li, X., Cheng, Q. and Wang, G., 2018. High-performance stretchable supercapacitors based on intrinsically stretchable acrylate rubber/MWCNTs@ conductive polymer composite electrodes. *Journal of Materials Chemistry A*, *6*(10), pp. 4432–4442.
32. Shown, I., Ganguly, A., Chen, L. C. and Chen, K. H., 2015. Conducting polymer-based flexible supercapacitor. *Energy Science & Engineering*, *3*(1), pp. 2–26.
33. Lokhande, P. E., Chavan, U. S. and Pandey, A., 2020. Materials and fabrication methods for electrochemical supercapacitors: Overview. *Electrochemical Energy Reviews*, *3*(1), pp. 155–186.
34. Dong, X., Wang, J., Wang, J., Chan-Park, M. B., Li, X., Wang, L., Huang, W. and Chen, P., 2012. Supercapacitor electrode based on three-dimensional graphene—polyaniline hybrid. *Materials Chemistry and Physics*, *134*(2–3), pp. 576–580.
35. ur Rehman, S. and Bi, H., 2021. Electrodes for flexible integrated supercapacitors. *Flexible Supercapacitor Nanoarchitectonics*, pp. 1–26.
36. Lee, M., Wee, B. H. and Hong, J. D., 2015. High performance flexible supercapacitor electrodes composed of ultralarge graphene sheets and vanadium dioxide. *Advanced Energy Materials*, *5*(7), p. 1401890.
37. Park, S. and Kim, S., 2013. Effect of carbon blacks filler addition on electrochemical behaviors of Co$_3$O$_4$/graphene nanosheets as a supercapacitor electrodes. *Electrochimica Acta*, *89*, pp. 516–522.
38. Wang, Y., Ding, Y., Guo, X. and Yu, G., 2019. Conductive polymers for stretchable supercapacitors. *Nano Research*, pp. 1–10.
39. Ali, S. W. and Bairagi, S., 2020. Conductive polymer based flexible supercapacitor. In *Self-standing substrates* (pp. 211–233). Springer, Cham.
40. Bu, X., Ge, Y., Wang, L., Wu, L., Ma, X. and Lu, D., 2021. Design of highly stretchable deep eutectic solvent-based ionic gel electrolyte with high ionic conductivity by the addition of zwitterion ion dissociators for flexible supercapacitor. *Polymer Engineering & Science*, *61*(1), pp. 154–166.
41. Li, W., Lu, H., Zhang, N. and Ma, M., 2017. Enhancing the properties of conductive polymer hydrogels by freeze—thaw cycles for high-performance flexible supercapacitors. *ACS Applied Materials & Interfaces*, *9*(23), pp. 20142–20149.
42. Xie, Y., Liu, Y., Zhao, Y., Tsang, Y. H., Lau, S. P., Huang, H. and Chai, Y., 2014. Stretchable all-solid-state supercapacitor with wavy shaped polyaniline/graphene electrode. *Journal of Materials Chemistry A*, *2*(24), pp. 9142–9149.
43. Wang, B., Song, W., Gu, P., Fan, L., Yin, Y. and Wang, C., 2019. A stretchable and hydrophobic polypyrrole/knitted cotton fabric electrode for all-solid-state supercapacitor with excellent strain capacitance. *Electrochimica Acta*, *297*, pp. 794–804.
44. Kandula, S., Kim, N. H., Lee, J. H. and Lee, J. H., 2019. Polymer-based flexible electrodes for supercapacitor applications. *Nanomaterials for Electrochemical Energy Storage Devices*, pp. 573–624.

45. Zhu, Y., Murali, S., Stoller, M. D., Ganesh, K. J., Cai, W., Ferreira, P. J., Pirkle, A., Wallace, R. M., Cychosz, K. A., Thommes, M. and Su, D., 2011. Carbon-based supercapacitors produced by activation of graphene. *Science*, *332*(6037), pp. 1537–1541.

46. Tiwari, S. K., Sahoo, S., Wang, N. and Huczko, A., 2020. Graphene research and their outputs: Status and prospect. *Journal of Science: Advanced Materials and Devices*, *5*(1), pp. 10–29.

47. Chang, P., Mei, H., Tan, Y., Zhao, Y., Huang, W. and Cheng, L., 2020. A 3D-printed stretchable structural supercapacitor with active stretchability/flexibility and remarkable volumetric capacitance. *Journal of Materials Chemistry A*, *8*(27), pp. 13646–13658.

48. Wu, S., Ladani, R. B., Zhang, J., Ghorbani, K., Zhang, X., Mouritz, A. P., Kinloch, A. J. and Wang, C. H., 2016. Strain sensors with adjustable sensitivity by tailoring the microstructure of graphene aerogel/PDMS nanocomposites. *ACS Applied Materials & Interfaces*, *8*(37), pp. 24853–24861.

49. Yun, T. G., Hwang, B. I., Kim, D., Hyun, S. and Han, S. M., 2015. Polypyrrole—MnO_2-coated textile-based flexible-stretchable supercapacitor with high electrochemical and mechanical reliability. *ACS Applied Materials & Interfaces*, *7*(17), pp. 9228–9234.

50. Wang, M., Chen, Q., Li, H., Ma, M. and Zhang, N., 2020. Stretchable and shelf-stable all-polymer supercapacitors based on sealed conductive hydrogels. *ACS Applied Energy Materials*, *3*(9), pp. 8850–8857.

51. Meng, C., Liu, C., Chen, L., Hu, C. and Fan, S., 2010. Highly flexible and all-solid-state paperlike polymer supercapacitors. *Nano Letters*, *10*(10), pp. 4025–4031.

52. Saborío, M. C. G., Lanzalaco, S., Fabregat, G., Puiggalí, J., Estrany, F. and Alemán, C., 2018. Flexible electrodes for supercapacitors based on the supramolecular assembly of biohydrogel and conducting polymer. *The Journal of Physical Chemistry C*, *122*(2), pp. 1078–1090.

53. Sundriyal, P. and Bhattacharya, S., 2020. Textile-based supercapacitors for flexible and wearable electronic applications. *Scientific Reports*, *10*(1), pp. 1–15.

54. Huang, Z., Ji, Z., Feng, Y., Wang, P. and Huang, Y., 2021. Flexible and stretchable polyaniline supercapacitor with a high rate capability. *Polymer International*, *70*(4), pp. 437–442.

55. Wang, Z., Carlsson, D. O., Tammela, P., Hua, K., Zhang, P., Nyholm, L. and Strømme, M., 2015. Surface modified nanocellulose fibers yield conducting polymer-based flexible supercapacitors with enhanced capacitances. *ACS Nano*, *9*(7), pp. 7563–7571.

56. Kang, S. W. and Bae, J., 2018. High-efficiency flexible and foldable paper-based supercapacitors using water-dispersible polyaniline-poly (2-acrylamido-2-methyl-1-propanesulfonic acid) and poly (vinyl alcohol) as conducting agent and polymer matrix. *Macromolecular Research*, *26*(3), pp. 226–232.

57. Maksoud, M. A., Fahim, R. A., Shalan, A. E., Abd Elkodous, M., Olojede, S. O., Osman, A. I., Farrell, C., Ala'a, H., Awed, A. S., Ashour, A. H. and Rooney, D. W., 2021. Advanced materials and technologies for supercapacitors used in energy conversion and storage: A review. *Environmental Chemistry Letters*, *19*(1), pp. 375–439.

6 Halloysite Filled Fluoropolymer Nanocomposites

Deepalekshmi Ponnamma and Igor Krupa

CONTENTS

6.1 INTRODUCTION

Portable electronic devices demand the utilization of supercapacitors with the high-power density and fast charge-discharge rates [1–2]. The uninterrupted power supply for a long period is highly targeted by supercapacitors [3]. Conducting polymers, metal oxides and carbon are the generally used materials for fabricating supercapacitors [3–4]. However, lightweight and flexible designs of polymers always trigger the manufacturing industry to focus on such materials and their composites. Polymers are notable in generating dielectric capacitors because of their intrinsic self-healing nature and large electric breakdown [1]. Additionally, their low elongation and brittleness can be resolved by employing a copolymerization process or composites/nanocomposites fabrication. Such capacitors made of polymer composites would significantly reduce the weight, volume, and cost of electrical energy storage systems and exhibit outstanding energy capability [5–7].

High-k nanofillers which include carbon-based materials, metal particles, and ferroelectric ceramic materials are added to polymers for increasing their dielectric permittivity [8–9]. Though such conducting fillers find promising enhancements in the dielectric performance of the polymer composites, their threshold limit needs considerable study, since all physical properties abruptly change around this concentration level [1]. The agglomeration tendency of high-k nanofillers is often resolved by modifying their surface by functionalization processes [8]. However, the high cost and complex chemical reactions involved challenge the mass quality production of composite materials. Halloysite nanotubes (HNTs) are one-dimensional hollow tubes of dioctahedral 1:1 clay minerals belonging to the kaolin group [10]. The low cost, non-toxicity, presence of silicon hydroxyl functional groups, hollow nature, and porosity cause HNTs to find applications in numerous areas such as catalysis, drug delivery, adsorption, and energy storage [11–12]. HNTs achieve a chemical form of positive internal surface and negative external surface, due to their pH variations at 2–8 [13]. This hydrated aluminum silicate, $(Al_2Si_2O_5(OH)_4 \cdot nH_2O)$ is extensively available in many countries and has excellent thermal, mechanical, and electrical characteristics [14]. Economically viable complex composites are reported using HNT nanomaterials. HNTs have a high aspect ratio with respective internal and external diameters of 15 nm and 60 nm and with a length of 1000 nm [15]. High-level dispersibility in polymers can be ensured by the adsorption behavior of HNTs by the cationic and anionic surfactants in the medium. When dispersed in polymers, HNTs uniformly align in the medium by entrapping mechanism, which improves the thermal and mechanical stability of the composites [16]. In addition, polar functional groups on the HNTs can be connected with certain polymers through hydrogen bonding interactions as well.

DOI: 10.1201/9781003174646-6

Fluoropolymers are fluorocarbon-based polymers with multiple C-F bonds. They are stable chemically due to the presence of strong C-F bonds [17–18]. Such polymers can be homopolymers such as polyvinylidene fluoride (PVDF), polytetrafluoroethylene (PTFE or Teflon), and polychlorotrifluoroethylene (PCTFE), and copolymers such as fluorinated ethylene-propylene (FEP), polyethylene tetrafluoroethylene (ETFE), and polyethylene chlorotrifluoroethylene (ECTFE) [18]. PVDF-based copolymers are also notable for their high dielectric property and breakdown strength with low dielectric loss value [19]. The additional significance of fluoropolymer composites containing HNTs is that it achieves large polarized response under electric field, contributing to the high-energy capability for the final composite [1].

This chapter addresses the dielectric properties of the HNTs and their fluoropolymer nanocomposites. The bonding between the polymer and the HNTs in regulating the interfacial polarization and thus the dielectric performance are addressed concerning the concentration and modifications of the HNTs. The chapter will provide information on the energy density of various fluoropolymer nanocomposites containing HNTs and how the electroactive phases and polarization effects of the polymers influence the dielectric capacity of the composites. This study will add to the information on flexible polymer/inorganic filler nanocomposites applicable for energy storage purposes.

6.2   DIELECTRIC PROPERTIES OF HNTs

The tubular architecture of HNTs is formed as the curvature of the octahedral and tetrahedral layers in halloysite due to their lateral mismatch and are arranged as rolling skeletons [1]. Figure 6.1 shows the tubular structure of HNT by the TEM image and the outer and inner surfaces of HNT with chemical groups similar to SiO_2 and Al_2O_3 [20]. The presence of hydroxyl functional groups on the external surfaces of HNTs enables high-level dispersion in polar polymers and thus a compatible interface. For core-shell structured conductive nanofillers, the outer insulating shell prevents current leakage and thereby reduces the probable dielectric loss, whereas the non-conductive shell suppresses the charge mobility [21]. Therefore, the dielectric constant of the whole composite only comes from the strengthened average electric field according to Maxwell-Wagner theory [22]. This decreases the dielectric performance of the sample when compared to the composites containing conductive fillers without a core-shell structure. In addition, a few nanomaterials will have a ceramic coating on the shell that deteriorates the compatibility with polymers, mechanical stability and defects reduce the dielectric properties. However, for HNTs, these effects are less due to their structural integrity and used for modifying conducting nanofillers and polymers [23].

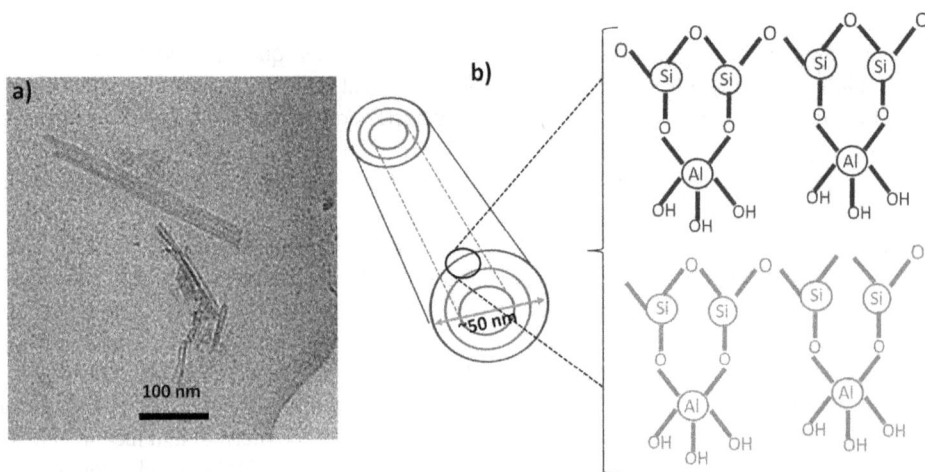

FIGURE 6.1   TEM image a) and schematic representation b) of HNT.

Zhang et al. [4] introduced MnO_2 on the surface of carbon-coated HNTs to develop electrode materials for supercapacitors. It is observed that the HNTs help to prevent the agglomerations in MnO_2, and its well-defined coaxial tubular structure facilitates the Carbon-MnO_2 ion diffusion. For the hybrid composite, the specific capacitance reached up to 274 F/g with symmetric charge-discharge curves, due to the high reversibility of the coaxial tubular composites. The electrochemical behavior was also improved for the HNT containing composite, making it suitable as the electrode for supercapacitors. A similar system was also identified by Yang and coworkers from the self-assembled monolayer amine-functionalized HNTs or SAM-HNTs [24]. Composites of polypyrrole (PPy) were made using the functionalized HNTs with PPy particles on its surface. In this case, the interface is strengthened by the Lewis acid-base type interactions between the basic amino group and the acidic N-H bonds. This generates a well-defined coaxial tubular morphology for the material and thus noticeable electrochemical properties. However, the specific capacitance depends on the concentration of the modified HNTs, for instance, the maximum value of 522 F/g is achieved at 15 wt.% concentration whereas it decreases to half of its value when the concentration reaches 75 wt.% (illustrated as inset of Figure 6.2a). This indicates the low utilization of PPy as the excess PPy does not favor its dispersion on HNT surfaces. The ability of HNTs in regulating the electrochemical capacitive behavior was also clearly explained by the authors using the cyclic voltammetry curves. The electrodes were charged and discharged at a pseudo constant rate and with scan rate, the ion/electrode effective interaction decreased (charge discharge curves are given in Figure 6.2a). Figure 6.2b again compares the power properties of PPy/HNT composite with that of PPy, from the calculated values of high rate discharge ability. It is the percentage of discharge capacity of an electrode at a certain current density to the discharge capacity at 1 mA cm^{-2}. The better high rate discharge ability of the composite when compared to PPy is clear from the figure. Moreover, the HNTs in the composite improve the conductivity, chemical durability, and rate capability, by lowering the charge transfer resistance, thus making the PPy/HNT composite ideal as supercapacitor electrodes.

A very recent study by Pandi et al. [13] describes the dielectric stability and high efficiency of HNT-polyaniline (PANI) supercapacitor electrodes. The PANI was polymerized in situ in the presence of HNTs to ensure good structural bonding between the materials, and the reaction did not affect the structural stability of HNT. However, polymerization made the HNT surface rough, which contributed to the enhanced electrical conductivity for the HNT-PANI composite. The specific surface area of the HNT-PANI composite was 38.49 m^2/g due to the high entanglement density of HNTs and the intermingling of the HNT-PANI network. The specific capacitance was at its maximum of 98.5 F/g at a 5 mV/s scan rate. In addition to the cyclic voltammetry study, galvanostatic charge-discharge measurements were also applied to derive the specific capacitance, and a value of 282.5 F/g at a 0.5 A/g current density was obtained for the HNT-PANI composite. At higher current density values, the specific capacitance decreased and the electrode utilization became lower. This was due to the diffusion of electrolytic ions at slower speeds indicating the partial doping and re-doping for the composites [25]. However, the obtained value of 282.5 F/g, was observed to be higher when compared to the previously reported values of 37 F/g and 137 F/g at a 0.5 A/g current density for HNT-PANI and HNT-PANI-PSS respectively [26]. The composite also showed very good cyclic stability, and at 5000 cycles of charge-discharge, the capacitance retention was 115%.

HNTs were also used to decorate reduced graphene oxide microstructures by controllable electrostatic self-assembly to generate micro capacitors [23]. The influence of HNTs in regulating the dielectric behavior was by i) preventing the direct contact between the rGO and ii) formation of dielectric interface within the capacitors. The specific modification method using functionalizing agent generated electrostatic interaction between the HNT and rGO to strengthen the interfacial bonding. The dielectric constant and loss values for the cold-pressed HNT pellet were respectively obtained as 25 and 0.2 at 10^3 Hz. The authors also noticed the relaxation of aluminosilicate groups within the HNT structure. The HNTs when combined with polyvinyl butyral (PVB) in the composites, interfacial polarization occurs, and applied field-induced electrical charge restrains at the interface. However, electrical conductivity was not improved for the high HNT containing

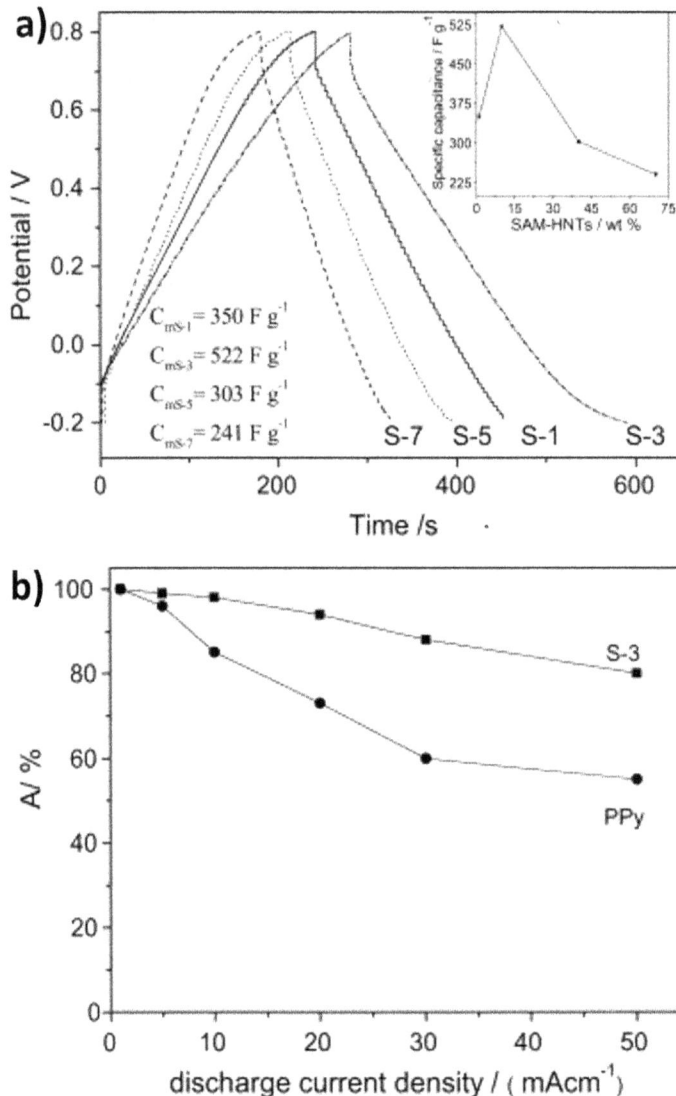

FIGURE 6.2 a) charge-discharge measurements at 5 mA cm^{-2} current density and b) the high rate discharge ability for HNT/PPy nanocomposites [24].

composites, due to the insulating nature of the HNTs. The 10 wt%HNTs@5wt%rGO/PVB composite showed a very high dielectric constant of 150 at 10^3 Hz with a low dielectric loss of 0.12. This can be explained using a micro capacitor model in which the capacitance of each micro capacitor contributes to the dielectric property. The capacitance values depend on the microelectrode areas according to the maximum dispersion of nanofillers. The conductivity of microelectrode determines the charge storage capacity of the micro capacitor. In addition, the dielectric layer between the microelectrodes is very thin that also contributes to enhanced capacitance by prohibiting the direct contact of microelectrode [27]. The synergistic effect of the HNT and rGO for enhancing the capacitance is schematically represented in Figure 6.3. There are three kinds of dielectric layers in the middle of two microelectrodes. In the first type, HNTs are sandwiched between the rGO microsheets, in which the HNTs prevent the direct contact between rGO sheets and act as a thinner dielectric layer to enhance the capacitance. In the second type, HNTs network between rGO sheets

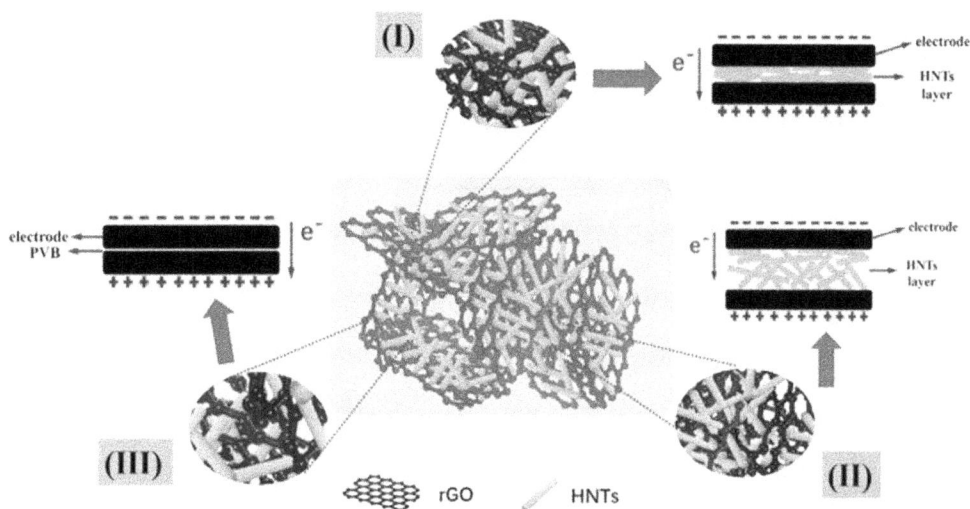

FIGURE 6.3 Schematic representation of HNT/rGO synergistic influence in improving dielectric properties and the energy conversion mechanism in HNTs@rGO [23].

with remote distance, forming thicker equivalent dielectric layers and thus low capacitance. In the third type, thin insulation of PVB layers is found between the rGO layers, which cannot completely restrain the charge migration through rGOs due to tunneling effect [28], and thus the mild percolation leads to high enhancement of dielectric constant and slight improvement in the dielectric loss. The microstructure theory illustrates a larger capacitance value under the external electric field for the type (I) and type (III) micro capacitors.

6.3 DIELECTRIC PROPERTIES OF HNT/FLUOROPOLYMER NANOCOMPOSITES

Xu et al. demonstrated [1] the enhanced polar response for the HNTs/poly(vinylidene fluoride-chlorotrifluoroethylene) (P(VDF-CTFE)) nanocomposite since the tubular architecture of HNTs serves for interior diffusion. The large free space inside the lumen allows the free motion of electrons and ions so that large polarization is reached with a very high energy density of 6.5 J/cm^3 at 350 MV/m for 1 wt.% of the HNT. The larger enhancement in the dielectric property can be due to several reasons such as i) high interaction between the uniformly distributed HNTs and the polar P(VDF-CTFE) segments, and thus compatible interface ii) increased electroactive phase transition due to matching clay crystal lattice and decreased activate energy barrier during crystallization and iii) efficient dipole response under high electric field. The static electric interaction between the nanoparticles and the CH$_2$ groups of the polymer is the prime factor for increasing the β-phase nucleation. This interaction with the negative charges of HNTs causes the polymer chains to assemble in the extended TTTT conformation as the β-phase. This in turn slows down the diffusion coefficient of dipoles and ions at the interface when compared to the bulk [29].

The dielectric constant showed a 300% enhancement for 3 wt.% HNT composite when compared to neat P(VDF-CTFE), however, the value decreased at 5 wt.% HNT concentration. The dielectric loss of the nanocomposite showed an interesting influence on the frequency. At lower frequencies, the charge mobilization within the matrix caused higher loss values, and at higher frequencies, dynamic polymer chain relaxation improved the loss [1]. Other than the dielectric constant and loss values, the electrical displacements, and electrical conductivity also change with the concentration of HNTs. This is because of the HNTs influence on the electroactive phase transformations in the

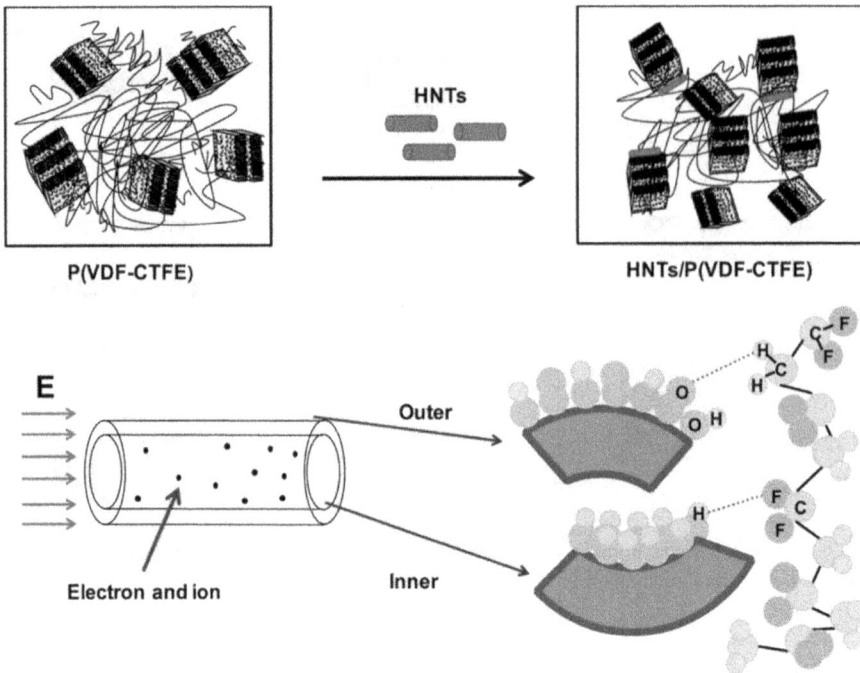

FIGURE 6.4   Schematic illustration of interfacial interaction between HNTs and P(VDF-CTFE) [1].

fluoropolymer. In addition, the high concentrations of HNTs result in a decrease in breakdown strength values for the nanocomposite. Since the outer and inner surfaces of HNTs respectively contain SiO_2 and Al_2O_3 groups, they offer strong hydrogen-bonding interactions with the fluoropolymer. This is represented in Figure 6.4. In addition, the free lumen space in HNTs allows the transport of energy units, causing the dielectric constant to enhance. The large area of HNT-polymer contact surface causes interfacial polarization and HNTs act as heterogeneous nucleation agents for the crystallization of P(VDF-CTFE). It also decreases the diffusion pathways and allows efficient ion/electron delivery [30]. At 1 wt.% HNT, the composite showed a permittivity and loss of $\varepsilon' = 15.1$, $\varepsilon''=0.06$ respectively at 100 Hz, energy density 6.5 J/cm^3 and charge discharge efficiency 41%.

PVDF and its copolymers contain different electroactive phases of which the polar β-phase is kinetically stable at room temperature and pressure and thus notable for high piezoelectric, pyroelectric, ferroelectric, and dielectric properties [31–33]. Many fillers such as carbon-based materials, metal oxides, and ceramic particles are widely used to improve the PVDF electroactive phases and thus the energy storage performance. Thakur et al. [34] explored the β-phase formation and dielectric performance mechanism of PVDF polymer composites containing 1–15 wt.% of HNT. The dielectric constant increased linearly with the HNT concentration reaching a maximum of 57 at 15 wt.%. Figure 6.5 demonstrates the frequency-dependent dielectric property variation of the nanocomposites in which the Maxwell—Wagner—Sillars (MWS) interfacial polarization mechanism plays a significant role in enhancing the dielectric constant values. With an external electric field, the charge carriers move and accumulate at the clay-polymer interface due to conductivity difference and causes large polarization. The charge confinement within the nanocomposite also causes the interfacial polarization to decrease with increased frequency, in a relatively long time. When HNTs are compared with kaolinite, their dc electrical conductivity and surface area are higher and this results in high interfacial polarization and a large dielectric constant. As illustrated in Figure 6.5b, the tangent loss for HNT composites decreases rapidly due to the dipolar relaxations happening within the system. In addition, the conductivities also show good enhancement attributed

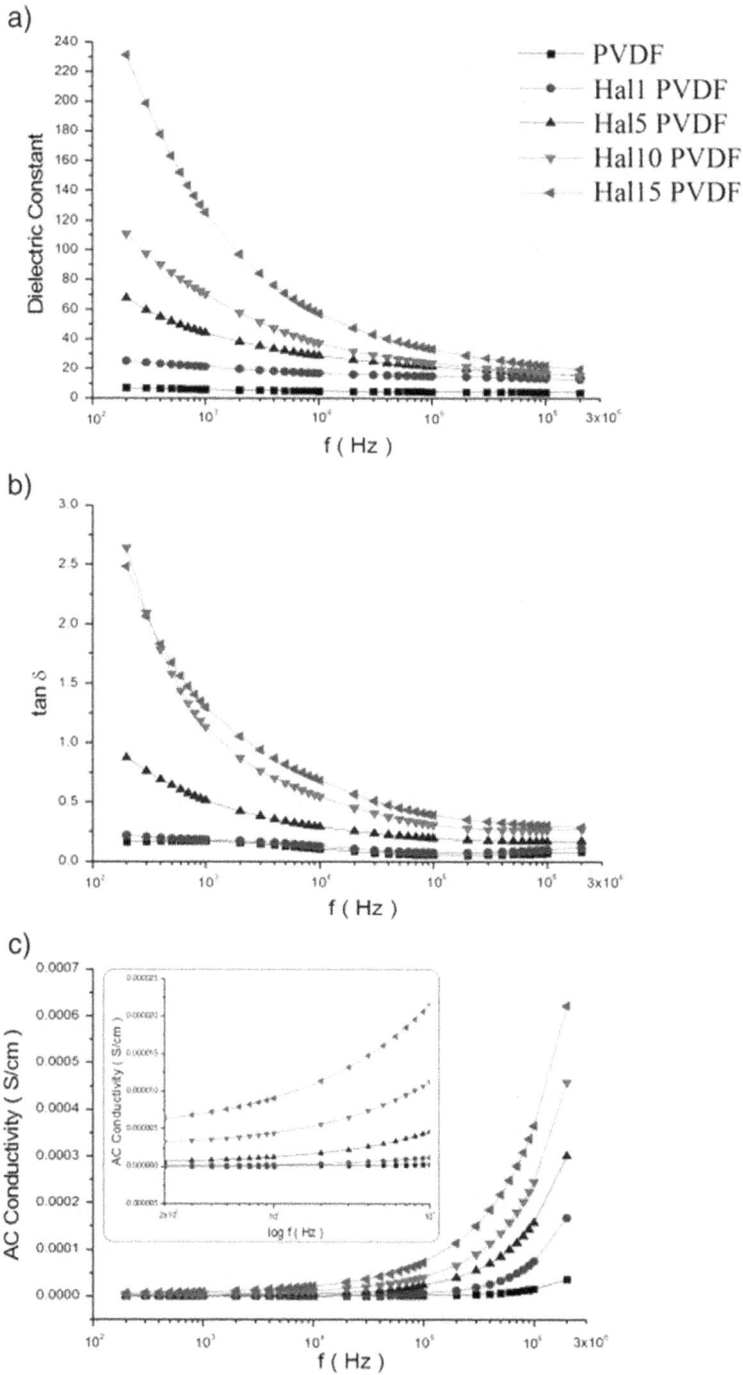

FIGURE 6.5 Variation of a) dielectric constant, b) tan δ, and c) AC conductivity of PVDF and its composites [34].

to the influence of HNTs. The negatively charged HNT surfaces attract the partially positive CH_2 dipoles of PVDF in the composites and cause β-phase nucleation. This aligns the PVDF in TTTT conformation and enhances the electroactive β-phase formation.

Thermoresponsive dielectric materials are reported from PVDF/polyethylene glycol (PEG)/ HNT composites [35] with relative dielectric constants. When the temperature is raised from 20 to 30 °C, PEG melting takes place and the effective dipole moment associated with orientational polarization enhances. For PVDF/HNT composite, the slight increase in dielectric constant noticed beyond 50 °C is related to slightly increased segmental motion. However, the crystalline regions in PVDF/HNT restrict the molecular motion and the effective dipole moment relatively decreases at other temperatures. The electrical conductivity of both PVDF/HNT and PVDF/PEG/HNT linearly increases with an observable variation in the low-frequency region. More temperature-dependent behavior of the PVDF/PEG/HNT is also due to the enhanced polarization effect. In short, the large dielectric constant variation in narrow temperature span makes the composite applicable in dielectric temperature-responsive applications such as smart sensors and switches.

Tian and coworkers designed a simple method of poly(dopamine) (PDA) functionalization to HNTs in tris-buffer solution to enhance the compatibility with PVDF-TrFE and observed high energy capability and cycle efficiency for the composite [36]. Figure 6.6 demonstrates the hysteresis displacement curves for the composites at various electric fields, with the loops obtained for neat PVDF-TrFE and the composite with 2 wt.% PDA-HNTs respectively at Figure 6.6a and 6.6b. Further, the displacement of modified composite is quantitatively analyzed from the maximum and remanent polarizations as shown in Figure 6.6 c and d. At 2 wt.% PDA-HNT, the displacement value becomes $6.1\,\mu C/cm^2$ at 250 MV/m, due to the interfacial polarization. The authors also studied the feature breakdown strength of the composite from Weibull expression, in which the possibility of breakdown failure is calculated based on the experimental breakdown value, shape parameter, and the breakdown strength with 63.2% possibility for the sample to be catastrophic. From Figure 6.6e, it is clear that the possibility of sample catastrophe decreases with the introduction of PDA-HNTs in the composite.

Abbasipour and coworkers [37] illustrated the higher polar phase formation for PVDF composites with a small concentration of HNTs (<0.1 wt.%), and the rod-like morphology of HNTs generated oriented and finer nanofibers when compared to graphene oxide and graphene containing nanocomposites. PVDF nanocomposite fibers showed typical percolation transition behavior in their dielectric constant values when nanomaterials are added. The increase in the dielectric constant near the percolation level of the fillers is due to the micro capacitor effect. Uniformly distributed fillers make mini capacitors when separated by polymer chains, and with the applied electric field charge carriers migrate and accumulate at the filler-polymer interface. With 0.4% HNT, the dielectric constant of the composite became 18, attributed also to the electrical conductivity difference at the interface and large surface area of HNTs. In the case of dielectric loss, the value increased up to 0.2 wt.% and thereafter decreases. This is mainly due to the decreased α-phase content of the PVDF as the nanofillers disrupt the polymer chain movement.

The significance of HNTs in reducing the average fiber diameter of PVDF and in influencing the Coulombic efficiency, energy capacity, and charge-discharge cycles is illustrated for the gel polymer electrolyte-based PVDF/HNT nanocomposite non-wovens by Khalifa and coworkers [20]. HNT influences the nanocomposite performance in many ways: i) overcomes the low mechanical strength and high crystallinity issues, ii) improves the thermal stability, iii) enhances the porosity, ionic conductivity, and electrochemical stability of gel polymer electrolyte, and iv) simplify the cost of fabrication. With HNT addition, the average fiber diameter decreases from 302 nm to 210 nm, and the space between the interconnected nanofibers enhances. This hinders the lithium dendritic structural growth and provides longer life for the electrolyte uptake. Impedance measurements showed lower bulk resistance of 0.9 Ω for the HNT composite when compared to the 2.3 Ω for the PVDF. The composite has a higher ionic conductivity (1.77 mScm^{-1}), which along with the cationic mobility improves the cycle performance and suppresses the dendritic structural nucleation. The interfacial resistance is lowered also with HNT addition (114 Ω/cm^2 compared to 189 Ω/cm^2 for PVDF), which is beneficial for C-rate performance for the Lithium batteries. The mechanism of HNT interaction with the electrolyte is explained based on a well-oriented structure along the PVDF fiber axis. But at some places, the HNTs are protruding out, which suggests the formation of H-bonds between the hydroxyl groups of HNTs and the fluorine atoms of $LiPF_6$ and dipole-dipole attraction between the

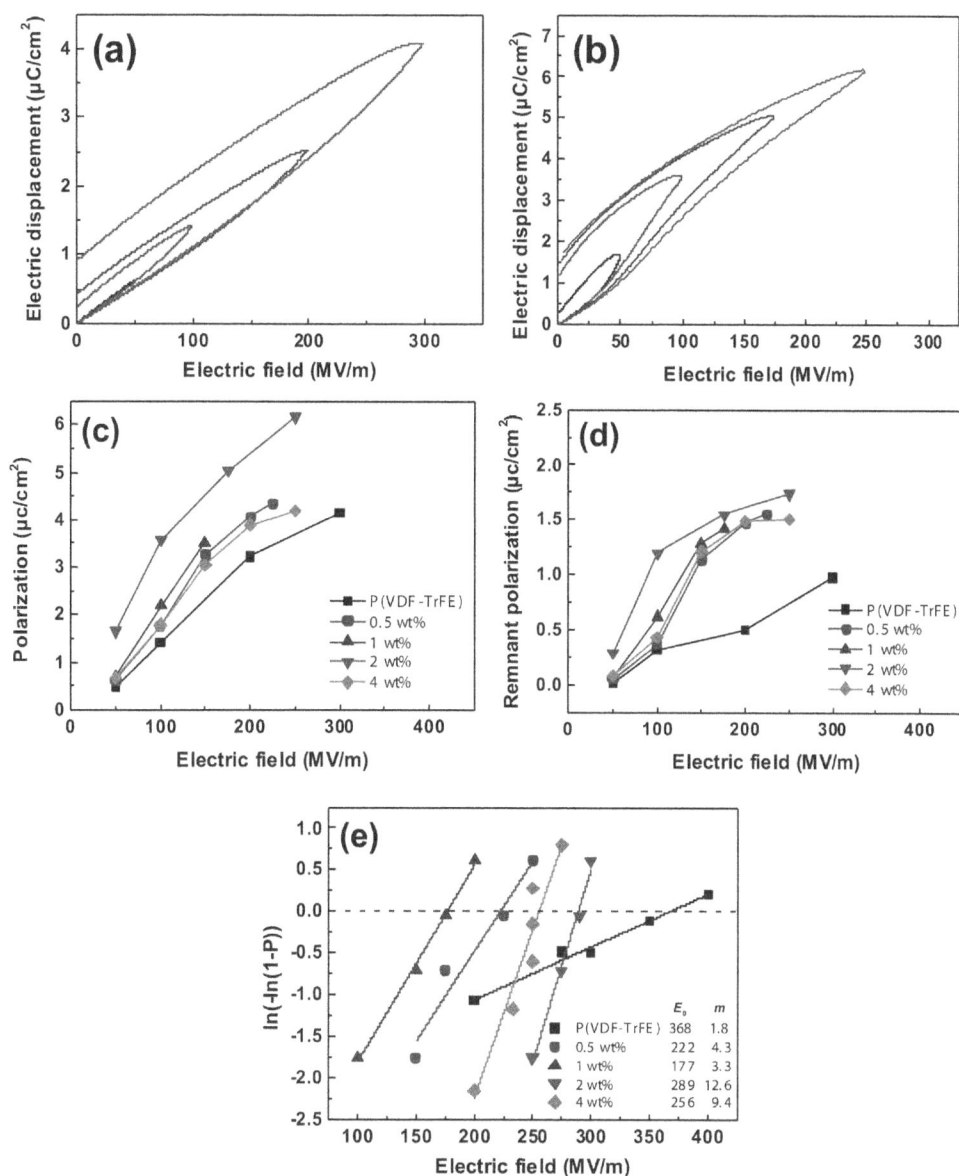

FIGURE 6.6 Electric hysteresis displacements of PDA-HNTs/P(VDF-TrFE) composites: (a) P-E loops for P(VDF-TrFE) film, (b) P-E loops for 2 wt% composite, (c) maximum displacement, (d) the remnant polarization, and (f) Weibull distribution [36].

oxygen atoms of HNTs and Li[+] of LiPF$_6$. This weakens the Li bonds and allows the migration of free Li[+] ions to the opposite electrodes. In this way, the nanocomposite separator at different current densities show remarkable performance towards repeated charge discharge cycles and thus could enhance the lithium battery performance.

HNT is also applied to resolve the safety issues in Li-ion batteries due to dendrite accumulation on the anode surface [38]. In this direction, Shaik et al. developed HNT-poly(vinylidene fluoride-co-hexafluoropropylene) (PVDF-HFP) hybrid composite with excellent interfacial stability during long term cycling. Figure 6.7 represents the performance analysis of the composite coating on Li-metal after 300th cycle of charge discharge process. While Figure 6.7a directly illustrates

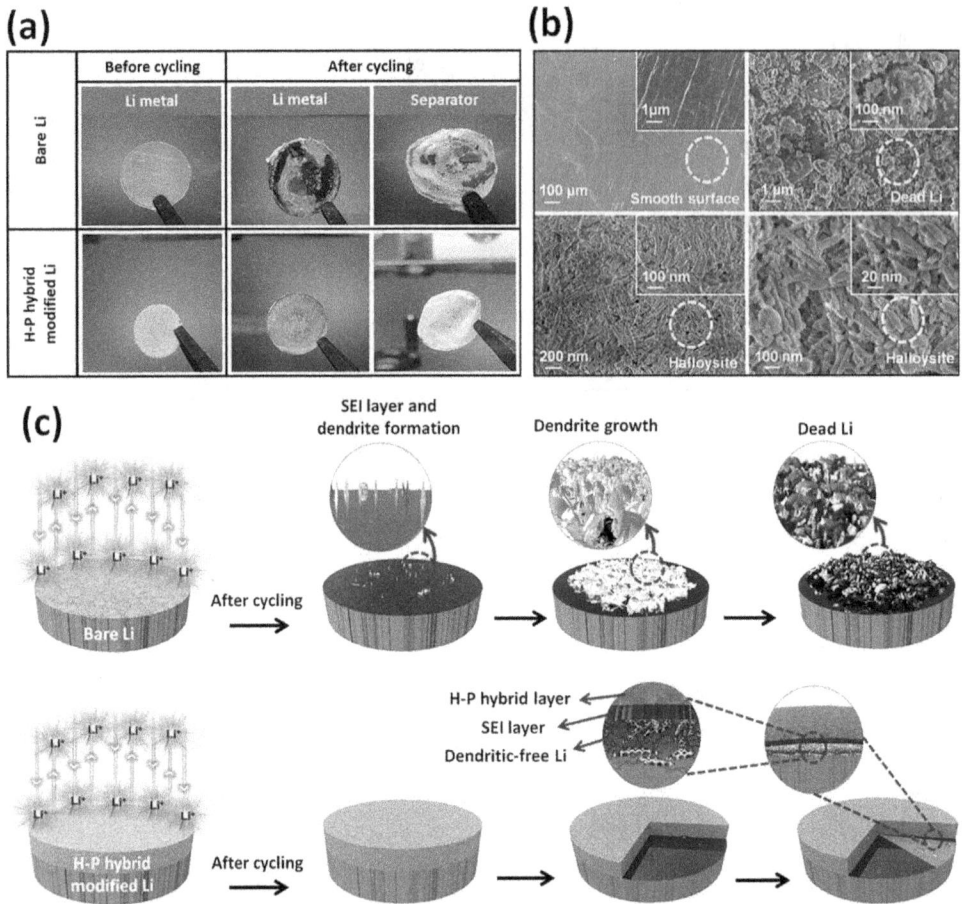

FIGURE 6.7 (a) Photographs of bare and HNT composite modified Li-electrode, and separator before and after 300th cycle, (b) SEM images and (c) schematic diagram showing the dendrite-free HNT modified Li- and bare Li- electrodes [38].

the clear difference between bare Li and the composite modified Li in dendrite formation through photographs, Figure 6.7b evidences the same by SEM images. Non-uniform Li deposition and dendrite formation make the bare Li metal surface rough. The influence of HNT composite protective coating in preventing dead Li formation on the anode surface before and after the charge discharge process is also represented schematically in Figure 6.7c.

Modification of HNT surfaces by the poly(3,4-ethylenedioxythiophene) (PEDOT) by chemical oxidative polymerization enhanced the conductivity around 100 times (up to 12.89 S/cm) when compared to the PEDOT, typically for 50 wt.% HNT composition [39]. The modification significantly improves the dispersibility of HNTs in PVDF and triggers the electroactive phase transformation from α to β-phase. The frequency-dependent dielectric performance of the composites is illustrated in Figure 6.8. The dielectric constant of 7.54 for pure PVDF at 1000 Hz reaches up to 790.94 with 13% of PEDOT-HNT. While the increase in the dielectric constant of the composites is due to the formation of mini capacitor networks, the increase in dielectric loss is due to the electrical conductivity. A concentration of 11% is marked as the percolation level at which both constant and loss values become high. The variation in AC conductivity also shows the transformation of the system from the insulating to conducting mode.

A very recent study by Li et al. correlates the dielectric performance of the HNTs/PVDF-HFP composite with triboelectric power generation [40]. Li or H_2O was used to modify the clay

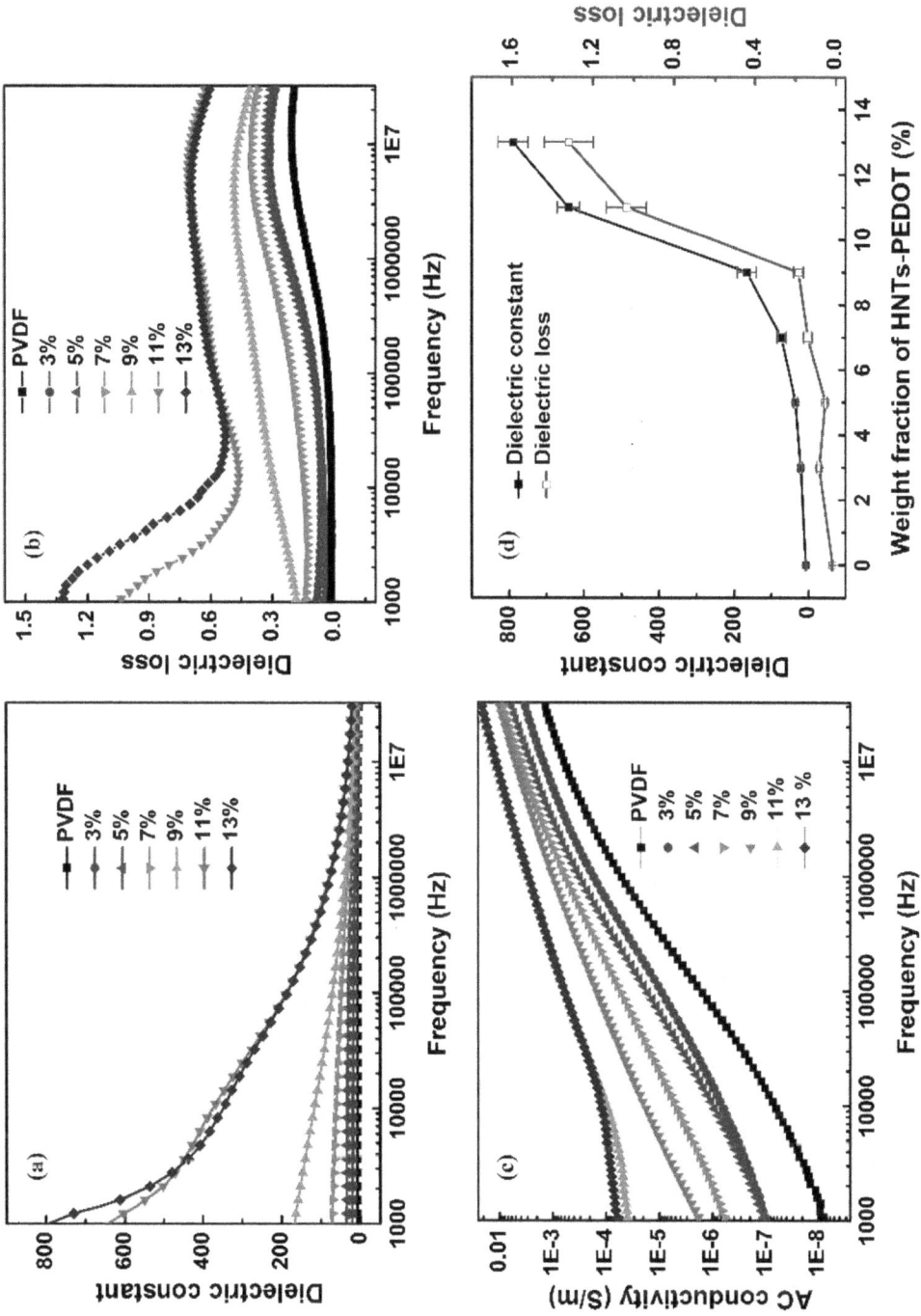

FIGURE 6.8 Variation of (a) dielectric constant, (b) dielectric loss, and (c) AC conductivity with frequency; d) comparison of dielectric constant and loss values for PVDF/HNT-PEDOT composites [39].

surface before making the polymer composites. Various cation exchange capacity modified clays were analyzed for their dielectric performance and observed high values for dielectric constant and loss with Li^+ modification. H_2O modification also improved the dielectric performance due to the increased ionic polarization. With 1 and 2 wt% H_2O, the loss values decreased since the partly ordered β-phase decreased the free electrons/ions movement in the PVDF-HFP and thus increased the density. However, above 3 wt% H_2O, the loss was mainly dependent on the ion in/on the clay material. It is concluded that the polarization of Li+ in clay with adsorbed H_2O improved the dielectric constant of composites film effectively and the adsorbed H_2O enhanced the β-phase and d33 by the hydrogen bonds.

6.4 CONCLUSIONS

Fluoropolymer HNT nanocomposites are good dielectric materials with notable dielectric performance and charge-discharge behavior. The reasons for the excellent energy storage capability of these materials are attributed to the typical structural features of the HNTs and their influence on the electroactive phases of fluoropolymer. With a good distribution of HNTs within the polymeric medium, the polarization enhances and thus the dielectric constant. However, in some cases, the dielectric loss values and conductivity varies depending on the concentration of the filler, temperature, and frequency differences other than the particle aggregation. The dielectric mechanism is mainly due to the microcapacitors influence as the HNT/fluoropolymer units interacting by the hydrogen bonds and other interfacial interactions act as tiny capacitor units. Such composites are ideal materials for microelectronic devices and smart gadgets.

ACKNOWLEDGMENT

This work is carried out by the NPRP grant NPRP10-0205-170349 from the Qatar National Research Fund (a member of the Qatar Foundation). The statements made herein are solely the responsibility of the authors.

REFERENCES

[1] Ye, H., Wang, Q., Sun, Q. and Xu, L. (2020) 'High energy density and interfacial polarization in poly (vinylidene fluoride-chlorotrifluoroethylene) nanocomposite incorporated with halloysite nanotube architecture', *Colloids and Surfaces A: Physicochemical and Engineering Aspects*, 606, pp. 125495.

[2] Ponnamma, D. and Al-Maadeed, M. A. (2017) '3D architectures of titania nanotubes and graphene with efficient nanosynergy for supercapacitors', *Materials & Design*, 117, pp. 203–12.

[3] Satapathy, K. D. *et al.* (2017) 'High-quality factor poly (vinylidenefluoride) based novel nanocomposites filled with graphene nanoplatelets and vanadium pentoxide for high-Q capacitor applications', *Advanced Materials Letters*, 8(3), pp. 288–94.

[4] Zhang, W., Mu, B. and Wang, A. (2015) 'Halloysite nanotubes induced synthesis of carbon/manganese dioxide coaxial tubular nanocomposites as electrode materials for supercapacitors', *Journal of Solid State Electrochemistry*, 19(5), pp. 1257–63.

[5] Ribeiro, C. *et al.* (2018) 'Electroactive poly (vinylidene fluoride)-based structures for advanced applications', *Nature protocols*, 13(4), p. 681.

[6] Deshmukh, K. *et al.* (2017) 'Fumed SiO_2 nanoparticle reinforced biopolymer blend nanocomposites with high dielectric constant and low dielectric loss for flexible organic electronics', *Journal of Applied Polymer Science*, 134(5).

[7] Li, M. *et al.* (2013) 'Revisiting the δ-phase of poly (vinylidene fluoride) for solution-processed ferroelectric thin films', *Nature materials*, 12(5), pp. 433–8.

[8] Deshmukh, K. *et al.* (2016) 'Eco-friendly synthesis of graphene oxide reinforced hydroxypropyl methylcellulose/polyvinyl alcohol blend nanocomposites filled with zinc oxide nanoparticles for high-k capacitor applications', *Polymer-Plastics Technology and Engineering*, 55(12), pp. 1240–53.

[9] Satapathy, K. D. *et al.* (2017) 'High-quality factor poly (vinylidenefluoride) based novel nanocomposites filled with graphene nanoplatelets and vanadium pentoxide for high-Q capacitor applications', *Advanced Materials Letters*, 8(3), pp. 288–94.

[10] Joussein, E. *et al.* (2005) 'Halloysite clay minerals—a review', *Clay Minerals*, 40(4), pp. 383–426.

[11] Machado G. S. *et al.* (2008) 'Immobilization of metalloporphyrins into nanotubes of natural halloysite toward selective catalysts for oxidation reactions', *Journal of Molecular Catalysis A: Chemical*, 283 (1–2), 99–107.

[12] Shchukin, D. G., Sukhorukov, G. B., Price, R. R. and Lvov, Y. M. (2005) 'Halloysite nanotubes as biomimetic nanoreactors', *Small*, 1(5), pp. 510–3.

[13] Pandi, N. *et al.* (2021) 'Halloysite nanotubes-based supercapacitor: Preparation using sonochemical approach and its electrochemical performance', *Energy, Ecology and Environment*, 6(1), pp. 13–25.

[14] Zhi, C. *et al.* (2005) 'Covalent functionalization: Towards soluble multiwalled boron nitride nanotubes', *Angewandte Chemie International Edition*, 44(48), pp. 7932–5.

[15] Wan, X., Zhan, Y., Zeng, G. and He, Y. (2017) 'Nitrile functionalized halloysite nanotubes/poly (arylene ether nitrile) nanocomposites: Interface control, characterization, and improved properties', *Applied Surface Science*, 393, pp. 1–0.

[16] Chamakh, M. M., Ponnamma, D. and Al-Maadeed, M. A. (2018) 'Vapor sensing performances of PVDF nanocomposites containing titanium dioxide nanotubes decorated multi-walled carbon nanotubes', *Journal of Materials Science: Materials in Electronics*, 29(6), pp. 4402–12.

[17] Cui, Z., Drioli, E. and Lee, Y. M. (2014) 'Recent progress in fluoropolymers for membranes', *Progress in Polymer Science*, 39(1), pp. 164–98.

[18] Hougham, G. G., Cassidy, P. E., Johns, K. and Davidson, T. (1999) 'Fluoropolymers 2', in *Springer Science & Business Media*.

[19] Ponnamma, D. *et al.* (2018) 'Stretchable quaternary phasic PVDF-HFP nanocomposite films containing graphene-titania-SrTiO$_3$ for mechanical energy harvesting', *Emergent Materials*, 1(1), pp. 55–65.

[20] Khalifa, M. *et al.* (2019) 'PVDF/halloysite nanocomposite-based non-wovens as gel polymer electrolyte for high safety lithium ion battery', *Polymer Composites*, 40(6), pp. 2320–34.

[21] Nair, N. and Sankapal, B. R. (2017) 'Nested CdS@ HgS core—shell nanowires as supercapacitive faradaic electrode through simple solution chemistry', *Nano-Structures & Nano-Objects*, 10, 159–66.

[22] Lopes, A. C. *et al.* (2013) 'Dielectric relaxation, ac conductivity and electric modulus in poly (vinylidene fluoride)/NaY zeolite composites', *Solid State Ionics*, 235, pp. 42–50.

[23] Su, Y. *et al.* (2020) 'Polyvinyl butyral composites containing halloysite nanotubes/reduced graphene oxide with high dielectric constant and low loss', *Chemical Engineering Journal*, 394, pp. 124910.

[24] Yang, C., Liu, P. and Zhao, Y. (2010) 'Preparation and characterization of coaxial halloysite/polypyrrole tubular nanocomposites for electrochemical energy storage', *Electrochimica Acta*, 55(22), pp. 6857–64.

[25] Huang, H. *et al.* (2014) 'Reinforced conducting hydrogels prepared from the in situ polymerization of aniline in an aqueous solution of sodium alginate', *Journal of Materials Chemistry A*, 2(39), pp. 16516–22.

[26] Huang, H. *et al.* (2016) 'Facile preparation of halloysite/polyaniline nanocomposites via in situ polymerization and layer-by-layer assembly with good supercapacitor performance', *Journal of materials science*, 51(8), pp. 4047–54.

[27] Tonkoshkur, A. S., Glot, A. B. and Ivanchenko, A. V. (2017) 'Basic models in dielectric spectroscopy of heterogeneous materials with semiconductor inclusions', *Multidiscipline Modeling in Materials and Structures*, 13(1), pp. 36–57.

[28] Elshurafa, A. M., Almadhoun, M. N., Salama, K. N. and Alshareef, H. N. (2013) 'Microscale electrostatic fractional capacitors using reduced graphene oxide percolated polymer composites', *Applied Physics Letters*, 102(23), p. 232901.

[29] Belovickis, J. *et al.* (2018) 'Dielectric, Ferroelectric, and Piezoelectric Investigation of Polymer-Based P (VDF-TrFE) Composites', *Physica Status Solidi (B)*, 255(3), p. 1700196.

[30] Yuksel, R. *et al.* (2020) 'Necklace-like nitrogen-doped tubular carbon 3D frameworks for electrochemical energy storage', *Advanced Functional Materials*, 30(10), p. 1909725.

[31] AlAhzm, A. M. *et al.* (2021) 'Piezoelectric properties of zinc oxide/iron oxide filled polyvinylidene fluoride nanocomposite fibers', *Journal of Materials Science: Materials in Electronics*, 1–3.

[32] Ponnamma, D., Aljarod, O., Parangusan, H. and Al-Maadeed, M. A. (2020) 'Electrospun nanofibers of PVDF-HFP composites containing magnetic nickel ferrite for energy harvesting application', *Materials Chemistry and Physics*, 239, p. 122257.

[33] Parangusan, H., Ponnamma, D. and AlMaadeed, M. A. (2019) 'Toward high power generating piezo-electric nanofibers: Influence of particle size and surface electrostatic interaction of Ce—Fe2O3 and Ce—Co3O4 on PVDF', *ACS Omega*, 4(4), pp. 6312–23.

[34] Thakur, P. *et al.* (2014) 'Enhancement of β phase crystallization and dielectric behavior of kaolinite/halloysite modified poly (vinylidene fluoride) thin films', *Applied Clay Science*, 99, pp. 149–59.

[35] Wang, W. *et al.* (2020) 'Dielectric response triggered by a non-ferroelectric phase transition in poly (vinylidene fluoride)/poly (ethylene glycol)/halloysite composites', *Journal of Macromolecular Science, Part B*, 59(12), pp. 867–77.

[36] Tian, Y. *et al.* (2021) 'High energy density in poly (vinylidene fluoride-trifluoroethylene) composite incorporated with modified halloysite nanotubular architecture', *Colloids and Surfaces A: Physicochemical and Engineering Aspects*, pp. 126993.

[37] Abbasipour, M. *et al.* (2017) 'The piezoelectric response of electrospun PVDF nanofibers with graphene oxide, graphene, and halloysite nanofillers: A comparative study', *Journal of Materials Science: Materials in Electronics*, Nov;28(21), pp. 15942–52.

[38] Shaik, M. R. *et al.* (2021) 'Soft, robust, Li-ion friendly halloysite-based hybrid protective layer for dendrite-free Li metal anode', *Chemical Engineering Journal*, 424, p. 130326.

[39] Wang, F. *et al.* (2018) 'Conductive HNTs-PEDOT hybrid preparation and its application in enhancing the dielectric permittivity of HNTs-PEDOT/PVDF composites', *Applied Surface Science*, 458, pp. 924–30.

[40] Li, Y. *et al.* (2020) 'Enhanced dielectric properties of halloysite/PVDF-HFP modified by Li-ion realizing superior energy conversion ability', *Chemical Physics Letters*, 761, p. 138089.

7 Carbon-Based Supercapacitors

Suganthi Nachimuthu

CONTENTS

7.1 INTRODUCTION

7.1.1 NANOSTRUCTURED SUPERCAPACITORS

The progress of green and renewable energy storage strategies is the focus of numerous researchers concerned with environmental challenges, the energy crisis, and rural evolution. Nowadays lithium-ion batteries (LIBs) and supercapacitors (SCs) are the foremost applicants for the storage of energy. LIBs have high energy density, however, they have a small power density and low poor cycle life. SCs have high power density (Pd), rapid charge-discharge, and lengthy cycle lives compared with batteries and traditional capacitors [1], which cause LIBs for energy storage applicants. Many reported carbon-based electrode materials such as graphene, CNTs, carbon fiber, carbon-derived carbon, etc., are used for supercapacitor applications. The reported electrode material mostly consists of transition metal oxides (TMOs) and conducting polymers (CPs). TMOs such as RuO_2, Fe_3O_4, MnO_2, NiO, Co_3O_4, IrO_2 including V_2O_5 and CuO wherein its charge storing process is dependent on the faradaic action [2]. However, the pseudocapacitors' execution is lesser compared to EDLC

DOI: 10.1201/9781003174646-7

due to the intrinsically lagging faradaic charge storing process. This leads to bad life cycle, energy density (Ed), and mechanical stability [3]. CPs utilized for pseudocapacitors can be unsteady on the nanoscale range, and ruthenium oxide is also enormously expensive. Hybrid capacitors reside in the middle position among batteries and capacitors. They have highter energy/power densities than EDLCs and pseudocapacitors that have high operative temperatures between −55 °C to 125 °C. The most commonly employed materials are carbon-coated CPs, metal oxides (like nickel and manganese-based oxides), and graphene oxides. Fiber-type flexible supercapacitors are produced from carbon fibers as a substrate.

Carbon-based supercapacitors (CSs) are auspicious high-power devices that are marketable and having the instant rapid pulses of energy, millions of charge-discharge cycles, widespread functioning temperatures, and extreme Coulombic efficacies. CSs are comprised of two carbonous electrodes immersed in an aqueous/nonaqueous electrolyte, and a permeable membrane separator permitting electrolyte penetration. Their energy storing process is derived from reversible charge split-up at every interface of the electrolyte through the cathode/anode surface creating two densely spaced charge zones. This interaction results in completely charged/discharged CSs within seconds and a large power distribution/retort. The marketable CSs is composed of activated carbons having energy densities (< 10 W h kg−1) in the organic electrolytes. To compensate for this drawback, substantial attempts have been applied for intensifying the energy storage of CSs by the two main factors such as high capacitance electrodes and large potential (V) electrolytes. Concerning the carbon-based electrodes, the design of large-surface-area carbon materials with doping for optimizing the electrochemical action, surface polarization, as well as electrical conduction has be considered in exhaustive research. Moreover, for advanced CSs, the desirable features are aqueous electrolytes having broader potential windows and nonaqueous electrolytes exhibiting extreme ionic conduction, electrochemical inactivity, and operation safety. Innovative fabrication approaches of CSs are emphasized by adapting the morphologies, pore assemblies, and surface functioning in detection of high capacitance electrodes.

The porous carbon and its products having larger specific surface areas are mostly employed as a widespread electrode material in EDLCs [4]. Non-faradaic electrostatic interactions are the basis for charge storage process in EDLCs, whereas in pseudocapacitors the process is situated on faradaic redox activity. The intention of the researchers is to advance porous carbon-based EDLCs that show excellent electrochemical activity. Several carbon-based electrode substances varying from 0D to 3D substances like fullerenes, activated carbon, carbon nanotube, graphene, and metal organic framework derived carbon, etc., have been investigated [5]. Becker established the first EDLC using carbon-based constituents in 1975 [6]. Because of enhanced features like high surface area, low cost, eco-friendliness, and synthesis approaches the carbonaceous substances are employed as electrode material for EDLCs.

Various fabrication methods have been employed for improving the functioning of graphene-based electrodes. The stacking properties of graphene substance impedes the charge-discharge action attributable to the thickness of the electrode [9]. Various template-based techniques and chemical functionalization approaches are executed for preventing stacking [10]. But these approaches are costly and difficult to apply on an industrialized zone. Presently, poisonous fluorine-based binderies and harmful solvents are employed to fabricate economical SCs [11]. Thus, eco-friendly bio solvents are examined for the industrialized production of SCs [12]. Garakani et al. established a spraying "green" ink for SCs production. This ink consists of activated carbon and graphene layer, and it performed like active substance for SC [13]. This spray method is suitable for industrialized appliances and creates an extremely uniform coating that is easily reachable by the electrolyte. The wire-type electrode is produced from conductive carbon fibers, which is fabricated using the carbonized and graphitized procedure employing the polyimide (PI) as a carbon fiber precursor [14]. Different synthesis methods have been narrated for fabricating the 3DGN such as hydrothermal, chemical cross linking, and CVD technique [15].

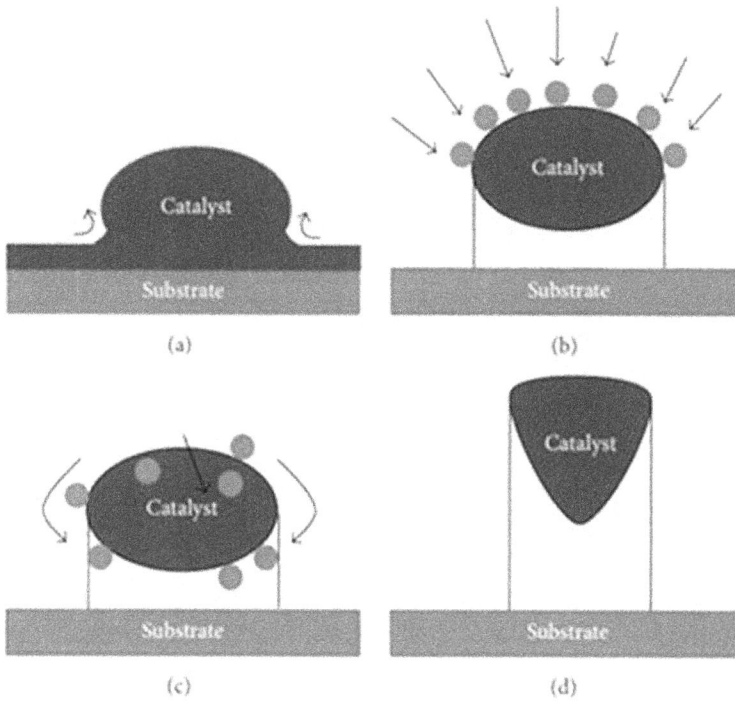

FIGURE 7.1   Growth mechanism of carbon nanofiber [8].

7.2   FACTORS INFLUENCING THE PERFORMANCE OF SUPERCAPACITORS

The capacitance of EDLC SCs primarily originates from the accumulated charges on the boundary among the electrodes and electrolyte. Thus, the capacitive functioning of an EDLC SCs is influenced by the pore sizes, pore dispersal and SSAs of the electrode substance [16].

7.2.1   STRUCTURE AND MORPHOLOGY

The surface morphology aids in releasing the blockade to surface availability for enhancing storage ability. Spherical carbon nano structures have the lowest surface-to-volume ratio. The porous carbon nanosheets have extreme versions of 2D sheet features, such as structures allowing quick ion transmission ways. Moreover, the porous structures efficiently avoid the overlying of the precious interfacing available sites [17]. Graphene integrating surface and their interfacial features are obtained by covalently coupling arylamine derivates having various reactive terminal groups (-SH, -OH, -NH$_2$, -COOH and -SO$_3$H). The amine performing assembly existing in the their derivate for creating a covalent bond among amine and graphene through thiol, hydroxy, amine, carboxyl, and sulfonate groups. The anodic oxidation of primary amine creates an amine radical cation that exhausts a proton and establishes an amine radical. Subsequently, the produced amine radical sticks the sp^2 carbon of the graphene and creates a C-N covalent bond among the amine and graphene. This process has been effectively formed among other carbon allotropes including fullerenes, carbon nanotubes, carbon fibers, and graphite [18]. Furthermore, 1D CNFs and nanorods could provide shortcut routes for electron movement, and which offers the suitable geometries for producing innovative SCs. Usually, external-templating approaches gives accurate influence on various morphologies which aid the active boundaries into a single structure. Yu et al. analyzed a

FIGURE 7.2 FE-SEM images of (a) graphene (b) thiolated graphene, (c) hydroxylated graphene, (d) aminated graphene, (e) carboxylated graphene, and (f) sulfonated graphene [18].

rigid-interface-induced outward contraction method for preparing a hollow mesoporous CNTs and MOF for the predecessor [18].

7.2.2 PORE SIZE AND SHAPE

Specific capacitance is influenced by surface area, as well as other vital factors like morphology, pore size dispersals, pore shape, and their availability in the electrolyte. According to the optimal pore size theory for wide pores, the energy density decreases with increasing pore width (at an applied voltage) due to an electro-neutral zone appearances interior, a pore does not participate in the storage energy, but nevertheless it supports the total volume [19, 20]. Altering the pore size and pore shape of the electrode substance enhances the transport of electrolyte ions. The physicochemical features of the carbon-based material are based upon the pore size, which is less than 1 nm [16]. The electrode material with smaller pores offers higher capacitance and energy density. Furthermore, appropriate size dispersal may enhance retention ability and it offers high power density in SCs. The effectual dispersal of micro/mesopores of the electrode substance could offer a rapid mass and ion transfer, and it improves the approachability of the electrolyte, which is essential for SC application [6]. The electrode pore size is lesser compared to the of electrolyte ion, the electrode unfavored for a charge storing purpose besides large pore sizing (mesopores) remain essential to SCs appliances [21].

7.2.3 SURFACE AREA

Surface area (SA) performs an important factor for creating an EDL. Generally, high specific surface area (SSA) delivers superior capacitance. Carbon-based materials such as carbon nanofibers, activated carbon, and few-layer graphene are some carbon allotropes have potential for EDLC supercapacitor electrodes due to its large SA. By raising the temperature, numerous appropriate pore sizings, i.e., 30–50 Å, were obtained for raising the capacity of EDL. The existence of a porous diameter increases the SA then reduces the resistance of CNT electrodes as a result of the transmission of hydrated ions in pores [21]. Porous activated carbon materials having high SSA have been prepared by KOH activation using a variety of carbon precursors, namely, pollen, wood powder, and other biomasses. Porous carbon derived from melamine-formaldehyde followed by KOH-activation gives an SSA of 3193 m^2 g^{-1} [22]. Activated reduced graphene oxide (a-rGO) is a material having a rigid 3D porous structure. High SSA by post-synthesis mechanical treatment gives the value of 1000–3000 m^2/g [23], and SSA of activated graphene grains interconnected by carbon nanotubes is ~1700 m^2/g [24].

7.2.4 FUNCTIONAL GROUPS

Functionalization on the electrode surface is a successful process for enhancing the capacitance of carbon substances [25]. By introducing heteroatom impurity, for instance oxygen and nitrogen, carbon substances hydrophilicity can upsurge. Additionally, faradaic reactivity because of the existing functional groups creates faradaic capacity, which enhances the total capacity of the substance. Therefore, addition of precise functional sets on the surface fabricates an effectual capacitor.

At lower temperatures, nitrogen is situated outside of the aromatic ring and includes a localized charge like amides and protonated amides. Functional grouping of oxygen, for example quinone, carbonyl, and carboxyl groupings, are able to boost functioning by accelerating the ion transport on high current density. Quinone groupings are singularly promising due to their excellent redox reaction and electrochemical reversions [26]. Oxygen functional groupings are unfavorable to organic electrolytes because they could blockade the pores due to augmentation of electrode wetness. Nitrogen- and oxygen-based functional groupings like pyridinic nitrogen, pyrrolic nitrogen, and quinone oxygen have larger porous around 10 Å and reveal pseudocapacitance within acid electrolyte. However, smaller porous around 5–6 Å exhibited EDLC SCs [25, 27].

7.2.5 ELECTROLYTES

The operation of electrolytes is an important consideration in EDL creation and in faradaic redox reactivity, which determines the electrochemical energy storage capacity. Owing to their smaller ion diameter and exceptional conductivity, aqueous electrolytes such as KOH and H$_2$SO$_4$ exhibit higher specific capacitance [28]. The device construction and electrochemical functioning is affected by hydrogen/oxygen development ensues at the anode/cathode potential about 0/1.23 V. The higher potential of neutral electrolytes such as Na$_2$SO$_4$, Li$_2$SO$_4$, etc. are used for reducing the H+/OH accessibility [29]. WIS electrolytes denote the super-concentrated salt mixtures, and it has commonly been used in energy-associated purposes of high-potential batteries as well as SCs [27]. Nowadays ionic solutions such as solvent-free electrolytes command attention because of their broad working potential window (43 V), wide temperature applicability, small volatility, and low flammability. It is an organic salt mixture of larger organic cations and inorganic/organic anions, with extremely asymmetric unification that proceeds to its aqueous state at room temperature [30]. Furthermore, a high SSA, pore size, and pore geometry of the carbon materials are main factors, as the latter has a direct consequence on the electric double layer. It was detected that micropores (< 2 nm) could raise the specific capacitance, when its size matches the size of electrolyte ions [31].

7.3 TYPES OF CARBON-BASED ELECTRODE MATERIALS FOR SUPERCAPACITORS

7.3.1 ACTIVATED CARBONS

Owing to the high porosity and SA, activated carbon (AC) has achieved significant interest. Recently there is growing attention for fabricating AC from biowaste leads to ecological development. The biowaste used for fabrication of activated carbon is animal, mineral, plant, and vegetable etc. as an electrode substance in electrochemical energy techniques. The SC electrode material from biowaste

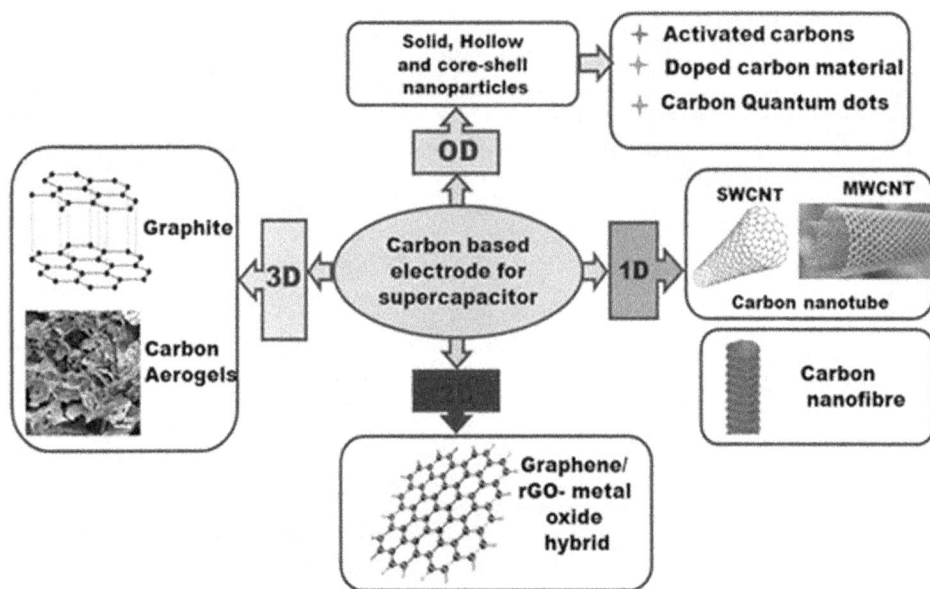

FIGURE 7.3   Illustrative representation of the carbon-based materials for SC electrodes.

has two advantages: (1) waste is converted into a useful product, and (2) it offers an inexpensive for SC technologies. Hierarchical porous activated carbon (HPAC) at very high specific surface area (greater than 1000 m^2/g) is used as the active electrode resource in SCs in electrochemical double layer for energy storing operations [32]. The porous structure determines the energy storage abilities of HPAC. For example, the macropores (> 50 nm) perform as the ion-buffering storage, mesopores (2–50 nm) assist as the electrolyte ions transportation paths, and micropores (< 2 nm) generally act like the charge storing locations [33].

7.3.2   DOPED CARBON MATERIALS

Doping processes are widely utilized for enhancing the electrochemical features of semiconductors. These processes have been developed for improving the conductivity features of carbon substances [34]. Integration of heteroatom within carbon-based substance can affect the properties such as thermal steadiness, localized electronic state, fermi level, and charge transportation. The doping of various single or many heteroatoms simultaneously within carbon-based substances in smaller density enhances the electrochemical features, which is due to the redox efficient functional grouping [35].

Nitrogen (N) doped in the carbon-based materials is a powerful method, and electron abundant nitrogen produces distinctive attribute alterations when it is substituted in a carbon lattice. The n-type and p-type doped reactions are detected upon nitrogen doping. In n-type material, nitrogen contributes the electron, while in p-type doping electrons are lacking and it extracts the electrons from a carbon lattice. The simple nitrogen doping illustrates enhanced SC properties attributable to the tunableness of electron localization. Consequently, wide research has been done in nitrogen doping compared to other heteroatoms within carbon. Nitrogen abundant resources like melamine, urea, polyacrylamide, polyaniline, human hair etc. are inspected to produce the nitrogen-doped carbon [21]. High surface area carbons (HSCAs) of the activated carbons under a nitrogen gas atmosphere under different pre-carbonization temperature is shown in Figure 9.4. Thus, the capacitance decreases with the carbon materials HSAC-300-5 and HSAC-400-5 [31].

Boron has less electronegativity than carbon, which is an extremely necessary property. It produces smaller binding energy among boron and carbon bonds than the bond among carbon atoms.

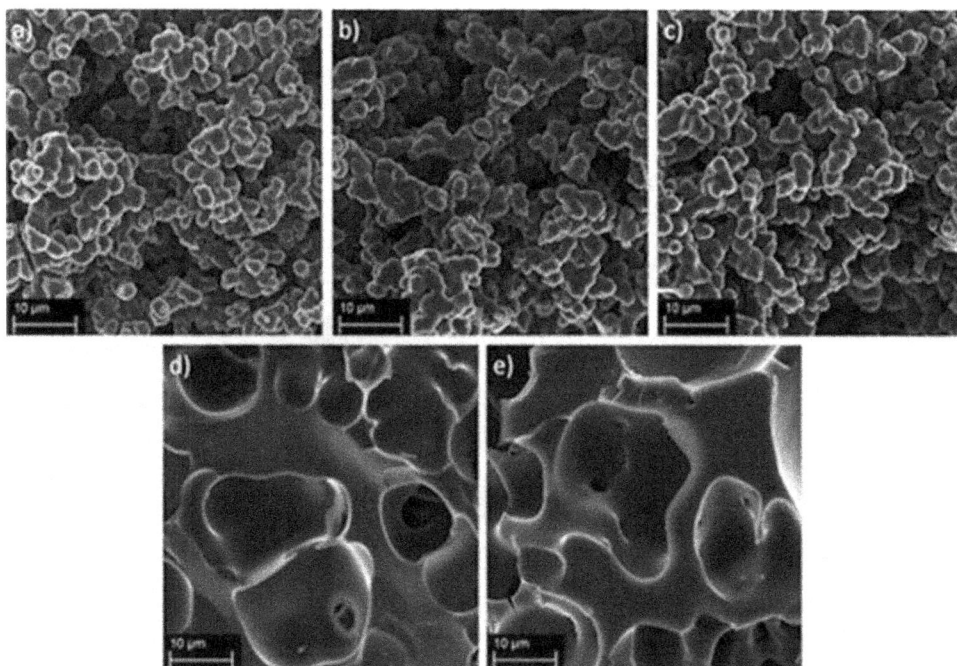

FIGURE 7.4 SEM images of the activated carbons under a nitrogen gas atmosphere with pre-carbonization temperature and the mass ratio of KOH to carbon a) HSAC-800-5, b) HSAC-7000-5, c) HSAC-600-5, d) HSAC-400-5, and e) HSAC-300-5[31].

Certainly, it is comparatively easy to incorporate boron within carbonous substances [36]. Generally boric acid (H3BO3) is employed like the predecessor of boron.

The sulfur doping could produce considerable distraction in the carbon lattice because of its large structure. The bond distance amid sulfur and carbon (C) is greater in comparison to carbon-carbon bonds. Sulfur doping enhanced the surface area and disclosed higher specific capacitance as well as high cycle stableness compared to the parent substance with no sulfur [37]. Phosphorus has low electronegativity than N and C. The polarity of phosphorous and carbon bonds is entirely contrary to a larger bond distance compared to the N-C bond. Theoretically verified graphene bandgap features are altered by phosphorus doping compared to sulfur doping at same dilution. Computational analyses showed that phosphorous-doped material is actively beneficial [38].

7.3.3 Carbon-Based Quantum Dots

Carbon-based quantum dots (CQDs) have significant electrochemical properties that can be used for various energy storage appliances. They refer to a zero-dimensional section in the nanoscale range. On the whole, CQDs show the capacitive actions of an EDLC mechanism. Though, lately, the usage of CQDs has been applied for pseudocapacitors. Carbon-based substances possess size-dependent properties. According to creation and configurations, Cayuela et al. suggest the classification of CQDs as CDs, CQDs, and graphene quantum dots (GQDs) [39]. They are created by the collection of C atoms, the crystalline configuration, and dimensionality. CDs or carbon nanodots are identification of the amorphous quasi-spherical nanodots and an lack of quantum confinement [40] and their diameter is in nanometer-sized, around 1–20 nm. Consequently, the bandgap of CDs does not depend on their size.

CQDs denote the spherical shaped carbon in the nanoscale possessing a crystalline nature and exhibit the quantum confinement effect. It is a quasi-spherical particle having lateral and height

FIGURE 7.5 Carbon quantum dots assembly powder of Optical image and their Inset image shows sponge-like CQDs-800 at100 µm scale bar [42]

sizing around 1–20 nm [40]. The lattice remains constant among graphene and graphite lattices, and because of the low crystalline sp2 carbon, the CQDs exhibit lesser crystallinity compared to GQDs [41]. In CQDs, the p-electron in the sp2 hybridization might act as electron (e-) acceptors/donors, like a conduction channel for e-transportation [40].

GQDs is a zero-dimensional graphene sheets at nanoscale dimension which show effective quantum confinement and excellent crystallinity including graphene lattice constant. GQDs are composed of π-link and sp2 carbon construction, an impression of polycyclic aromatic hydrocarbon fragments. GQDs include some layers of graphene sheets of 1–10 nm sizing compared to the pristine graphene. CQDs are produced through rupturing C_{60} via KOH activation. Annealed CQD construction exhibits high density (1.23 gcm^{-3}) electrodes in SCs, and also they exhibit high volumetric and real capacitance [42].

7.3.4 CARBON NANOTUBES

Carbon nanotubes (CNTs) possess electrically conductive networks and can also store energy at low-cost, since they are durable materials having high surface areas. Owing to their unique properties CNTs are used in SC electrode applications. The excellent operation of CNTs is attribute to the usage of extreme SA for unceasing charge dispersal [36]. Also, the mesoporous features permit electrolytes to transport freely, which decreases the equivalent series resistances and hence enhance power production [43]. CNTs are categorized as single-walled CNTs (SWNTs) and multi-walled CNTs (MWNTs). CNTs proposes a large available pore SA and which could ease the efficient transport for electrolyte ions. Owing to this flexibleness, SWNTs and MWNTs are utilized to EDLC electrode appliances to obtain extreme power of the electrode [44]. SWCNTs were coated on the SBS (solution blow spinning) nanofibers using a simple dipping and drying process. Supercapacitor cells were prepared by sandwiching a SBS nanofiber mat separator immersed in an ionic liquid between two SWCNT/SBS electrodes attached on Ecoflex substrates as shown in Figure 7.6c [45].

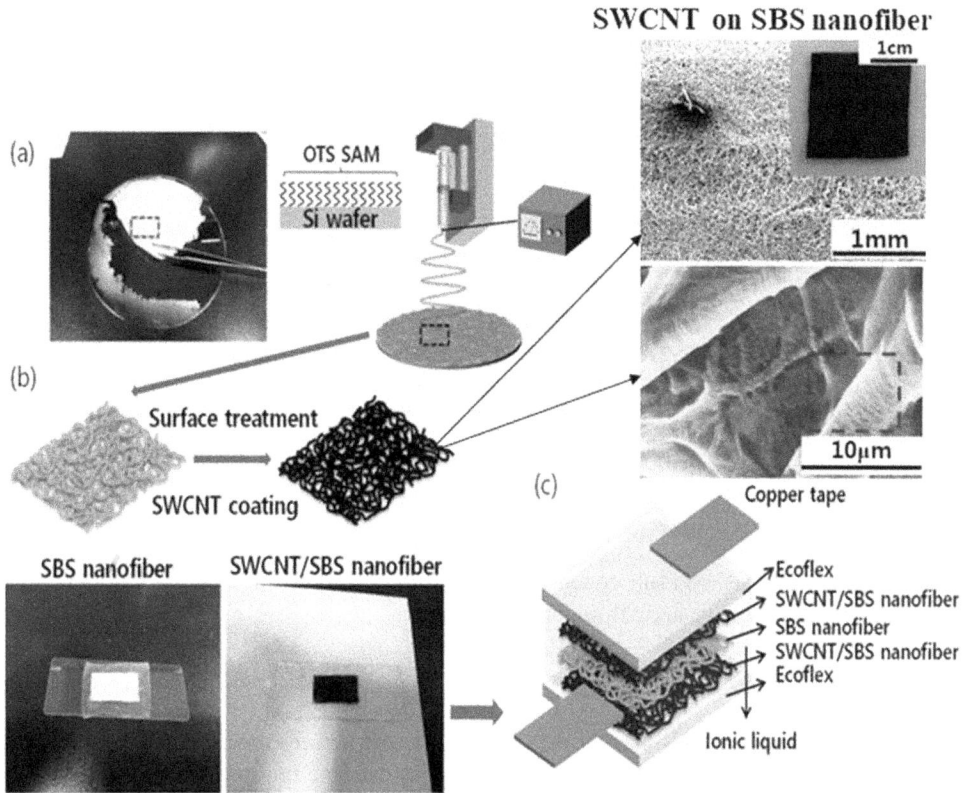

FIGURE 7.6 Schematic diagram for the stretchable supercapacitor fabrication process (SWCNT coating on SBS nanofibrous mat) [45].

7.3.5 CARBON-BASED FIBERS

The fibrous carbon materials have good conduction with high power density, while the high SA can deliver good energy density. CVD grown CNFs are produced through curled graphite layers stacking on one another creating cone formed coatings. Figure 7.7c shows the packed cone constructions, also called herringbone and bamboo type CNFs [46]. A week interplane van der Waals joining among cones constitutes CNFs weakly compared to CNTs. Moreover, there has been increasing attention on fibrous substances for precise usage in energy storing. Thus, carbonous fibrous substances are used in EDLC electrodes. Shirshova et al. confirmed the feasibility of using high conduction, nonactivated carbon fiber as electrodes in SCs appliances [47].

7.3.6 GRAPHENE

Graphene is the preeminent 2D single-layer carbon arranged in sp^2-hybridized carbons comprising several fascinating features like lightweightness, high electrical and thermal conduction, extreme tunefulness SA (upto 2675 m^{2}g^{-1}), tough mechanical potency, and chemical steadiness [48]. This unique combination of excellent features creates graphene-based substances developing in electrochemical energy storing and ecological energy production [49–50]. But, in reality, the capacity performance of graphene is less than the estimated value because of severe agglomeration through both fabrication and application methods. Thus, enhancing the complete electrochemical functioning of graphene-based materials is still debatable. Generally, the chemical exfoliation of graphite

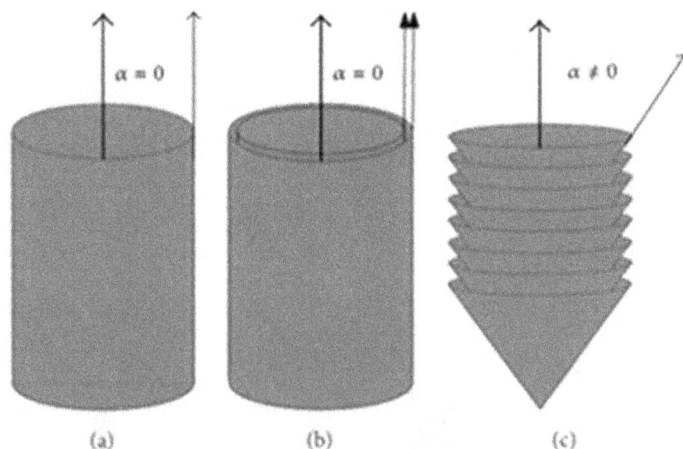

FIGURE 7.7   Illustrative figure of (a) SWNTs, (b) MWNTs and (c) carbon nanofibers.

into graphite oxides (GOs) and subsequent controllability of diminishing the GOs into graphene is considered the most effective and affordable technique.

7.3.7 Graphite

3D graphene networks (3DGNs) having penetrating microstructures revealed a large potential for hybrid SCs applications. The exclusive 3D structure has extreme porous construction and exceptional SA. It offers an approachable area to the electrolyte transport and charge movement for active substances [15]. Recently several composites of 3DGNs with MO/hydroxide and polymers, like $MnO2$, Co_3O_4, Ni (OH)$_2$, Co (OH)$_2$, and PANI, were reported [51]. Dong et al. studied the Co_3O_4 nanowires on 3D graphene foam as an electrode for SC and they exhibited a higher specific capacitance of ~1100 F g^{-1} at 10 A g^{-1} together with the outstanding cycle steadiness [52].

7.3.8 Carbon Aerogels

Research has been accomplished over the years on multifunctional anisotropic carbon aerogels (CAs). CAs can be produced from kraft lignin (KL) and TEMPO-oxidized cellulose nanofibers (TOCNFs), which perform as a template through the carbonization procedure, and produce micropores and mesopores in CA configuration. Hierarchically porous anisotropic carbon aerogels exhibited a specific capacitance of 124 F g^{-1} at a current density of 0.2 A g^{-1}, demonstrating the probabilities for utilization of lignin/CNF as a prospective carbon resource for preparing SC electrodes [53]. Bony Thomas et al. studied carbon aerogel produced from sustainable resources attaining an exceptional EDLC. Two different green, rich, and carbon-rich lignins obtained from several biomasses as raw materials, i.e., kraft and soda lignins, led to distinctive physical and structural features – including electrochemical features – of CAs after carbonization [54].

7.3.9 Hybrid Carbon Materials

7.3.9.1 Carbon Nanomaterial Metal-Oxide-Based Supercapacitors

Fabrication of composite materials with carbon can increases the electrochemical usage of pseudoactive sites and improve electron transference and mechanical stableness. Also, hybrid carbon capacitance and pseudocapacitive/battery-type electrodes could connect the voltage variance among the two electrodes to increase device capacity. MO-adapted carbons serve as the typical surface redox pseudocapacitive substance for symmetric and asymmetric SCs [16].

7.3.9.2   Graphene/Reduced Graphene Oxide Metal-Oxide-Based Supercapacitors

Graphene-based substance have been widely explored as a conducting system for enhancing the redox activity of TMOs, hydroxides (HOs), and CPs. The nanohybrid electrodes consist of graphene, and nanoparticles of TMO/HOs or CPs showed the higher electrochemical operation. The synergetic result is that graphene layers accelerate the diffusion of MOs/HOs nanoparticles, and perform like high conductive medium to improve the electrical conduction, and the MOs/HOs/CPs propose the required pseudocapacitance. Recently, Zhao et al. investigated the developments in graphene-based hybrid substances for LIBs and SC utilizations [55]. To enhance operation, graphene-based systems are mixed with redox-active oxides and polymers. Several graphene oxide composites are employed like electrode substances in the SC.

7.3.9.3   Carbon Fiber Metal-Oxide-Based Supercapacitors

By comparing 2D planar structural substance, 1D fibrous (like CNTs, graphene fibers, carbon fibers (CFs), etc.) are broadly utilized for producing flexible electrodes because of their lightweightness, great electrical conduction, and higher flexibility [56]. The exclusive mechanical features of CFs make them a great option for performing as flexible electrodes. The acid subjected carbon fiber paper (A-CFP) electrode affords more energetic locations than does carbon fiber paper. The combination of CNFs and CNTs were coated over the A-CFP by vacuum-filtration because of the higher hydrophilicity of A-CFP enhanced by acidic action (as displayed in Figure 7.8a) [57]. The recycled CFs (RCFs) strengthened PANI/MnO$_2$ flexible incorporated electrode was fabricated using an electrodeposition method as shown in Figure 7.9a. When electropolymerization of aniline and electrodeposition of

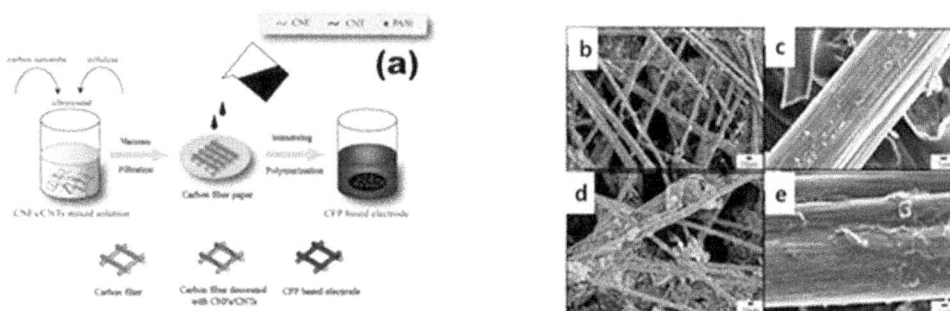

FIGURE 7.8   Experimental process for producing (a) CNFs and CNTs deposit over the A-CFP by vacuum-filtration and (b) SEM images of low enlargement of raw CFPs, (c) high amplification of raw CFPs, (d) low amplification of A-CFP and (e) high enlargement of A-CFP [57].

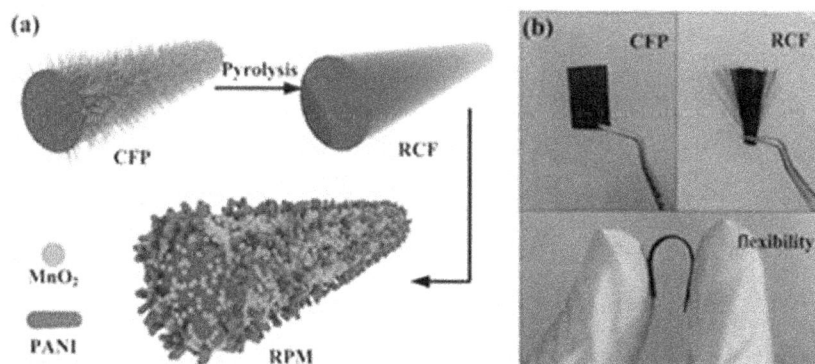

FIGURE 7.9   (a) Fabrication process of RCF-supported PANI/MnO$_2$ and (b) snapshots of carbon fiber plates (CFP) and RCF [58].

TABLE 7.1
A Comparison of the Carbon-Based Electrode Substance Used in Several Kinds of SC Devices

Material	Method	Morphology	Specific capacitance	Energy density	Power density/ current density	% of specific capacitance at no. of cycles	Ref
Zinc chloride activation of lotus seed powder	Carbonization	Hierarchically porous structure	317.5 F g^{-1}		50 A g^{-1}	99.2% after 10,000 cycles	[59]
Zinc chloride activation of Washnut seed	Carbonization	Hierarchical micro- and meso-pore architectures	225.1 F g^{-1}		1 A g^{-1}	98% at 10,000 cycles	[60]
Graphitic Carbon from Lapsi Seed	Carbonization and activation	Mesoporous architecture	284 F g^{-1}		1 A g^{-1}.	99% at 10,000	[61]
CNF/MnOx	Carbonization	Hierarchical 3D porous aerogel	269.7 F g^{-1}		37.5 Wh kg^{-1}	80% after 1000	[62]
MnO$_2$/rGO	Chemical method	Nanoscrolls	223.2 F/g	105.3 Wh/kg	308.1 W/kg	92% at 10,000 cycles	[63]
Recycle carbon fiber/ polyaniline/MnO$_2$	Electrodeposition	Nanofiber	475.1 F·g^{-1}		1 A·g^{-1}	86.1% at 5000 cycles	[58]
CoMnO$_2$-Polyimide- Carbon Fiber Electrodes		Porous hierarchical interconnected nanosheet	221 F g^{-1}	60.2 Wh kg^{-1}	490 W kg^{-1}	95% up to 3000 cycles	[14]
N-doped carbon aerogel	Carbonization	3D network structure	185 F g^{-1}		1 A g^{-1}		[64]
Activated carbon fiber papers with CNT	Vacuum–Filtration	Carbon Fiber	626 mF·cm^{-2}	87 μWh·cm^{-2}			[57]
Green CAs based on kraft lignin	Carbonization	Aerogels	163 F g^{-1}	5.67 Wh kg^{-1}	50 W kg^{-1}		[54]

The various carbon-based electrode substances are used in several kinds of SC devices for example with reference 59 to 64. The various materials included zinc chloride activation of lotus seed powder, zinc chloride activation of washnut seed, graphitic carbon from lapsi Seed, carbon nanofiber/MnOx, MnO$_2$/rGO, Recycle carbon fiber/polyaniline/MnO$_2$, CoMnO$_2$-Polyimide- Carbon Fiber Electrodes, N-doped carbon aerogel, Activated carbon fiber papers with CNT and green CAs based on kraft lignin via carbonization, chemical method, electrodeposition and vacuum-filtration method. The different morphologies yielded specific capacitance and power densities: 163–475 F g^{-1}, 1 A g^{-1} to 490 W kg^{-1} at 99.2% to 80% of specific capacitance after 1000–10,000 cycles.

MnO_2 happen concurrently, the produced MnO_2 could perform as an oxidant and make chemical polymerization of aniline, and as a result form fluffy structures of the hybrid film [58].

7.4 CONCLUSION

Carbon based materials are auspicious for providing high specific surface area and promising chemical and thermal stableness with less electrical resistance. Applications of biomass-based sustainable resources represent an alternative to fabricating doped carbon substances for SC appliances. Various factors including diverse morphologies, dopants, composites, and functionalized substances determine the electrochemical capacitors. Recent research developments on carbonous resources show that poisonous materials might be prevented in upcoming SCs with no deficiency of efficacy. The chief advantages of using carbonous substances are low cost and that several types of eco-friendly carbon exist in nature. Easily modifying electronic features can be achieved by altering the porosity, dopant strength, and functional grouping. The approaches reviewed for enhancing the SC function and their merits and restrictions may be helpful in future research for constructing SC devices with novel approaches and materials.

REFERENCES

[1] Reis, G.S.; Larsson, S.H.; Oliveira, H.P.; Thyrel, M.; Lima, E.C. Sustainable biomass activated carbons as electrodes for battery and supercapacitors—a mini-review. *Nanomaterials* 2020, 10, 1398.

[2] Forouzandeh, P.; Kumaravel, V.; Pillai, S.C. Electrode materials for supercapacitors: A review of recent advances. *Catalysts* 2020, 10, 969.

[3] Zhu, Z.; Wang, G.; Sun, M.; Li, X.; Li, C. Fabrication and electrochemical characterization of polyaniline nanorods modified with sulfonated carbon nanotubes for supercapacitor applications. *Electrochim. Acta* 2011, 56, 1366–1372.

[4] Fitri Aulia Permatasari; Muhammad Alief Irham; Satria Zulkarnaen Bisri; Ferry Iskandar. Carbon-based quantum dots for supercapacitors: Recent advances and future challenges. *Nanomaterials* 2021, 11(1), 91.

[5] Gopalakrishnan, A.; Badhulika, S. Effect of self-doped heteroatoms on the performance of biomass-derived carbon for supercapacitor applications. *Journal of Power Sources* 2020, 480, 228830.

[6] Becker, H.I.; General Electric Company. Low voltage electrolytic capacitor. U.S. Patent 2,800,616, 1957.

[7] M.; Feng, X. Thin-film electrode-based supercapacitors. *Joule* 2019, 3, 338–360.

[8] Hasegawa, G.; Aoki, M.; Kanamori, K.; Nakanishi, K.; Hanada, T.; Tadanaga, K. Monolithic electrode for electric double-layer capacitors based on macro/meso/microporous S-Containing activated carbon with high surface area. *Journal of Materials Chemistry* 2011, 21, 2060–2063.

[9] Song, Y.; Liu, T.-Y.; Xu, G.-L.; Feng, D.-Y.; Yao, B.; Kou, T.-Y.; Liu, X.-X.; Li, Y. Tri-layered graphite foil for electrochemical capacitors. *Journal of Materials Chemistry A* 2016, 4, 7683–7688.

[10] Ping, Y.; Gong, Y.; Fu, Q.; Pan, C. Preparation of three-dimensional graphene foam for high performance supercapacitors. *Progress in Natural Science: Materials International* 2017, 27, 177–181.

[11] Wood, D.L., III; Li, J.; Daniel, C. Prospects for reducing the processing cost of lithium ion batteries. *Journal of Power Sources* 2015, 275, 234–242.

[12] Garakani, M.A.; Bellani, S.; Pellegrini, V.; Oropesa-Nuñez, R.; Castillo, A.E.D.R.; Abouali, S.; Najafi, L.; Martín-García, B.; Ansaldo, A.; Bondavalli, P. Scalable spray-coated graphene-based electrodes for high-power electrochemical double-layer capacitors operating over a wide range of temperature. *Energy Storage Mater* 2020, 34, 1–11.

[13] Young-Hun Cho; Jae-Gyoung Seong; Jae-Hyun Noh; Da-Young Kim; Yong-Sik Chung; Tae Hoon Ko; Byoung-Suhk Kim. $CoMnO_2$-decorated polyimide-based carbon fiber electrodes for wire-type asymmetric supercapacitor applications. *Molecules* 2020, 25(24), 5863.

[14] Li, C.; Zhang, X.; Wang, K.; Zhang, H.; Sun, X.; Ma, Y. Three dimensional graphene networks for supercapacitor electrode materials. *Journal of New Carbon Materials* 2015, 30(3), 193–206.

[15] Parnia Forouzandeh; Vignesh Kumaravel; Suresh C. Pillai. Electrode materials for supercapacitors: A review of recent advances. *Catalysts* 2020, 10(9), 969.

[16] He, Y.; Zhuang, X.; Lei, C.; Lei, L.; Hou, Y.; Mai, Y.; Feng, X. Porous carbon nanosheets: Synthetic strategies and electrochemical energy related applications. *Nano Today* 2019, 24, 103–119.

[17] Chiranjeevi Srinivasa Rao Vusa; Manju Venkatesan; Aneesh K; Sheela Berchmans; Palaniappan Arumugam. Tactical tuning of the surface and interfacial properties of graphene: A Versatile and rational electrochemical approach. *Scientific Reports* 2017, 7, 8354.

[18] Liu, C.; Huang, X.; Wang, J.; Song, H.; Yang, Y.; Liu, Y.; Li, J.; Wang, L.; Yu, C. Hollow mesoporous carbon nanocubes: Rigid-interface-induced outward contraction of metal-organic frameworks. *Advanced Functional Materials* 2018, 28, 1705253.

[19] Kondrat, S.; Perez, C. R.; Presser, V.; Gogotsib, Y.; Kornyshev, A. A. Effect of pore size and its dispersity on the energy storage in nanoporous supercapacitors. *Energy & Environmental Science* 2012, 5, 6474–6479.

[20] Noureen Siraj; Samantha Macchi; Brian Berry; Tito Viswanathan. Metal-free carbon-based supercapacitors—a comprehensive review. *Electrochem* 2020, 1, 410–438.

[21] Zheng Yue; Hamza Dunya; Maziar Ashuri; Kamil Kucuk; Shankar Aryal; Stoichko Antonov; Bader Alabbad; Carlo U. Segre; Braja K. Mandal. Synthesis of a very high specific surface area active carbon and its electrical double-layer capacitor properties in organic electrolytes. *ChemEngineering* 2020, 4, 43.

[22] Artem Iakunkov; Vasyl Skrypnychuk; Andreas Nordenstrom; Elizaveta A. Shilayeva; Mikhail Korobov; Mariana Prodana; Marius Enachescu; Sylvia H. Larssond; Alexandr V. Talyzin. Activated graphene as a material for supercapacitor electrodes: Effects of surface area, pore size distribution and hydrophilicity. *Physical Chemistry Chemical Physics* 2019, 21, 17901–17912.

[23] Vasyl Skrypnychuk; Nicolas Boulanger; Andreas Nordenstrom; Alexandr Talyzin. Aqueous activated graphene dispersions for deposition of high-surface area supercapacitor electrodes. *Journal of Physical Chemistry Letters* 2020, 11(8), 3032–3038.

[24] Bresser, D.; Buchholz, D.; Moretti, A.; Varzi, A.; Passerini, S. Alternative binders for sustainable electrochemical energy storage—the transition to aqueous electrode processing and bio-derived polymers. *Energy & Environmental Science* 2018, 11, 3096–3127.

[25] Hulicova-Jurcakova, D.; Seredych, M.; Lu, G.Q.; Bandosz, T.J. Combined effect of nitrogen- and oxygen-containing functional groups of microporous activated carbon on its electrochemical performance in supercapacitors. *Advanced Functional Materials* 2009, 19, 438–447.

[26] Zhou, C.; Gao, T.; Liu, Q.; Wang, Y.; Xiao, D. Preparation of quinone modified graphene-based fiber electrodes and its application in flexible asymmetrical supercapacitor. *Electrochim Acta* 2020, 336, 135628.

[27] Seredych, M.; Hulicova-Jurcakova, D.; Lu, G.Q.; Bandosz, T.J. Surface functional groups of carbons and the effects of their chemical character, density and accessibility to ions on electrochemical performance. *Carbon* 2008, 46, 1475–1488.

[28] Ling Miao; Ziyang Song; Dazhang Zhu; Liangchun Li; Lihua Gan; Mingxian Liu. Recent advances in carbon-based supercapacitors. *Materials Advances* 2020, 1, 945–966.

[29] Zhou, Z.; Miao, L.; Duan, H.; Wang, Z.; Lv, Y.; Xiong, W.; Zhu, D.; Li, L.; Liu, M.; Gan, L. Highly active N, O-doped hierarchical porous carbons for high-energy supercapacitors. *Chinese Chemical Letters* 2020, 31, 1226–1230.

[30] Watanabe, M.; Thomas, M. L.; Zhang, S.; Ueno, K.; Yasuda, T.; Dokko, K. Application of ionic liquids to energy storage and conversion materials and devices. *Chemical Reviews* 2017, 117, 7190–7239.

[31] Ruben Heimbockel; Frank Hoffmann; Michael Froba. Insights into the influence of the pore size and surface area of activated carbons on the energy storage of electric double layer capacitors with a new potentially universally applicable capacitor model. *Physical Chemistry Chemical Physics* 2019, 21, 3122–3133.

[32] Bai, X.; Wang, Z.; Luo, J.; Wu, W.; Liang, Y.; Tong, X.; Zhao, Z. Hierarchical porous carbon with interconnected ordered pores from biowaste for high-performance supercapacitor electrodes. *Nanoscale Research Letters* 2020, 15, 88.

[33] Tai-Feng Hung; Tzu-Hsien Hsieh; Feng-Shun Tseng; Lu-Yu Wang; Chang-Chung Yang; Chun-Chen Yang. High-mass loading hierarchically porous activated carbon electrode for pouch-type supercapacitors with propylene carbonate-based electrolyte. *Nanomaterials* 2021, 11(3), 785.

[34] Abbas, Q.; Raza, R.; Shabbir, I.; Olabi, A. Heteroatom doped high porosity carbon nanomaterials as electrodes for energy storage in electrochemical capacitors: A review. *Journal of Science: Advanced Materials and Devices* 2019, 4, 341–352.

[35] Wang, X.; Sun, G.; Routh, P.; Kim, D.-H.; Huang, W.; Chen, P. Heteroatom-doped graphene materials: Syntheses, properties and applications. *Chemical Society Reviews* 2014, 43, 7067–7098.

[36] Panchakarla, L.S.; Subrahmanyam, K.S.; Saha, S.K.; Govindaraj, A.; Krishnamurthy, H.R.; Waghmare, U.V.; Rao, C.N.R. Synthesis, structure, and properties of boron- and nitrogen-doped graphene. *Advanced Materials* 2009, 21, 4726–4730.

[37] Desmaris, V.; Saleem, M. A.; Shafiee, S. Examining carbon nanofibers: Properties, growth, and applications. *IEEE Nanotechnology Magazine* 2015, 9(2), 33–38.

[38] Denis, P.A. Concentration dependence of the band gaps of phosphorus and sulfur doped graphene. *Computational Materials Science* 2013, 67, 203–206.

[39] Ilani, S.; Donev, L.A.K.; Kindermann, M.; McEuen, P.L. Measurement of the quantum capacitance of interacting electrons in carbon nanotubes. *Nature Physics* 2006, 2, 687–691

[40] Essner, J.B.; Baker, G.A. The emerging roles of carbon dots in solar photovoltaics: A critical review. *Environmental Science: Nano* 2017, 4, 1216–1263.

[41] Li, M.; Chen, T.; Gooding, J.J.; Liu, J. Review of carbon and graphene quantum dots for sensing. *ACS Sensors* 2019, 4, 1732–1748.

[42] Guanxiong Chen; Shuilin Wu; Liwei Hui; Yuan Zhao; JianglinYe; Ziqi Tan; Wencong Zeng; Zhuchen Tao; LihuaYang; Yanwu Zhu. Assembling carbon quantum dots to a layered carbon for high-density supercapacitor electrodes. *Scientific Reports* 2016, 6, 19028.

[43] An, K.H.; Kim, W.S.; Park, Y.S.; Choi, Y.C.; Lee, S.M.; Chung, D.C.; Bae, D.J.; Lim, S.C.; Lee, Y.H. Supercapacitors using single-walled carbon nanotube electrodes. *Advanced Materials* 2001, 13, 497–500.

[44] Frackowiak, E.; Beguin, F. Carbon materials for the electrochemical storage of energy in capacitors. *Carbon* 2001, 39, 937–950.

[45] Juyeon Yoon; Joonhyung Lee; Jaehyun Hur. Stretchable supercapacitors based on carbon nanotubes-deposited rubber polymer nanofibers electrodes with high tolerance against strain. *Nanomaterials* 2018, 8, 541.

[46] Nishino, A. Capacitors: Operating principles, current market and technical trends. *Journal of Power Sources* 1996, 60(2), 137–147.

[47] Shirshova, N.; Greenhalgh, E.; Shaffer, M.; Steinke, J. H. G.; Curtis, P.; Bismarck, A. Structural polymer composites for energy storage devices. In *Proceedings of the 17th International Conference on Composite Materials (ICCM'09)*, Edinburgh, UK, 2009.

[48] Xia, J. L.; Chen, F.; Li, J. H; Tao, N. J. Measurement of the quantum capacitance of graphene. *Nature Nanotechnology* 2009, 4, 505–509.

[49] Simon, P.; Gogotsi, Y. Materials for electrochemical capacitors. *Nature Materials* 2008, 7, 845–854.

[50] Xia, J. L.; Chen, F.; Li, J. H.; Tao, N. J. Measurement of the quantum capacitance of graphene. *Nature Nanotechnology* 2009, 4, 505–509.

[51] Sachin Kumar; Ghuzanfar Saeed; Ling Zhua; Kwun Nam Huic; Nam Hoon Kimb; Joong Hee Lee. 0D to 3D carbon-based networks combined with pseudocapacitive electrode material for high energy density supercapacitor: A review. *Chemical Engineering Journal* 2021, 403, 126352.

[52] Dong, X. C.; Xu, H.; Wang, X. W. 3D graphene-cobalt oxide electrode for high-performance supercapacitor and enzymeless glucose detection. *ACS Nano* 2012, 6, 3206–3213.

[53] Geng, S.; Wei, J.; Jonasson, S.; Hedlund, J.; Oksman, K. Multifunctional carbon aerogels with hierarchical anisotropic structure derived from lignin and cellulose nanofibers for CO_2 capture and energy storage. *ACS Applied Materials & Interfaces* 2020, 12, 7432–7441.

[54] Bony Thomas; Shiyu Geng; Mohini Sain; Kristiina Oksman. Hetero-porous, high-surface area green carbon aerogels for the next-generation energy storage applications. *Nanomaterials* 2021, 11, 653.

[55] Yao, X.; Zhao, Y. Three-dimensional porous graphene networks and hybrids for lithium-ion batteries and supercapacitors. *Chem* 2017, 2, 171–200.

[56] Hatzell, K.; Fan, L.; Beidaghi, M.; Boota, M.; Pomerantseva, E.; Kumbur, E.; Gogotsi, Y. Composite manganese oxide percolating networks as a suspension electrode for an asymmetric flow capacitor. *ACS Applied Materials & Interfaces* 2014, 6, 8886–8893.

[57] Sicong Tan; Jiajia Li; Lijie Zhou; Peng Chen; Jiangtao Shi; Zhaoyang Xu. Modified carbon fiber paper-based electrodes wrapped by conducting polymers with enhanced electrochemical performance for supercapacitors. *Polymers* 2018, 10(10), 1072.

[58] Xiaoning Wang; Hongli Wei; Wei Du; Xueqin Sun; Litao Kang; Yuping Zhang; Xiangjin Zhao; Fuyi Jiang. Recycled carbon fiber-supported polyaniline/manganese dioxide prepared via one-step electrodeposition for flexible supercapacitor integrated electrodes. *Polymers* 2018, 10(10), 1152.

[59] Ram Lal Shrestha; Rashma Chaudhary; Timila Shrestha; Birendra Man Tamrakar; Rekha Goswami Shrestha; Subrata Maji; Jonathan P. Hill; Katsuhiko Ariga; Lok Kumar Shrestha. Nanoarchitectonics of lotus seed derived nanoporous carbon materials for supercapacitor applications. *Materials* 2020, 13(23), 5434.

[60] Ram Lal Shrestha; Timila Shrestha; Birendra Man Tamrakar; Rekha Goswami Shrestha; Subrata Maji; Katsuhiko Ariga; Lok Kumar Shrestha. Nanoporous carbon materials derived from washnut seed with enhanced supercapacitance. *Materials* 2020, 13(10), 2371.

[61] Lok Kumar Shrestha; Rekha Goswami Shrestha; Subrata Maji; Bhadra P. Pokharel; Rinita Rajbhandari; Ram Lal Shrestha; Raja Ram Pradhananga; Jonathan P. Hill; Katsuhiko Ariga. High surface area nanoporous graphitic carbon materials derived from Lapsi seed with enhanced supercapacitance. *Nanomaterials* 2020, 10(4), 728.

[62] Xiaoyu Guo; Qi Zhang; Qing Li; Haipeng Yu; Yixing Liu. Composite aerogels of carbon nanocellulose fibers and mixed-valent manganese oxides as renewable supercapacitor electrodes. *Polymers* 2019, 11, 129.

[63] Janardhanan R. Rani; Ranjith Thangavel; Minjae Kim; Yun Sung Lee 2; Jae-Hyung Jang. Ultra-high energy density hybrid supercapacitors using MnO_2/reduced graphene oxide hybrid nanoscrolls. *Nanomaterials* 2020, 10(10), 2049.

[64] Lei, E.; Wei, Li; Jiaming Sun; Zhenwei Wu; Shouxin Liu. N-doped carbon aerogels obtained from APMP fiber aerogels saturated with rhodamine dye and their application as supercapacitor electrodes. *Applied Sciences* 2019, 9, 618.

8 Bionanocomposite Systems for Supercapacitor Applications

Greeshma Kuzhipalli Perayikode and
Sella Muthulingam

CONTENTS

8.1 NANOCOMPOSITES

The combination of inorganic solids with polymers at a nanometric scale is often referred to as nanocomposites; the structure of nanocomposites is more complicated than normal microcomposites due to their composition and individual properties [1]. Normally nanocomposites are multiphasic materials of nanoscale morphology and one phase should have dimensions in the range of nanoparticles (10–100 nm). To improve the properties of nanocomposites, usually reinforcing fillers may be incorporated with organic or inorganic materials. Nowadays, different alternatives in nanocomposites are introduced to minimize limitations observed in engineering materials and also many novel approaches have been developed to synthesize existing nanocomposites with multifunctional applications [2]. The properties of these nanocomposites not only depended on the parent constituent but also its morphological and interfacial behaviors.

The utilization of polymers has increased rapidly due to their ductility and lightweight properties. But when compared with metals and ceramics they exhibit some drawbacks in strength, stability, and mechanical properties [3]. Polymer nanocomposites are considered a leading engineering material for current research development. Similarly, polymer nanocomposites with carbon-based nanoparticles can improve these properties to some extent [4]. These newly derived multifunctional nanocomposites with improved mechanical strength, heat resistance, and biodegradability are applicable in many engineering fields such as sports, aerospace components, solar cells, and supercapacitors.

Nanocomposites are generally obtained by combining more than one material like fullerenes, metal oxides, nanoclusters, organometallic compounds, biological molecules, and enzymes. Due to this combination of two or more constituents, they exhibit unique properties, which arise from their small size, interfacial interaction between phases, and, of course large surface area. These

DOI: 10.1201/9781003174646-8

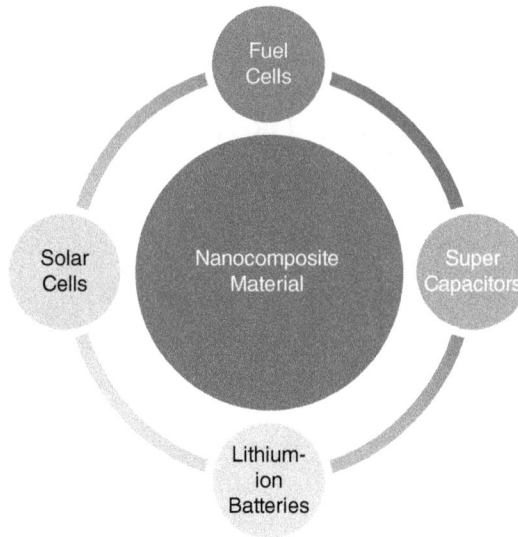

FIGURE 8.1 Applications of nanocomposites materials.

TABLE 8.1
Different Varieties of Nanocomposites [6]

Sl.No	Classification	Examples
1	Metal Matrix Nanocomposites (MMNC)	Co/Cr, Fe/MgO, Fe-Cr/Al$_2$O$_3$ and Mg/CNT
2	Ceramic Matrix Nanocomposites (CMNC)	Iron (Fe)-Chromium (Cr)/Al$_2$O$_3$ & Ni/Al$_2$O$_3$
3	Polymer Matrix Nanocomposites (PMNC)	Polymer/CNT & Polymer/layered double hydroxide.

extraordinary properties have been utilized in many biomaterials, sensors, supercapacitors, and value-added products [5].

It has been observed that when the size of the particle is at the nanometer level, phase interface interactions become appreciable and that can enhance the material properties. Furthermore, these structural properties could change due to their altered surface. The introduction of carbon nanotubes in 1996 and their unique properties added discoveries in nanocomposite-based research. Recently the synergic effect on carbon nanotube-based nanocomposites has made them widely applicable across all industries. Based on their matrix phase, nanocomposites are classified in Table 8.1 [6].

A combination of flexible material with molecular-level reinforcing components are called metal matrix nanocomposites. These have properties of both metals and ceramics and show excellent properties in strength, ductility, and toughness. Due to these properties, this type of material can be applicable in high-temperature conditions and also for the production of industrial materials with compression processes. This extraordinary property is applicable in many areas like aerospace, automotive, and the production of structural materials. Nilhara et al. reported promising strength of Al_2O_3/SiC ceramic matrix nanocomposites and their toughening mechanism using nanosized reinforcing materials. Similar studies revealed incorporation of nanosized reinforcing materials with ceramic matrix and development of advanced matrix nanocomposites [7].

Polymer materials are considered to have many disadvantages due to their low modulus and strength, but the addition of fibers, platelets, or materials essential to improve mechanical properties lead to the development of polymer matrix nanocomposites [8]. To improve mechanical strength, heat resistance, flame retardantness, polymers are usually filled with materials like inorganic compounds. So polymer matrix nanocomposites are widely applicable in industries [9–10].

8.2 GENERAL METHODS OF SYNTHESIS

During the synthesis of nanocomposites, naturally and commercially available materials are used. In addition to this, carbon nanotubes, biodegradable molecules, and graphene oxide are also used in the nanometer range [11]. Energy storage devices and electronic components are developed by using conducting polymers. Polymer nanocomposites can be prepared by in situ polymerization, melt processing, and solution blending. Nanomaterials are prepared either by green synthesis or by chemical method by ball milling, chemical vapor deposition, and solution method. To get an efficient interfacial bonding between polymers and fillers, surface modification and noncovalent functionalization is required [12]. This type of modification can be achieved by grafting to and grafting from methods. Both include the bonding of monomer and polymer by a polymerization reaction. Radiation grafting is also applicable for the surface modification process, which uses an electron beam or plasma beam as a radiation source [13].

8.2.1 SOLUTION BLENDING

Nanocomposites are usually fabricated by the solution blending method. This method polymer solution and fillers are mixed with suitable solvents like chloroform, water, toluene, and dimethylformamide, and finally, nanocomposites are obtained by filtration and precipitation method [14]. Usage of toxic solvents and solvent removal are the main drawbacks of this solution blending method. Polymer incorporated graphene nanocomposites with polymethyl methacrylate, polyacrylamide, and polyamides have been synthesized by this method [15]. This type of nanocomposite is widely used for water purification. The solution blending method is also applicable for Silica polymer nanocomposites. Silica sources like TMSO, silicate, sodium silicate on polycondensation in presence of non-ionic or ionic surfactants lead to the synthesis of mesoporous silica [16]. To improve interfacial linkage and to get well-dispersed nanocomposites, sufficient surface modifications need to be generated using silane by co-condensation post-synthetic functionalization [17–18].

8.2.2 MELT PROCESSING

The melt processing method is an eco-friendly one; this avoids the usage of harmful chemicals. This method mainly focuses on processing methods, surface modification of fillers, and compatibility of polymer matrix [19–20]. During this process, nanocomposites are obtained by the direct dispersion of fillers and melted polymer. Injection molding or extrusion can be applied for the blending process. Due to the high viscosity of thermosetting plastics, filler distribution becomes difficult in this method and also requires high temperature for polymer melting [21–22]. In the melt processing method, nanoparticles are agglomerated easily, so close monitoring is required during the reaction. Biopolymers and natural polymers show very close melting and degradation temperature [23].

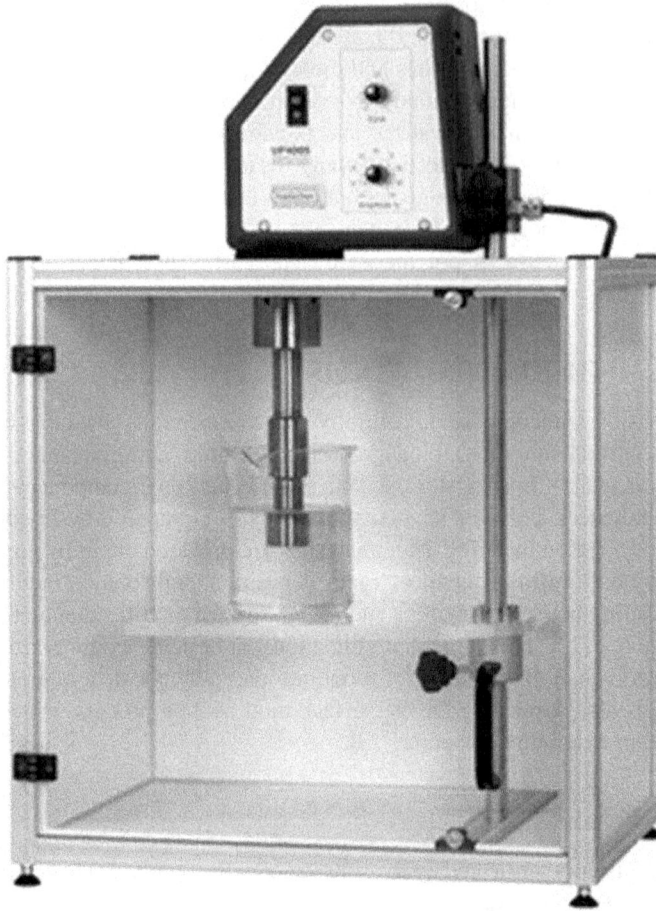

FIGURE 8.2 Schematic representation of solution blending process.

8.3 BIONANOCOMPOSITES

Non-degradable polymers like polyethylene, polypropylene, and polystyrene are disposed of either through landfilling or by incineration; both methods are undesirable to the environment and emit a huge amount of carbon dioxide into the atmosphere [24]. Bionanocomposites are generally composed of inorganic solids and biopolymers. They exhibit a size in the range of a nanometer scale. In other words, it consists of two or more phases and forms a single material with enhanced performance over individual components [25]. These types of materials include hair protein, cellulose acetate, starch, and polylactic acid. These materials can be applied in medical science as artificial bone tissue shells, in drug delivery, and in in vitro bone regeneration. Bionanocomposites are eco-friendly and show the most significant applications in biomedical sciences and medicine [26]. Enhanced surface reactivity of the bionanocomposites provides an ideal environment for the antimicrobial system. This ability makes them inactivate microorganisms and produce better results than micro or macro scale components [27]. Bionanocomposites exhibit an important application in medicine as it does not cause any side effects and is also able to generate a desirable effect on the human body. In addition to this, green synthesized nanocomposites are excellent in other applications like vaccinations, drug discovery, and supercapacitors.

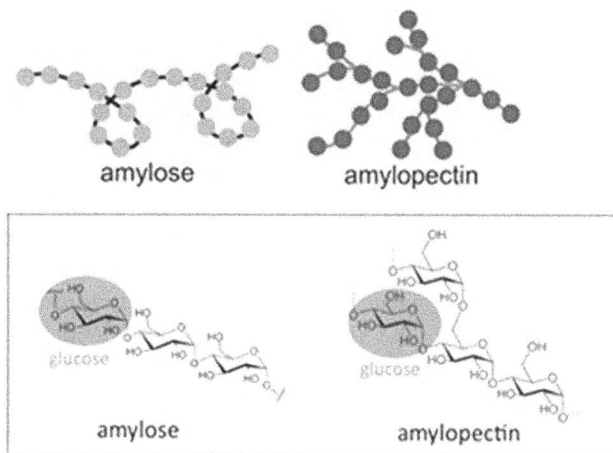

FIGURE 8.3   The structure of amylose and amylopectin.

8.3.1 STARCH-DERIVED NANOCOMPOSITES

Starch, a polysaccharide considered as the major requirement for photosynthesis and comprises of two different types of polymers namely amylopectin and amylase which are in the composition of 70%–90% granule and 10%–30% granule respectively [28]. In this amylopectin consist of cross-linked α (1–6) bonds and a high molecular weight polymer while amylase comprises of α (1–4) linked D-glucose units linearly. Starch-derived nanocomposites with multi-walled carbon nanotubes (MWCNT) show amazing properties in bone regeneration and tissue engineering [29]. Starch-based hydrogels are commonly applied with antimicrobial properties; these are synthesized using gamma radiation polymerization technique [30].

8.4   HUMAN HAIR PROTEIN-DERIVED NANOCOMPOSITES

8.4.1   SCOPE OF HUMAN HAIR PROTEIN

Nowadays human hair is one of the major wastes and it creates many environmental problems due to its low degradability. But its biocompatibility, antibacterial properties, and high carbon content makes it an efficient material for researchers and industrialists [31]. According to elemental analysis, human hair fiber mainly consists of 50 wt.% carbon, 16 wt.% nitrogen, 7 wt.% hydrogen, 22 wt.% oxygen, and a small portion of sulfur (5 wt.%). In addition to these, proteins, carbohydrates, lipids, and inorganic compounds enhance their biological activities [32]. The presence of disulfide bonds in keratin and keratin-associated proteins are the major contributors to the mechanical properties of the hair fiber. Many investigations have been made on the biological importance of hair fiber, which includes its application as a scaffold made up of composites for tissue engineering. Currently investigated composites for bone scaffolds for tissue engineering are silk and hydroxyapatite, nano hydroxyapatite, wool keratin, collagen, and polyamide. Recently biological fiber such as human hair can be used as a better alternative for fiber-reinforced polymers [33].

8.5   SUPERCAPACITORS

Recently, engineering technology has required materials with multifunctional properties that can satisfy both mechanical properties and electrical energy storing capacity [34]. In the early stages of the research, fiber-reinforced nanocomposites structures performed both roles simultaneously. Based on the energy and power density, electrical energy devices are mainly classified as batteries, super-capacitors, and dielectric capacitors [35–36]. Due to short charging time and safe operation, super-capacitors facilitate intensive attention in energy storage devices [37]. These are mainly designed for electric vehicles, conversion systems, continuous power supply, and in modern cellular devices, supercapacitors can minimize the gap between capacitors and batteries and provide fast-charging energy storage devices for the current energy crisis. Supercapacitors differ from capacitors in many ways [38]. Different types of polymer nanocomposites can provide enhanced applications in super-capacitors. Carbon-based materials, especially carbon nanotubes, graphene, carbon dots, activated carbon, because of their high mechanical strength, surface volume ratio and outstanding thermal and electrical properties make them an efficient candidate for electroactive material in supercapacitors [39]. Electric double-layer capacitors (EDLCs) and pseudocapacitors are the major class of supercapacitors based on energy storage mechanisms. In any supercapacitor some factors affect their final electrochemical performance, these include surface to volume ratio, geometry, electrical conductivity, and wet nature. Recently many materials have been developed to satisfy all these factors, these carbon-based materials have a significant role in supercapacitors [40]. These types of Supercapacitors have sufficient stability and high specific capacitance. Supercapacitors are eco-friendly and easy to dispose of to meet basic environmental standards. In supercapacitors, energy is stored electrostatically in an electrical double layer (EDL) and it is directly relative to the area of the electrode surface, hence materials with a high surface area can provide high specific energy [41]. Normally carbonaceous materials like carbon black or carbon nanotubes can be used for the synthesis of an electrode; these can enhance technological advancement in supercapacitors [42].

Supercapacitors are made up of two electrodes, an electrolyte, and a separator. From these two electrodes are constructed by activated carbon with high surface area and boundary layers between these two electrodes are usually conducting ions [43]. The produced energy due to the accumulation of charges is stored between these two electrodes and electrolytes. The stored energy depends on the concentration of ions, electrolyte stability, and electrode surface. Gibson et al. reviewed multifunctional materials that instantly store electrical energy as well as carry mechanical load [44].

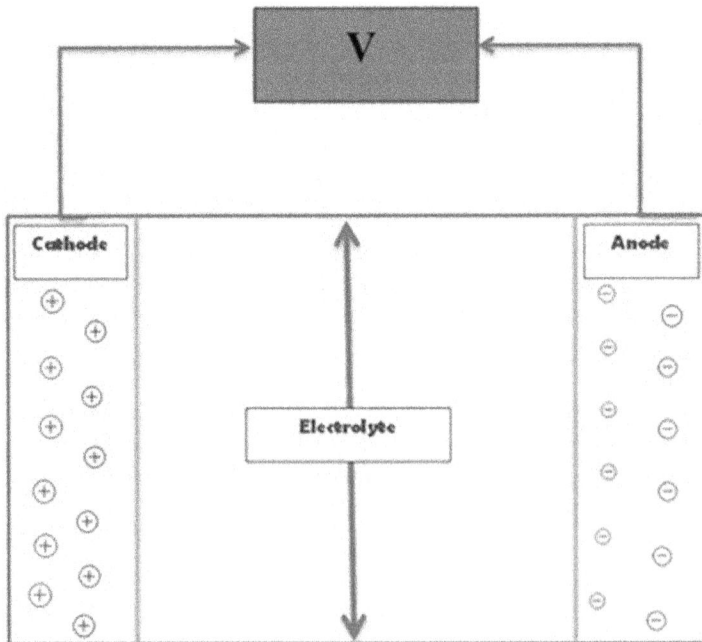

FIGURE 8.4 Schematic diagram for supercapacitors.

8.6   HUMAN HAIR PROTEINS AS AN ELECTRODE FOR SUPERCAPACITORS

Many researchers have reported activated carbon, fullerenes, carbon nanotubes, and carbon fibers as superior electrodes for supercapacitors; this activated carbon, due to its low cost and large surface area, is preferable to others [45]. Many eco-friendly materials are available as carbon sources. Among them, human hair is a low cost and eco-friendly biowaste, and it can be easily converted into activated carbon by the simple NaOH method [46]. Qian et al. reported carbon quantum dots from waste human hair for supercapacitor electrodes. These reported carbon flakes show appreciable current density and specific conductance [47].

Activated carbon from human hair can efficiently perform as electrode material in acid or base electrolyte, but it shows poor capacitance in neutral electrolytes like potassium sulfate and sodium sulfate. So the development of a supercapacitor electrode in acid, base, or neutral electrolyte is an important task for researchers. Introduction of metal oxides like RuOx, NiOx, CuOx, and MnOx with activated carbon can enhance the activity of supercapacitors. In these materials the most promising one is MnO_2 because it can give high capacitance values and energy density. Moreover, it is superior to others due to its low cost, nontoxic nature, and high theoretical surface area. Hence the combination of MnO_2 with graphene or any other activated carbon from biomass can demonstrate very high capacitance [48].

Gopiraman et al. reported the development of electrode material for supercapacitors from human hair-derived activated carbon with MnO_2 in different electrolytes like KOH, H_2SO_4 and Na2SO4. These are the first reported human hair-derived nanocomposites with versatile supercapacitor application. They are able to achieve capacitance, and are capable of utilizing hair derived carbon dots as a good electrode for supercapacitors [49].

8.7   GREEN ELECTRODES FOR SUPERCAPACITORS

Biologically activated polysaccharides chitin and chitosan with suitable modification using organic and inorganic nanofillers can able to contribute various applications in electrochemical devices [50]. These polymeric compact biomaterials have high mechanical and chemical properties and can be used as excellent barrier properties. The utilization of organic and inorganic fillers with chitosan is possible only because of their hydroxyl and amine groups, which makes them form many inter- and intramolecular hydrogen bonds within the composites. This property can increase the functionalization of chitosan derived nanocomposites for energy storage devices [51].

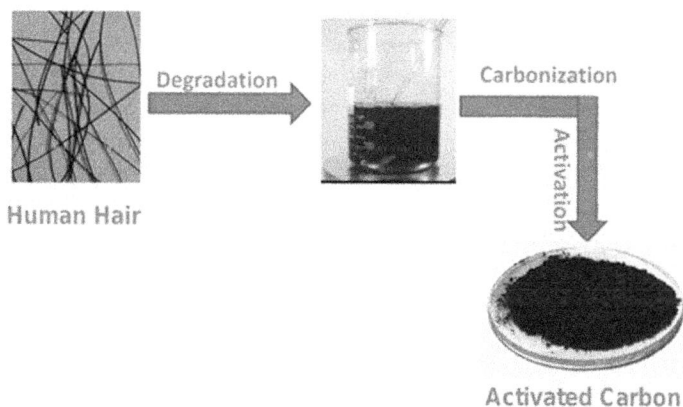

FIGURE 8.5   Application of hair derived nanocomposites in supercapacitors.

Similarly, graphene and graphene oxide doped chitosan-derived nanocomposites were also reported with enormous applications in supercapacitors. The synthesis of chitosan-derived nanocomposites for supercapacitors includes nitrogen self-doping aerogel synthesis and aerogel carbonization. These types of nanocomposite can produce excellent specific capacitance for supercapacitors. Similarly, the ultra-fast hydrogenation method was also used to prepare highly efficient supercapacitors with notable specific capacitance [52].

Recently Ciplak et al. reported the synthesis of nanoparticle incorporated graphene oxide polyaniline and reduced graphene oxide-gold polyaniline nanocomposites using a biological method. For this, initially, graphene oxide (GO) was converted into reduced graphene oxide (rGO) and then gold and aniline monomer was allowed to deposit on the surface of reduced graphene oxide, and in situ polymerization was used to prepare nanocomposites. Ciplak et al. synthesized various nanocomposites electrodes via green and reported noteworthy capacitance of 63%, 42% and 17% [53].

Arthisree et al. biogenically synthesized polyacrylonitrile and polyaniline graphene quantum dot sandwiched with sodium chloride and alumina. This combined PAN/PANI and GQD can produce a supercapacitor with an output of 1.5 V and working efficiency of about 60 minutes [54]. These reported values are higher than the traditional polyaniline-derived supercapacitors. Many inorganic and organic nanoparticles with biopolymer nanocomposites were reported with efficient supercapacitors. This type of styrene-maleic anhydride with zinc oxide nanoparticle comprised nanocomposites for supercapacitor application was reported by Chakraborty et al. [55].

In addition to the carbon nanotube and graphene, transition metal oxides are also used with conducting polymers for the development of supercapacitor electrodes. The introduction of regenerated cellulose aerogel with graphene oxide can able to prove the concept of green material for electrode preparation [56]. This graphene with regenerated cellulose aerosol supercapacitors with conducting polymers has shown noticeable specific capacitance in energy storage devices [57]. Similar green electrodes were prepared using porous cellulose with aniline by in situ polymerization method [58]. Zu et al. reported the synthesis of carbon and cellulose aerogel green electrodes with large surface area by the pyrolysis method [59]. Many green electrodes are reported with high specific capacitance and this includes bamboo cellulose-based green electrodes that reached a high specific capacitance of 382 Fg-1[60]. The nitrogen or sulfur-doped cellulose aerogel can enhance the efficiency of green electrodes [61–62].

Gao Feng et al. synthesized an eco-friendly by-product-free green electrode with highly effective specific conductance. In this paper, they reported reduced coal-derived graphene oxide with Mn_3O_4 for electrode material and K_2SO_4 as an electrolyte. During the synthesis, instead of traditional graphene flakes, they utilized coal-derived graphene and modified the Hummers method for the Mn_3O_4 synthesis. The main advantage of this whole process is to achieve an atom economy of 97% and a supercapacitance of nearly 261 Fg-1 [63].

Quantum dots due to their appreciable size and surface volume ratio, can be widely acceptable for electrochemical applications. These applications are limited due to some challenges like monodisperse of particles and stability. The digestive ripening method for the synthesis of quantum dots can able to minimize this limitation and such semiconductor material-based nanocomposites are suitable for energy storage devices. Nasser et al. reported solvent-free green synthesized carbon quantum dots with cobalt sulfide nanocomposites for supercapacitor electrode materials. In this work, they utilized a microwave-assisted method followed by a hydrothermal process for CQD/CoS_2 synthesis. The synthesized brick-like structured nanocomposites have shown unique surface area and large specific conductance suitable for new and advanced supercapacitors [64].

Among green synthesized carbon quantum dots, conducting polymers incorporated nanocomposites have of great interest. Arthisree et al. developed such a system for supercapacitor applications [65]. They introduced the application of polyvinyl butyral, polyaniline, and poly(3,4-ethylene dioxythiophene)-polystyrene sulfonate with green synthesized graphene quantum dot loaded nanocomposites with high specific capacitance. The solution casting method was used to synthesize polymer-carbon-based nanocomposites and was characterized by general electrochemical methods.

The invention of this type of green synthesized nanocomposites will be a promising one for super-capacitor applications.

8.8 CONCLUSION

The introduction of biodegradable electrodes for supercapacitors could contribute novel materi-als that enhance energy storage, reduce environmental issues, and replace usage of non-renewable energy sources for sustainable development. This type of green initiative can minimize challenges like global warming and the energy crisis and lead to the development of efficient, clean, and eco-friendly energy storage devices. The emergence of supercapacitors introduces a new era in energy storage devices, with high power density, high capacitance, large surface area, chemical, and mechanical stability. Bionanocomposite-derived supercapacitors proved the scope of biodegrad-able materials with electrochemical applications. Novel nanomaterials from natural resources can overcome recent technical issues and enable the development of economic growth and resource utilization. Carbon-based nanomaterials are considered a promising material for ongoing research and development. This chapter depicts the development, utilization, and importance of bionano-composites in energy storage devices.

REFERENCES

1. Wang, X, Zhuang, J, Peng, Q & Yadong, L 2005, 'A general strategy for nanocrystal synthesis', *Nature*, no. 437, pp. 121–124. https://doi.org/10.1038/nature03968.
2. Schmidt, D, Shah, D & Giannelis, EP 2002, 'New advances in polymer/layered silicate nanocompos-ites', *Current Opinion in Solid State & Materials Science*, vol. 6, no. 3, pp. 205–212. DOI:10.1016/S1359-0286(02)00049-9
3. Gleiter, H 1992, 'Materials with ultrafine microstructures: Retrospectives and perspectives', *Nanostructured Materials*, vol. 1, no. 1, pp. 11–19. https://doi.org/10.1016/0965-9773(92)90045-Y.
4. Braun, T, Schubert, A & Sindelys, Z 1997, 'Nano composites and their application—A review', *Nanoscience and Nanotechnology on the Balance. Scientometric*, vol. 3, no. 1, pp. 321–325.
5. Iijima, S 2002, 'Helical microtubes of graphitic carbon', *Nature*, vol. 354, no. 6348, pp. 56–58. https://doi.org/10.1038/354056a0.
6. Biercuk, MJ, Llaguno, MC & Radosvljevicm, HJ 2002, 'Carbon nanotube, composites for thermal management', *Applied Physics Letters*. vol. 80, no. 15, pp. 2767–2769. http://repository.upenn.edu/mse_papers
7. Ounaies, Z, Park, C, Wise, KE, Siochi, EJ & Harrison, JS 2003, 'Electrical properties of single wall carbon nanotube reinforced polyimide composites', *Composites Science and Technology*, vol. 63, no. 1, pp. 11637-1646. DOI:10.1016/S0266-3538(03)00067-8.
8. Niihara, K 1991, 'New design concept of structural ceramics-ceramic nanocomposites', *Journal of the Ceramic Society of Japan*, vol. 99, no. 6, pp. 974–982. https://doi.org/10.2109/jcersj.99.974
9. Awaji, H, Choi, SM & Yagi, E 2002, 'Mechanisms of toughening and strengthening in cermaic-based nanocomposites', *Mechanics of Materials*, vol. 34, no. 7, pp. 411–422. https://doi.org/10.1590/S1516-14392009000100002.
10. Nakahira, A & Niihara, K 1992, 'Structural ceramics-ceramic nanocomposites by sintering method: Roles of nano-size particles', *Journal of the Ceramic Society of Japan*, vol. 100, no. 4, pp. 448–453.
11. Ferroni, LP, Pezzotti, G, Isshiki, T & Kleebe, HJ 2009, 'Determination of amorphous interfacial phases in Al2O3/SiC nanocomposites', *Nanocomposites: Synthesis, Structure, Properties and New Application Opportunities*, vol. 12, pp. 135–149.
12. Mittal, G, Dhand, V, Rhee, KY, Park, SJ & & Lee, WR 2015, 'A review on carbon nanotubes and graphene as fillers in reinforced polymer nanocomposites', *Journal of Industrial and Engineering Chemistry*, vol. 21, pp. 11–25. https://doi.org/10.1016/j.jiec.2014.03.022.
13. Robinette, E & Palmese, G 2005, 'Synthesis of polymer-polymer nanocomposites using radiation graft-ing techniques', *Nuclear Instruments and Methods in Physics Research*, B, vol. 236, no. 1–4, pp. 216–222. https://doi.org/10.1016/j.nimb.2005.04.032
14. Kumar, SK, Jouault, N, Benicewicz, B & Neely, T 2013, 'Nanocomposites with polymer grafted nanopar-ticles', *Macromolecules*, vol. 46, no. 9, pp. 3199–3214. https://doi.org/10.1021/ma4001385.

15. Potts, JR, Dreyer, DR, Bielawski, CW & Ruoff, RS 2011, 'Graphene-based polymer nanocomposites', *Polymer*, vol. 52, no. 1, pp. 5–25. https://doi.org/10.1016/j.polymer.2010.11.042

16. Camargo, PHC, Satyanarayana, KG & Wypych, F 2009, 'Nanocomposites: Synthesis, structure, properties and new application opportunities', *Materials Research*, vol. 12, pp. 1–39. https://doi.org/10.1590/S1516-14392009000100002

17. Wei, L, Hu, N & Zhang, Y 2010, 'Synthesis of polymer-mesoporous silica nanocomposites', *Materials*, vol. 3, pp. 4066–4079.

18. Godovsky, DY 2000, 'Device applications of polymer-nanocomposites in Biopolymers_ PVA hydrogels, anionic polymerization nanocomposites', *Springer*, pp. 163–205.

19. Samson, O, Adeosun, GI, Lawal, Sambo A, Balogun & Emmanuel, I 2012, 'Review of green polymer nanocomposites', *Journal of Minerals and Materials Characterization Engineering*, vol. 11, no. 4, pp. 385–416. DOI:10.4236/jmmce.2012.114028.

20. Jacob, A. LaNasa, Anastasia Neuman, Robert, A. Riggleman & Robert, J. Hickey 2021, 'Investigating nanoparticle organization in polymer matrices during reaction-induced phase transitions and material processing', *ACS Applied Materials Interfaces*, vol. 13, no. 35, pp. 42104–42113. https://doi.org/10.1021/acsami.1c14830

21. Shamim Ahmed Hira, Mohammad Yusuf, Dicky Annas, Saravanan Nagappan, Sehwan Song, Sungkyun Park & Kang Hyun Park 2021, 'Recent advances on conducting polymer-supported nanocomposites for nonenzymatic electrochemical sensing', *Industrial & Engineering Chemistry Research*, vol. 60, no. 37, pp. 13425–13437. https://doi.org/10.1021/acs.iecr.1c02043

22. Lettow, James H, Kaplan, Richard Y, Nealey, Paul, F & Rowan, Stuart J 2021, 'Enhanced Ion conductivity through hydrated, polyelectrolyte-grafted cellulose nanocrystal films', *Macromolecules*, vol. 54, no. 14, pp. 6925–6936. https://doi.org/10.1021/acs.macromol.1c01155

23. Sharon, D, Bennington, P, Dolejsi, M, Webb, MA, Dong, BX, De Pablo, JJ, Nealey, PF & Patel, SN 2020, 'Intrinsic Ion transport properties of block copolymer electrolytes' *ACS Nano*, vol. 14, no. 7, pp. 8902–8914. https://doi.org/10.1021/acsnano.0c03713.

24. Eichhorn, SJ, Dufresne, A, Aranguren, M, Marcovich, NE, Capadona, JR, Rowan, SJ, Weder, C, Thielemans, W & Roman, M 2010, 'Review: Current international research into cellulose nanofibres and nanocomposites', *Journal of Materials Science*, vol. 45, no. 1, pp. 1–33. https://doi.org/10.1007/s10853-009-3874-0.

25. Krishnasamy Ravinchandran, Prabhakaran Kala Praseetha, Thirumurugan Arun & Suyamprakam Gopalakrishnan 2018, 'Synthesis of nanocomposites', *Synthesis of Inorganic Nanomaterials*, pp. 141–168. https://doi.org/10.1016/B978-0-08-101975-7.00006-3

26. Manafi, M, Manafi, P, Agarwal, S, Bharti, AK, Asif, M & Gupta, VK 2017, 'Synthesis of nanocomposites from polyacrylamide and graphene oxide: Application as flocculants for water purification', *Journal of Colloid and Interface Science*, vol. 490, no. 15. pp. 505–510. https://doi.org/10.1016/j.jcis.2016.11.096

27. Rhim, JW, Park, HM & Hac, CS 2013, 'Bio-nanocomposites for food packaging applications', *Programme Polymer Science*, vol. 38, no. 10–11, pp. 1653–1689. https://doi.org/10.1016/j.progpolymsci.2013.05.008

28. Ratner, BD, Hoffman, AS, Schoeri, FJ & Lemons, JE 2004, 'An introduction to materials in medicine', in *Biomaterials Science*, Elsevier Academic Press, London, p. 864.

29. Hitzky, ER, Aranda, P & Darder, M 2008, 'Bionanocomposites', *Kirk-Othmer Encyclopedia of Chemical Technology*, pp. 1–28.

30. Zafar, R, Zia, KM, Tabasum, S, Jabeen, F, Noreen, A & Zubera, M 2016, 'Polysaccharide based bio nanocomposites, properties and applications: A review', *International Journal of Biological Macromolecules*, vol. 92, pp. 1012–1024.

31. Gonzalez, CR, Hernandez, ALM, Castano, VM, Kharissova, OV Ruoff, RS & Santos, CV 2012, 'Polysaccharide nanocomposites reinforced with graphene oxide and keratin-grafted graphene oxide', *Industrial & Engineering Chemistry Research*, vol. 5, pp. 13619–3629.

32. Ruan, D, Zhang, L, Zhang, Z & Xia, X 2003, 'Structure and properties of regenerated cellulose/tourmaline nano crystal composite films', *Journal of Polymer Science, Part B: Polymer Physics*, vol. 42, no. 3, pp. 367–373.

33. Wei, H, Rodriguez, K, Renneckar, S & Vikesland, PJ 2014, 'Environmental science and engineering applications of nanocellulose-based nanocomposites', *Environmental Science: Nano*, no. 4, pp. 302–316.

34. Popescu, Crisan & Hartwig, Hocker 2007, 'Hair- the most sophisticated biological composite material', *Chemical Society Reviews*, vol. 36, no. 8, pp. 1282–1290.

35. Ratner, BD 2013, 'An introduction to materials in medicine', *Elsevier Academic*, vol. 694.

36. Redepenning, Jody Guhan, Venkataraman, Jun Chen & Nathan, Stafford 2003, 'Electrochemical preparation of Chitosan/hydroxyapatite composite coatings on Titanium substrates', *Journal of Biomedical Materials Research*, vol. 66, no. 2, pp. 411–416.

37. Bonaccorso, F, Colombo, L, Yu, G, Stoller, M, Tozzini, V, Ferrari, AC & Ruoff, RSV 2015, 'Pellegrino', *Science*, vol. 347, p. 1246501.

38. Zhang, XY, Hou, LL, Ciesielski, A & Samorì, P 2016, 'Water splitting progress in tandem devices: Moving photolysis beyond electrolysis', *Advanced Energy Materials*, vol. 6, p. 1600671.

39. Winter, Martin & Brodd, Ralph J 2004, 'What are batteries, fuel cells, and supercapacitors', *Chemical Reviews*, vol. 104, no, 10, pp. 4245–4269.

40. Snook, Graeme A, Pon Kao & Best, Adam S 2011, 'Conducting-polymer-based supercapacitor devices and electrodes', *Journal of Power Sources*, vol. 196, no. 1, pp. 1–12. doi:10.1016/j.jpowsour.2010.06.084.

41. Carlson, T, Ordéus, D, Wysocki, M & Asp, LE 2010, 'Structural capacitor materials made from carbon fibre epoxy composites', *Composites Science and Technology*, vol. 70 no. 7, pp. 11–35.

42. Lin, Y & Sodano, HA 2009, 'Characterization of multifunctional structural capacitors for embedded energy storage', *Journal of Applied Physics*, vol. 106, p. 114108.

43. Chang, HH, Chang, CK, Tsai, YC & Liao, CS 2012, 'Electrochemically synthesized graphene/polypyrrole composites and their use in supercapacitor', *Carbon*, vol. 50, pp. 2331–2336.

44. Shi, W, Zhu, J, Sim, DH, Tay, YY, Lu, Z & Zhang, X 2011, 'Achieving high specific charge capacitances in Fe3O4/reduced graphene oxide nanocomposites', *Journal of Materials Chemistry*, vol. 21, pp. 3422–3427.

45. Gibson RF 2010, 'A review of recent research on mechanics of multifunctional composite materials and structures', *Composite Structures,* vol. 92, pp. 2793–2810.

46. Lota, K. Fic & Frackowiak, E 2011, 'Carbon nanotubes and their composites in electrochemical applications', *Energy & Environmental Science*, vol. 4, pp. 1592–1605.

47. Qian, W, Sun, F, Xu, Y, Qiu, L, Liu, C, Wang, S & Yan, F 2014, 'Human hair derived carbon flakes for electrochemical supercapacitors', *Energy & Environmental Science*, vol. 7, no. 379–386.

48. Biswal, M, Banerjee, A, Deo, M & Ogale, S 2013, 'From dead leaves to high energy density supercapacitors', *Energy & Environmental Science*, vol. 6, pp. 1249–1259.

49. Mayakrishnan Gopiraman, Dian Deng, Ke-Qin Zhang, Wei Kai, Ill-Min Chung, Ramasamy Karvembu & Ick Soo Kim 2017, 'Utilization of human hair as a synergistic support for Ag, Au, Cu, Ni, and Ru Nanoparticles: Application in catalysis', *Industrial & Engineering Chemistry Research*, vol. 56, no. 8, pp. 1926–1939.http//:doi. 10.1021/acs.iecr.6b04209.

50. Yuan, M, Zhang, Y, Niu, B, Jiang, F, Yang, DJ & Li, X 2019, 'Chitosan-derived hybrid porous carbon with the novel tangerine pith-like surface as supercapacitor electrode', *Journal of Materials Science,* vol. 54, pp. 14456–14468.

51. Chen, K, Weng, S, Lu, J, Gu, J, Chen, G, Hu, O, Jiang, X & Hou, L 2021, 'Facile synthesis of chitosan derived heteroatoms-doped hierarchical porous carbon for supercapacitors', *Microporous Mesoporous Mater*, vol. 320, pp 106–111.

52. Lin, CH Wang, PH, Lee, WN, Li, WC & Wen, TC 2021, 'Chitosan with various degrees of carboxylation as hydrogel electrolyte for pseudo solid-state Supercapacitors', *Journal of Power Sources*, vol. 494, pp. 229–236.

53. Çıplak, Z, Yıldız, A & Yıldız, N 2020, 'Green preparation of ternary reduced graphene oxide-au@ polyaniline nanocomposite for supercapacitor application', *Journal of Energy Storage*, vol. 32, pp101–112. DOI:10.1016/J.EST.2020.101846

54. Arthisree, D & Madhuri, W 2020, 'Optically active polymer nanocomposite composed of polyaniline, polyacrylonitrile and green-synthesized graphene quantum dot for supercapacitor application', *International Journal of Hydrogen Energy*, vol. 45, pp. 9317–9327. DOI:10.1016/j.ijhydenc.2020.01.179

55. Chakraborty, S & Mary, N 2020, 'Biocompatible supercapacitor electrodes using green synthesised ZnO/ Polymer nanocomposites for efficient energy storage applications', *Journal of Energy Storage*, vol. 28, pp. 101–275.

56. Zhong, C, Deng, Y, Hu, W, Qiao, J, Zhang, L & Zhang, J 2015, 'A review of electrolyte materials and compositions for electrochemical supercapacitors', *Chemical Society Reviews*, vol. 44, pp 7484–7539. https://doi.org/10.1039/D1CS00841B

57. Ren, F, Li, Z, Tan, WZ, Liu, XH, Sun, ZF, Ren, PG & Yan, DX 2018, 'Facile preparation of 3D regenerated cellulose/graphene oxide composite aerogel with high-efficiency adsorption towards methylene blue', *Journal of Colloid and Interface Science,* vol. 532, pp. 58–67. https://doi.org/10.1039/D1CS00841B

58. Tian, J, Peng, D, Wu, X, Li, W, Deng, H & Liu, S 2017, 'Electrodeposition of Ag nanoparticles on conductive polyaniline/cellulose aerogels with increased synergistic effect for energy storage', *Carbohydrate Polymers,* vol. 156, pp. 19–25. DOI:10.1016/j.carbpol.2016.09.005

59. Fan, H & Shen, W 2016, 'Gelatin-based microporous carbon nanosheets as high performance supercapacitor electrodes', *ACS Sustainable Chemistry & Engineering,* vol. 4, pp. 1328–1337. https://doi.org/10.1021/acssuschemeng.5b01354

60. Yang, X, Fei, B, Ma, J, Liu, X, Yang, S, Tian, G & Jiang, Z 2018, 'Porous nanoplatelets wrapped carbon aerogels by pyrolysis of regenerated bamboo cellulose aerogels as supercapacitor electrodes', *Carbohydrate Polymers.* vol. 180, pp. 385–392.

61. Tian, W, Gao, Q, Zhang, L, Yang, C, Li, Z, Tan, Y, Qian, W & Zhang, H 2016, 'Renewable graphene-like nitrogen-doped carbon nanosheets as supercapacitor electrodes with integrated high energy—power properties', *Journal of Materials Chemistry A,* vol. 4, pp. 8690–8699. https://doi.org/10.1016/j.enbenv.2019.09.002

62. Li, Y, Wang, G, Wei, T, Fan, Z & Yan, P 2016, 'Nitrogen and sulfur co-doped porous carbon nanosheets derived from willow catkin for supercapacitors', *Nano Energy,* vol. 19, pp. 165–175. DOI:10.1016/j.nanoen.2015.10.038.

63. GaoFeng, QuJiangying, Zhao Zongbin, Zhou Quan, Li Beibei & Qiu Jieshan 2014, 'A green strategy for the synthesis of graphene supported Mn_3O_4 nanocomposites from graphitized coal and their supercapacitor application', *Carbon,* vol. 80, pp. 640–650. https://doi.org/10.4236/sgre.2021.121001

64. Nasser Arsalani, Laleh Saleh Ghadimi, Iraj Ahadzadeh Amin Goljanian Tabrizi & Thomas Nann 2021, 'Green synthesized carbon quantum dots/cobalt sulfide nanocomposites as efficient electrode material for supercapacitors', *Energy Fuels,* vol. 35, pp. 9635–9645.

65. Arthisree, D & Madhuri, W 2021, 'A ternary polymer nanocomposite film composed of green-synthesized graphene quantum dots, polyaniline, polyvinyl butyral and poly (3,4-ethylenedioxythiophene) polystyrene sulfonate for supercapacitor application', *Journal of Energy Storage,* vol. 35, pp. 102–113. https://doi.org/10.1016/j.est.2021.102333.

9 Nanocellulose-Based Supercapacitors

Ekta Jagtiani and Manishkumar D. Yadav

CONTENTS

9.1 INTRODUCTION

As environmental concern about our reliance on limited, non-renewable, and depleting resources such as metals and petroleum has risen, sustainable, eco-friendly, and biodegradable materials have been developed extensively. Renewable materials, such as nanocellulose, can be used to replace metal and plastic in a range of applications, thereby decreasing the potential pollution-causing leftovers in the environment (J. Huang et al., 2013). By utilizing sophisticated and novel hybrid nanomaterials, the development and innovation of green technology has become a critical component of ecological sustainability. Customers are now seeking materials that are transparent, beneficial, durable, and flexible, in accordance with pre-industry standards.

Cellulose is a biodegradable and versatile raw material that is capable of replacing a broad variety of non-renewable resources (D. B. et al., 2011; Wegner & Jones, 2006). Advances in nanotechnology have reintroduced the debate about isolating nanocellulose from natural sources. Due to its low cost, higher Young's modulus, environmental-friendliness, higher mechanical strength, higher surface-to-volume ratio, low density, and excellent stability in most solvents, nanocellulose offers numerous advantages over various synthetic materials in myriad of applications, such as energy storage devices, water and oil filter membranes, as well as wound dressings (Kumar & Pizzi, 2019). Owing to its excellent properties for flexible, lightweight membrane electrode fabrication and loading conductive materials, nanocellulose has particularly bright future displaying high electrochemical performance too. Researchers are particularly interested in carbon compounds produced from nanocellulose for

DOI: 10.1201/9781003174646-9

long-term energy storage since they are abundant, cheap and possess high conductivity. The growing research interest in supercapacitors (SCs) may be attributed to a decade-long period of increased activity as a result of their increased capacity to store energy, enhanced delivery of power density and faster time for charging and discharging (Beidaghi & Gogotsi, 2014; G. Wang et al., 2012; Yu et al., 2015). Nanocellulose-based energy storage systems is an attractive alternative of renewable and "green" electronics, which is expected to expand the horizons for the next-generation dubbed as the "Battery-of-Things" era (W. Chen et al., 2018; Ling et al., 2018; Z. Wang, Tammela et al., 2017).

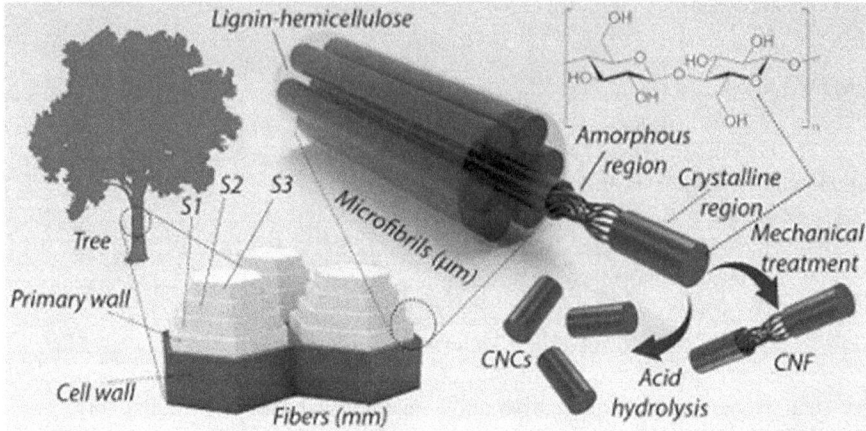

FIGURE 9.1 Hierarchical structure of cellulose depicted hierarchically along with the amorphous and crystalline regions in cellulose fiber.

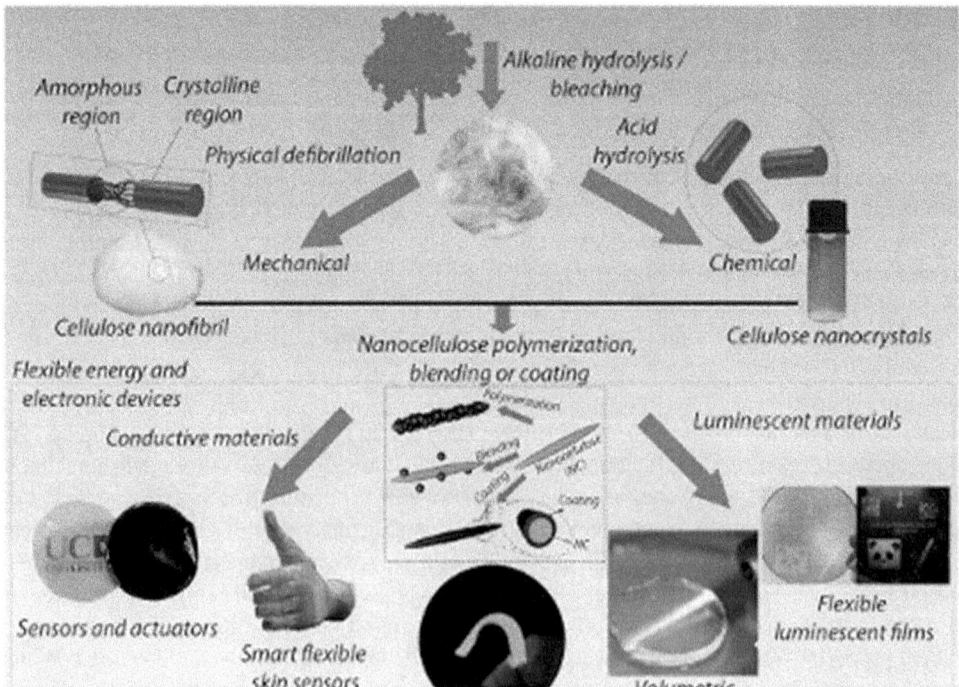

FIGURE 9.2 Extraction of CNF and CNC from cellulose, surface functionalization of nanocellulose for electronic materials and flexible energy applications.

9.1.1 HISTORICAL ASPECTS

Cellulose is the most abundant naturally occurring polymer on the planet (Postek et al., n.d.). Nanocellulose can be used or modified in a variety of ways, depending on the application. Nanocellulose can be used or modified in a variety of ways, depending on the application. During the development stage, cellulose nanofibers can be synthesized from a range of biomasses, including tall plants, agricultural wastes, algae, and bacteria (W. Chen et al., 2018). Since the 1980s, numerous strategies for synthesizing nanocellulose from plant cell walls, algae, and other organisms have been developed, including top-down methods such as acid hydrolysis and mechanical nanofibrillation, as well as bottom-up methods such as solvent dissolution and bacterial cellulose (BC) synthesis (Ling et al., 2018). Cellulose nanofibers are classified into three categories based on their structure and source material: cellulose nanocrystals (CNCs), nanofibrillated cellulose (NFC), and cellulose nanofibers (CNFs). While NFCs and CNCs are frequently created from plants, BCs are bacteria-derived high aspect ratio nanofibrils. NFCs have long lengths and high aspect ratios but a low crystallinity (both amorphous and crystalline areas), whereas CNCs have shorter lengths, lower aspect ratios, and a high crystallinity due to the various manufacturing intensities.

Cellulose (40–50 wt.%), hemicellulose (2035 wt.%), and lignin (2030 wt.%) are the principal components of the tree cell wall, interweaving and contributing to the wood's excellent mechanical properties (S. J et al., 2018). Cellulose bundles with diameters of several tens of nanometers can be further subdivided into nanocellulose fibers with diameters ranging from 5 to 20 nm, and on a finer scale, elementary fibrils with a diameter of 1.53.5 nm can be found in the nanocellulose fiber, acting as the fundamental building blocks of hierarchical cellulose fibers. The linear and rigid cellulose chain is formed when repeated units of D-glucose unite at the molecular level via covalent bonding and intra- and interchain hydrogen bonding (Z. Wang, Tammela et al., 2017). Due to the hydrophilic nature of nanocellulose, its ability to absorb water under specific circumstances is without doubt its most significant impediment in high-end goods.

9.1.2 NANOCELLULOSE-BASED MATERIALS FOR SUPERCAPACITORS

Lightweight and free-standing flexible membrane electrodes are usually fabricated using conductive substances due to its brittle nature, such as polymers, metallic particles and carbon, which are commonly applied in combination with a soft and flexible substrate (HH & W, 2019). Nanocellulose is an excellent substrate for loading a variety of conductive materials with extremely porous structures

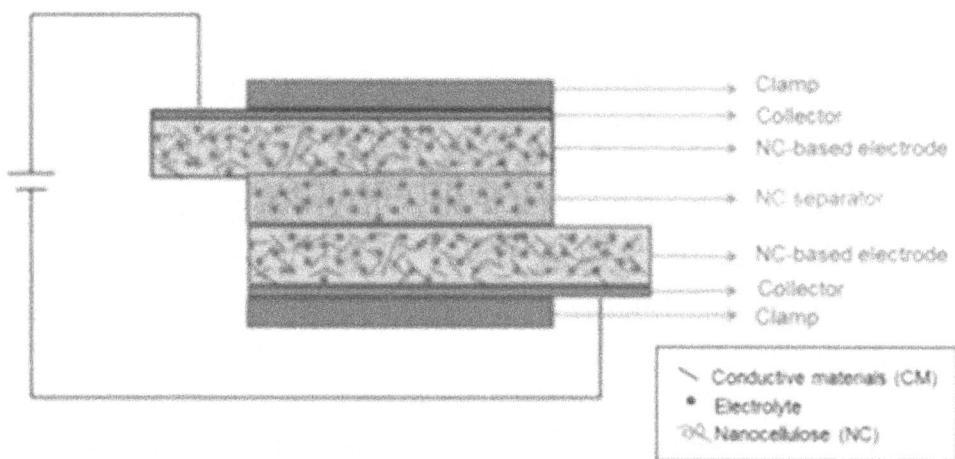

FIGURE 9.3 Fabrication of nanocellulose-based supercapacitors in usage of separator and electrode material. Reproduced with permission. Copyright 2013, Royal Society of Chemistry.

that permit transport of the ions. Additionally, the nanocellulose exhibits superior mechanical characteristics, with a Young's modulus of 100–130 GPa and tensile strength of 1.7 GPa, which is equivalent to that of aramid and glass fibers (Miloh et al., 2009). Apart from supercapacitors, membranes from nanocellulose have been extensively utilised as flexible substrates for various types of energy storage devices, such as and lithium ion batteries and solar cells (Du et al., 2017; Takagi et al., 2016). These energy storage technologies have demonstrated significant potential for huge energy storage systems, electric vehicles and wearable electronics powered by new energy sources. Recent advances in the supercapacitor design have enabled the development and use of pH-sensitive, shape memory, thermosensitive properties, self-healing and intelligent properties (Peng et al., 2017; X. Wang et al., 2016; H. Y et al., 2016; Y. Y et al., 2017).

Nanocellulose is extremely strong, modular and possess a high aspect ratio. It has a broad electrochemical window and is stable in a wide range of solvents. It is suitable for various applications such as a binder, electrolyte, and separator. Due to the presence of highly reactive hydroxyl surface groups, the surface can be chemically modified and the resulting nanocomposite's characteristics can be tuned to optimize its electrochemical performance for a particular application (W. Chen et al., 2018; Ling et al., 2018; W. Chen et al., 2018). Cellulose nanocrystals have a higher number of active sites than Bacterial cellulose or Cellulose nanofibers, making it their appropriate electrical alternative (W. Chen et al., 2018; Nie et al., 2019).

The interest in supercapacitors, or SCs, has peaked in recent decades due to their unique advantages: rapid charging/discharging rate, high energy density, great fundability, high power density and safe operation (Brodd., 2005; Y. Liu, Zhou, Tang et al., 2015; Nyholm et al., 2011). Energy density and power density denotes the energy and power stored respectively per unit volume/mass/ length/area. The energy density, power density and specific capacity indicate the electrochemical performance of a supercapacitor (Y.-Z. Zhang et al., 2015). Supercapacitors are classified into two categories according to their charging and discharging mechanisms: (i) electric two-layer condensers, which store electrochemical energy by desorption and (ii) ion adsorption pseudocapacitors, which use redox processes.

Supercapacitors are hailed as the future of energy storage due to their high power and energy density, long life, extended cycle life, wide range of working temperatures and quick rate of loading and low maintenance costs, among other advantages. The capacity of electrical double-layer supercapacitors (EDLCs) is determined by the surface area and conductivity of the electrode. The larger the surface area, the more ions are at the point of contacts and hence the greater the energy storage capacity. The pace of the reversible redox reaction between the electrolytes and functional electrode, as well as the conductivity of the electrode, affect the capacitive performance of pseudocapacitors (Yu et al., 2015; Y.-Z. Zhang et al., 2015). In a nutshell, EDLCs have better power densities and runnability than pseudocapacitors. As the hybrid supercapacitor incorporates the benefits of both, it is chosen over batteries and condensers.

9.2 FUNDAMENTALS: SYNTHESIS AND MODIFICATION

9.2.1 Preparation Techniques

The two primary forms of nanocellulose, NFCs and CNCs, have been produced by variety of methods (XW et al., 2010). CNFs have regions/domains that are both crystalline and amorphous. CNCs are crystalline structures which are synthesized by hydrolyzing the amorphous portions of CNFs. Purification and fibrillation processes are employed to synthesize CNFs. Mechanical fibrillation is utilized to remove CNFs from their underlying raw materials, thereby disrupting the intermolecular bond and enabling separation. On the other hand, purification removes the non-cellulose components from CNFs. Numerous factors, such as post-processing, ambient conditions, bacterial strains, might change the nanofiber structure. Furthermore, CNCs may also be made from Bacterial Cellulose.

Although NFCs may be recovered from plants by refining, grinding and high-pressure homogenization, CNCs are separated via acid hydrolysis methods. The formation of cellulose whiskers during sonication and acid hydrolysis causes the material to break transversely (Trinh & Mekonnen, 2018; Yoo & Youngblood, 2016). The internal fiber topologies, chemical structures, cell widths and microfibril angles of cellulose nanofibers generated from natural sources, are found to have major impact on their performance characteristics.

Decortication is a procedure used to create high-quality fibers by eliminating contaminants from plant fibers (Cheng et al., 2016; Wei et al., 2016). Firstly, the long bast fibers of the stems are separated using water or dew retting for about 20 days in order to eliminate lignin, hemicellulose, and pectin. Chemical methods are frequently associated with safety hazards and environmental pollution (Raghuwanshi & Garnier, 2019; Tao et al., 2020; H. Xu et al., 2020). To decorate the bast fibers derived from raw plants such as linseed, flax or hemp, a toothed roller is employed. Mechanical techniques usually require straws that are well-retted, dry and capable of breaking lengthy fibers (Cheng et al., 2016; Wei et al., 2016). Separation of nanocellulose from harvested and cleaned bast plants and fibers is possible through shearing methods without degrading the cellulose. Nanocellulose's aspect ratio and morphology can be determined by the defibrillation procedures such as homogenization, bleaching, grinding, refining and ultrasonication. Manufacturers have begun to employ high-pressure homogenization and refining techniques due to their increased overall output and efficiency. On the other hand, mechanical methods have a significant disadvantage owing to their considerable energy consumption. Cryocrushing is a process that begins by freezing the nanofibers in liquid nitrogen along with subjecting them to severe shear stresses. Shearing takes place in a refinery while crushing is undertaken in liquid nitrogen atmosphere. Following the processing, the fibers are suspended in distilled water or freeze-dried.

Enzymatic, alkaline, oxidation, acidic and other chemical treatments can significantly reduce the mechanical energy required to manufacture nanocellulose. Alkaline treatment of bast fibers enables the dissociation of lignin and carbohydrate bonds. Mild alkali treatments cause the dissolution of hemicellulose, pectin, and lignin. Then, the nanofibers were recovered from the source using dimethyl sulfoxide (DMSO) and an alkali treatment followed by acid hydrolysis. Recovery of nanocellulose was done from banana fibers by subjecting them to a steam explosion at temperatures ranging from 220 to 300 degrees Celsius, which results in the disassembly of glycosidic links in cellulose and thermal depolymerization of hemicellulose (Takagi et al., 2016). Another approach for optimizing nanocellulose production is enzymatic pre-treatment. For example, endoglucanase has been discovered to aid in the degradation of wood fiber pulp into nanocellulose. Additionally, pretreatment with enzymes further enhanced the structural uniformity of wood nanocellulose compared to acid hydrolysis (Habibi, 2014; Thakur, n.d.; Z et al., 2018; K. Zhang, Ketterle et al., 2020).

9.2.2 Surface Modification

The abundance and high surface area of hydroxyl groups in the structure of nanocellulose make these nanofibers an attractive substrate for surface modification via a variety of chemical procedures (Habibi, 2014). The characteristics of nanocellulose enable the incorporation of nearly any required surface functionality using attractive surface modification methods. Esterification methods are employed to hydrophobize the surface of cellulose and is then utilized in food packaging applications (Hofmann & Reid, 1929). The hydrophobicity of cellulose fibers, starch, chitin (G. N. K et al., 2003), cellulose fibrils (P. Huang et al., 2012) and xylan (X. W. et al., 2010; Zhao et al., 2014) has been studied thoroughly (P. Huang et al., 2012; Vaca-Garcia et al., 1998). Although nanocellulose's hydrophobicity has been extensively researched, efforts to enhance functionality for advanced applications have gained interest. Numerous long-chained aliphatic compounds are grafted onto CNC and CNF using a variety of methods to decrease moisture absorption (Sethi et al., 2017), increase thermal stability (Sharma & Deng, 2016) and the interfacial affinity between resins and nanofibers (Tan et al., 2015; Trinh & Mekonnen, 2018; Yoo & Youngblood, 2016). More advanced

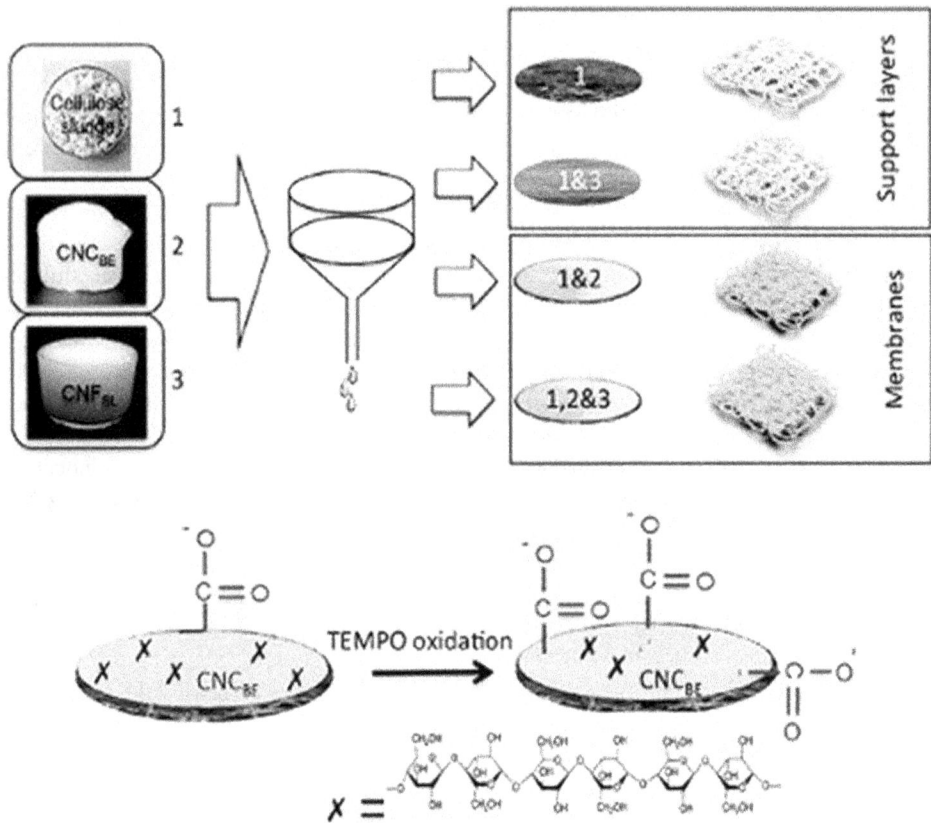

FIGURE 9.4 Schematic representation of TEMPO-mediated oxidation procedure.

cellulose materials have been researched in the context of next-generation technologies over the years. Recent emphasis has been focused on novel nanocellulose-based devices such as volumetric displays, biosensors, stretchable circuits, and artificial skin. To attain this aim, the structure of cellulose can be altered and new properties can be introduced.

The 2,2,6,6-Tetramethylpiperidin-1-yl)oxyl TEMPO-mediated oxidation methodology is employed to the surface of nanocellulose in order to create ion exchangeable carboxy groups (Cheng et al., 2016; Raghuwanshi & Garnier, 2019; Tao et al., 2020; Wei et al., 2016; H. Xu et al., 2020; Zeng et al., 2020; K. Zhang, Chen et al., 2020; C. Zhu et al., 2020) as it introduces aldehyde and carboxylate functional groups. The TEMPO-mediated oxidation of NFC films resulted in an increased transparency and reduced shrinkage compared to the other methods (Anirudhan & Rejeena, 2014) (Figure 9.4). It is found that the TEMPO-oxidized Nanocellulose (TONC) provides a promising emulsifier platform (Sharma & Deng, 2016) for conductive devices (K. Zhang, Ketterle et al., 2020), fluorescence sensors (H. Wang et al., 2020) and anchor carbon dot (Y. Jiang et al., 2016) in addition to assisting in material thermal expansion minimization (Fukui et al., 2018). Thus, the synthesis of TONCs paved the way for the creation of new functional materials (Zhao et al., 2014) Alternatively, nanocellulose can be altered or modified through one- or two-stage grafting process (Vadakkekara et al., 2020). Graft vinyl monomer polymerization (A & D, 2015; Anirudhan & Rejeena, 2014) offers active sites for attaching desired functional groups to cellulose structures (R et al., 2020; Thakur, n.d.). Physical and chemical bonding are used to enhance certain properties like the production novel materials like flexible aerogels and the moisture sensitivity of nanocellulose (S. M et al., 2020; B. Wu et al., 2019). Additionally, functional coatings on nanofiber surfaces were studied, in accordance with the nanofiber adhesion to a variety of nanoparticles (H. Wang et al., 2020). Two up-to-date methods for improving the functions of nanocellulose are CNF dopamide coatings and core-shell tannic acid (Nguyen et al., 2016; Y. Wang et al., 2017).

9.2.3 CONDUCTIVE MATERIALS

9.2.3.1 Metal Particles

Metallic particles possess an inherent conductivity of about 10^5 S/cm, that is five times that of conductors based on carbon (P. Huang et al., 2012; Zhao et al., 2014). In general, the technique of fabricating nanocellulose membrane conductive electrodes involves the direct surface coating of the nanocellulose substrate with carbon materials and metallic particles, which is an uncomplicated, simple approach. On the other hand, the conductivity of membrane electrodes is related to the thickness or loading of the metal layer (Vaca-Garcia et al., 1998). Metallic particles, on the other hand, are hefty and brittle, making uniform distribution difficult. Enhancing material dispersion by coating the substrate's surface with metallic particles is one of the techniques that is employed. Several metal oxides and hydroxides produced from nickel foam have also been deposited on and covered by nanocellulose sheets to provide flexible well-conducting, supercapacitor electron materials (Sethi et al., 2017). Polypyrrole and copper oxide were coated on bacterial cellulose substrates to construct a versatile range of supercapacitors (385 F/g) (Sharma & Deng, 2016). Carbon compounds or conductive polymers can often diminish the transparency of nanocellulose membranes because the layer's non-uniformity, resulting in substantial light dispersion.

Due to their exceptional conductivity, noble metals such as silver and gold are frequently used as conductive agents. Silver nanowires (AgNWs) and silver nanoparticles (AgNPs) can be directly inkjet printed on the surface to generate conductive items. A thin layer of silver, carbon nanotubes and indium oxide was applied on nanocellulose substrates to create conductive and transparent nanopapers with power conversion approximately 0.4% and high conductivity (25 ohmsq^{-1}) for use in optoelectronic devices such as displays applications, interactive paper and touchscreens (Tan et al., 2015).

Highly conductive, flexible, and lightweight circuits were created using CNF-based nanopapers via gold sputtering or printing with AgNP inks (see Figure 9.5) (Hsieh et al., 2013; Hu et al., 2013). Conductivities of up to 2.5 S cm^{-1} are possible with these circuits with straight and sharp

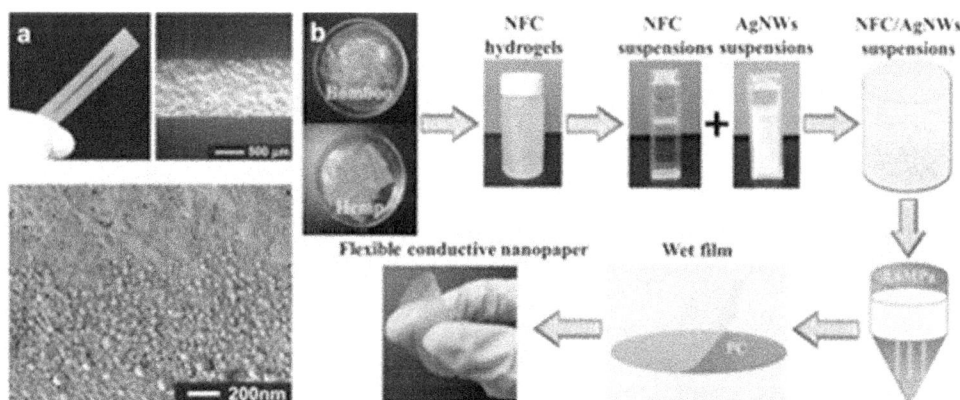

FIGURE 9.5 The routes to fabricate nanocellulose (NC) based conductive hybrid. Images of CNF, Reproduced with permission. Copyright 2012, American Chemical Society. Image of CNC, Reproduced with permission. Copyright 2011, Royal Society of Chemistry. Image of BC, Reproduced with permission. Copyright 2007, American Chemical Society.

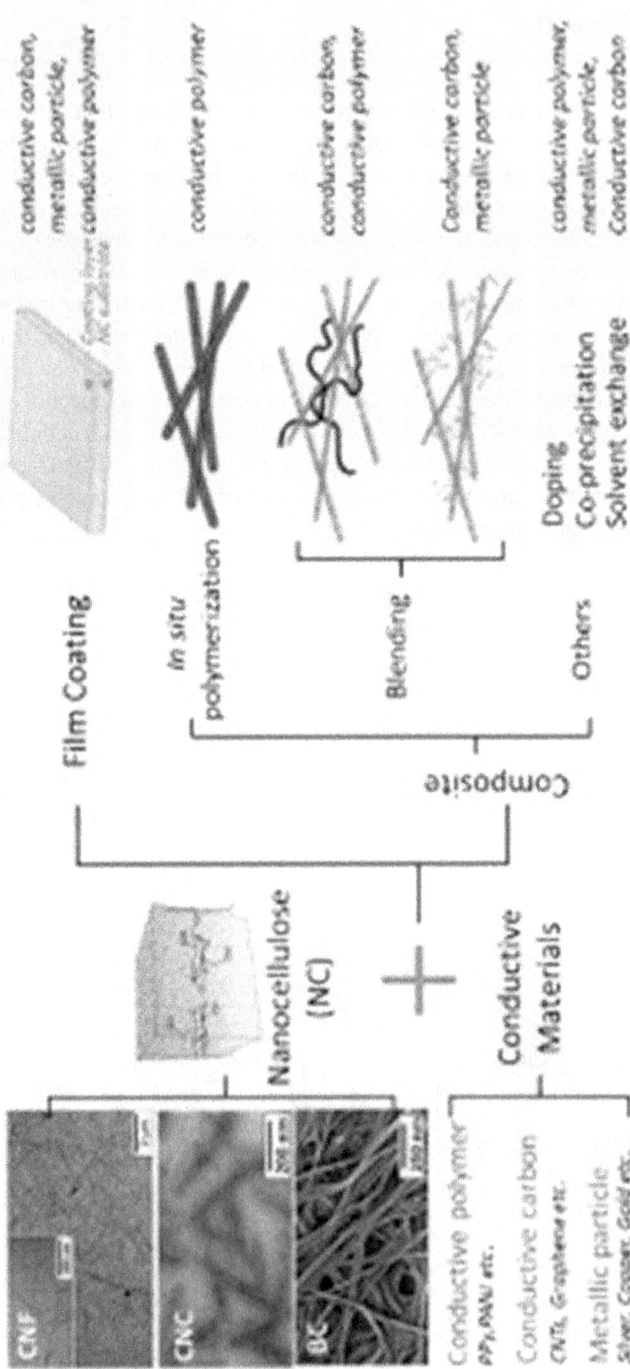

FIGURE 9.6 (a) Nanopaper consisting of silver nanoparticle lines. Reproduced with permission [64]. Copyright 2013, Royal Society of Chemistry. (b) NFC/AgNW nanopaper solution processing procedure. Reproduced with permission [65]. Copyright 2015, Royal Society of Chemistry.

edges. On the other hand, conventional papers produced conductive lines with extraordinarily high resistances (6,340 after gold sputtering, > 107 for AgNP) and irregular borders. This is due to the fact that ordinary pulp sheets contain large pores (about 20–60 m). These holes increased the width of their printing lines while decreasing the connections between adjacent metallic particles (Hsieh et al., 2013). Apart from noble metals, less expensive conductive metals such as copper (Cu) can be utilized to create conductive coatings. Cu's widespread availability and low cost make it a feasible option for functionalizing NC templates on a big scale (Pras et al., 2013). Tin has comparable economic benefits too. Hu et al. used radio frequency magnetron sputtering to generate flexible and transparent films from carbon nanotubes and indium tin oxide. This hybrid material demonstrated excellent conductivity with a sheet resistance of 12 sq^{-1}. If conductivity is the key consideration, low-cost metals are the logical choice. Many applications, however, such as fuel cells and batteries, would demand not only high conductivity but also additional functions such as chemical durability and catalysis. Doping (Z. Y. Wu et al., 2016), pressurized extrusion papermaking technique (Song et al., 2015), solvent exchange (G. J et al., 2012), and co-precipitation (L. K et al., 2015) are methods for embedding nanoscale metallic particles and derivatives in NC. As indicated in Figure 9.5, the conductivity of nanopaper was raised to 500 S cm^{-1} when AgNWs were introduced to bamboo/hemp CNF crosslinked with hydroxypropylmethy cellulose (Song et al., 2015). Additionally, nanoparticles of vanadium (G. J et al., 2012) and iron (II,III) oxide (Fe_3O_4) (L. K et al., 2015) were added to the NC matrix to generate a conductive composite.

In conclusion, since metallic particles have a high intrinsic conductivity, coating them on the surface of cellulose nanopapers or incorporating them into an NC matrix can yield more conductive sheets than other materials (e.g. carbon nanomaterials and conductive polymers). NC sheets augmented with metal nanoparticles have a conductivity of up to 103 S cm^{-1}.

9.2.3.2   Conductive Carbon Materials

Conductive carbon compounds consist of single, multiple wall nanotubes, reduced graphite graphite and reduced graphite oxide. In general, conductive carbon materials have high tensile strength and excellent electrical conductivity. Carbon nanotubes are well-known for their tensile strength, which is exceptionally high. Graphene is a two-dimensional material with remarkable mechanical characteristics, including an ultimate thickness of 130 GPa and a Young module of 1 TPa (H et al., 2015). Carbon materials can be deposited on the nanocellulose film's surface or mixed with it. Due to the possibility of carbon particles getting incorporated or trapped with the substratum, mixing or blending typically allows for the integration of more carbon particles into the nanocellulose substratum than coating the surface, resulting in a membrane with a higher conductivity (E. Feng et al., 2017; Hsu et al., 2019). Coating of carbon nanoparticles in nanocellulose membranes may diminish their transparency, limiting their application in advanced photosensitive conductors. During bending and stretching, these hybrids demonstrated outstanding electromechanical stability and conductivity. A recent research revealed that conductive nanocellulose-coated polymer and graphene flex electrodes possessed a capacity of 373 F/g (at 1 A/g) and a 85 percent cyclic stability after 1000 cycles (Liang et al., 2012).

NCs are utilized as a substrate paper for coating, also known as nanopaper and their surfaces are coated with conductive carbon compounds. Coating carbon nanomaterials on NC films frequently reduces optical transparency due to the coating materials' large-scale non-uniformity, which results in significant light scattering. An exceptional conductive composite, a CNF-based conductive nanopaper was demonstrated with a high degree of optical transparency (Hu et al., 2013). The nanopaper substrate is made of extremely porous CNF (40 m thickness). This is a significant advancement over conventional paper, which cannot be made with a clean conductive sheet. For CNT integration, CNF-based nanopaper substrates were used in place of conventional substrates such as plastic and glass. To include CNTs by Meyer rod coating, the resistivity of the nanopaper was lowered to 200 sq$^{-1.}$ The performance stated is comparable to that of the plastic substrate. This research enables a variety of optoelectronic applications, including displays, touch screens, and interactive papers. A similar investigation was conducted coating CNT on CNF-based nanopaper (Habibi, 2014; G. N. K et al., 2003) sing a technique called "filtration coating" (Figure 9.7) derived from the papermaking process, in which an aqueous solution of fiber-shaped carbon nanotubes was filtered through a wire mesh to create uniformly dispersed networks. The composite was then modified to achieve a high dielectric constant (k) for usage in the fabrication of miniature antennas. To enhance the thermal endurance and low coefficient of thermal expansion of the conductive nanopaper, the CNFs were chemically altered to lower their carboxylate content. Additionally, electricity usage was lowered during fabrication (H. et al., 2015). Although simple carbon nanomaterial coating is a more convenient method for producing more conductive carbon nanosheets than conductive polymer-based carbon nanosheets, increasing conductivity by increasing the thickness or amount of the coating layer is typically not possible due to the typically weak carbon-carbon particle bonding in the coating layer and loss of the layer is a concern. In comparison to surface coating, blending allows for the incorporation of more carbon particles into the substrate due to the large number of particles that can be physically confined within the CN substrate. Fabrication of the conductive composite is enabled by backfilling the NC template with an infinitely connected network of carbon

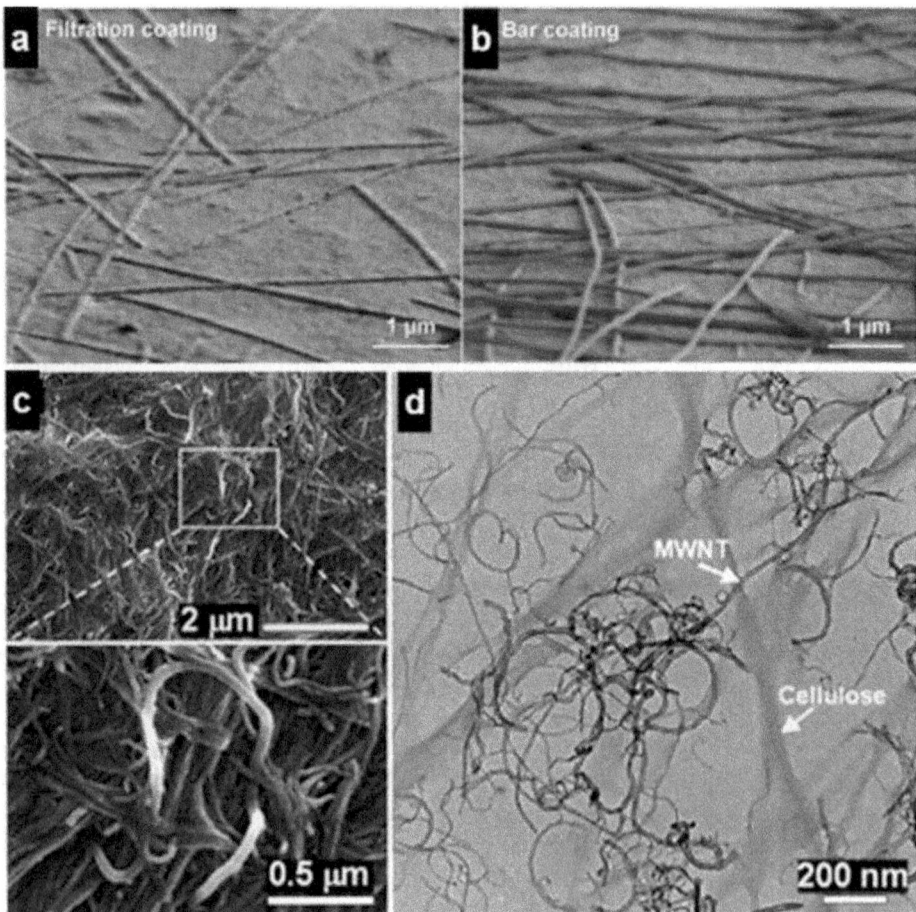

FIGURE 9.7 Illustration of FE-SEM images of AgNW nanopapers prepared by (a) filtration coating and (b) bar coating. (c) FE-SEM and (d) TEM images of BC pellicles incorporated with MWCNTs after sonication in water for 24 h. Reproduced with permission [46–47]. (a, b) Copyright 2014, *Nature*. (c, d) Copyright 2006, American Chemical Society.

components. When stretched and bent, these particular composites displayed reasonable electrome-chanical stability and electrical conductivity (Liang et al., 2012). Conductivity values of 0.14 (SH et al., 2006), 0.104 (C. J et al., 2012), and 1.2 S cm^{-1} have been reported for MWCNT/BC composites in different studies (Zhou et al., 2013) (Figure 9.7c,d). The conductivity of a single-walled carbon nanotube/carbon nanotube composite was significantly increased to 200 S cm^{-1} (MM et al., 2014). Graphene oxide (GO) and its derivative, reduced graphene oxide (RGO), were also employed as conductive agents in the construction of conductive composites (G. K et al., 2013; Nguyen Dang & Seppälä, 2015). The amount of graphene loaded could be utilized to control the conductivity level. When the RGO content was increased from 0.1 to 1%, the conductivity of RGO/CNFs nanopaper increased to $1.1*10^{-6}$ S cm^{-1} (Y. Feng et al., 2012). It was increased greatly from $7.3*10^{-4}$ S cm^{-1} with 1% RGO to 0.154 S cm^{-1} with 10% RGO (Nguyen Dang & Seppälä, 2015). Another outstand-ing graphite material, graphite nanoplatelets (GNPs), was employed to boost conductivity from 0.75 S cm^{-1} with 2% GNPs to 4.5 S cm^{-1} with 8.7% GNPs (Zhou et al., 2013). Conductive carbon particles can be easily placed on the surface of CN paper or blended with NC fibers to generate the conductive hybrid. Conductivity can be varied between 10^{-6} and 10^3 S cm^{-1} by adding conductive carbons in varying proportions.

9.2.3.3 Conductive Polymers

Conductive polymers were originally employed in the battery industry in the 1980s (Ma et al., 2014) as a substitute for metallic materials owing to their low cost, light weight and superior electrochemi-cal performance (Hu et al., 2013). The most frequently used conductive polymers for the fabrication of energy storage electrodes are poly(3,4-ethylene-dioxy-thiophene)(PEDOT), polyaniline (PANI), P-(phenylene sulfide), polyacetylene (PAC), poly(p-phenylic vinylene) (PPV), polypyrrole (PPy), poly(p-thiophene)s (PT). Due to its high claimed capacities, controlled conductivity and ease of manufacturing, PANI is regarded one of the most promising conductive polymers for application in membranes of battery electrodes or superconductor (Dong et al., 2017; Hu et al., 2013). The two major ways for incorporating conductive polymers into nanocellulose membranes are in situ polym-erization and mixing. As conductive polymers are combined with nanocellulose, easy manufac-turing processes with consistent three-dimensional (3D) network topologies and cheap production costs contribute to the final composite membranes' outstanding electromechanical performance. As a result, in situ polymerization is an effective approach for producing PANI on a nanocellulose substrate for the production of composite membranes. In general, monomer solution is impregnated on nanocellulose and then treated with an initiator such as ammonium persulfate to form a nano-composite of PANI and nanocellulose, developing in situ polymers.

Other conductors, such as PPV and PPy, can also be explored in case of situ polymerization as well. According to a recent research, in situ polymerization of polypyrrole and polystyrene sulfo-nate (PEDOT:PSS) resulted in the production of leading composite films composed of nanocel-lulose and poly (3,4-ethylenedioxythiophene). It has an electrical conductivity of 10.55 S/cm and a specific capacity of 315.5 F/g (Dong et al., 2017). Additionally, PEDOT:PSS-PPy nanopapers were more flexible than polypyrrole nanopapers (Dong et al., 2017). Numerous studies have been con-ducted to develop electric composites with superior electrochemical and mechanical characteristics by combining conductive polymers like PANI, PPy and PPV with nanocellulose (Du et al., 2017; F. Liu, Luo et al., 2017). Acidic environments are typically necessary for in situ polymerization of PPy and PANI in order to maximize their development and therefore generate well-dispersed con-ductive polymers in the nanocellulose substratum. Dodecylbenzenyl sulfonic acid, sulfuric acid and hydrochloric acid are frequently employed for these applications (Dubal et al., 2018; Khosrozadeh et al., 2016; F. Liu, Luo et al., 2017). However, when coating conductive nanocellulose films to cre-ate high-power conductive nanocellulose films, it is not easy to find a suitable solvent that is com-patible with the majority of the conductive polymers. At elevated temperatures, polymer thermal breakdown precludes the electrodeposition and coating of conduction polymers on nanocellulose films. Thus, the most often used technique for integrating conductive polymers into nanocellulose hybrids is in situ polymerization (E. Feng et al., 2017; Hu et al., 2013). Nonetheless, one disad-vantage of in situ polymerization is the associated environmental concerns and complexity of the procedures owing to the multiple stage reactions along with usage of hazardous solvents. A simple filtering procedure may also be used to produce nanocellulose-based composite membrane elec-trodes (Khosrozadeh, Darabi et al., 2015; Khosrozadeh et al., 2016; F. Liu, Luo et al., 2017) with conducting components added by in situ polymerization or mixing. After combining the conductive nanocellulose solvent component or suspension and transferring it to to a filtering system, where the liquids pass through the filter, a correctly mixed nanocellulose composite membrane or conductive material is left behind on it (Khosrozadeh, Xing et al., 2015).

9.3 ELECTRICAL PROPERTIES AND APPLICATION OF NANOCELLULOSE-BASED SUPERCAPACITORS

Nanocellulose is used to bind electroactive material to an electrode. The structural foundations for electrode sheets and supercapacitor systems were explored using aerogels and nanocellulose-derived films. Additionally, nanocellulose is also employed as a precursor for carbon compounds which can be obtained through pyrolysis.

9.3.1 ELECTRODE BINDERS

The majority of electrodes in smart SCs are composed of carbon-based materials and synthetic polymers that exhibit exceptional flexibility and mechanical strength allowing them to bend, stretch or twist in various shapes and then return to their original size and shape irrespective of stimula-tion (Z. Wang, Tammela et al., 2017). To enhance the electrochemical performance of the electrode membranes, a considerable amount of conductive material is applied on the substrate, compro-mising the membrane electrodes' flexibility and mechanical strength. The composite membrane electrodes encounter difficulties in striking an appropriate balance between mechanical and elec-trochemical performance. There are two primary methods for integrating conductive components into nanocellulose substrates to create composite membrane electrodes: one is to apply conductor materials on top of the nanocellulose matrix and the other is to include conducting agents within the nanocellulose substratum.

When employed as an electrode, nanocellulose gives such electrodes (aerogels or films) flex-ibility and mechanical strength, as well as a very large surface area that enhances capacity perfor-mance. By loading such an electrode with carbon-conductive materials such as graphite oxides and carbon nanotubes, its conductivity can be increased (Manthiram et al., 2014; H. W et al., 2014).

Nanocellulose is utilized as a substitute for synthetic polymer-based electrode binder in a wide range of electroactive materials (Manthiram et al., 2014; Pang et al., 2016) (Hu et al., 2013; Jabbour et al., 2010; KH et al., 2014; Leijonmarck et al., 2013). CNF has been widely investigated for this purpose due to its mechanical compliance and unique 1D fibrous function. Several nanocellulose-based composite electrodes were produced via super-critical drying, vacuum filtering, solvent exchange of Carbon Mixture Suspensions. Drying with supercritical carbon dioxide (CO_2) pro-duced a CNF/multiwall carbon nanotube (MWCNT) hybrid aerogel (Manthiram et al., 2014). To develop an all-solid, flexible, and symmetrical and supercapacitor, a hybrid aerogel is employed. The synthesized (Jabbour et al., 2010) hybrid aerogel composed of reduced graphene oxide (rGO) and carbon nanotubes. Here, the carbon nanotubes act as a nanospacer for the rGO. The pi-pi stacking interactions of rGO was significantly decreased by CNF and prevented rGOs from aggre-gating in this rGO and CNF hydrogel. CNF and single-walled carbon nanotube (SWCNT) mats were formed by extruding the solution of CNF and SWCNT in an ethanol coagulation bath fol-lowed by controlled drying. The CNF and SWCNT hybrid mats displayed a well formulated poros-ity structure, orienting the SWCNTs toward extrusion in a preferred manner. The CNF prevents SWCNTs from aggregating, enabling electrons to flow along the fibers' longitudinal direction. The

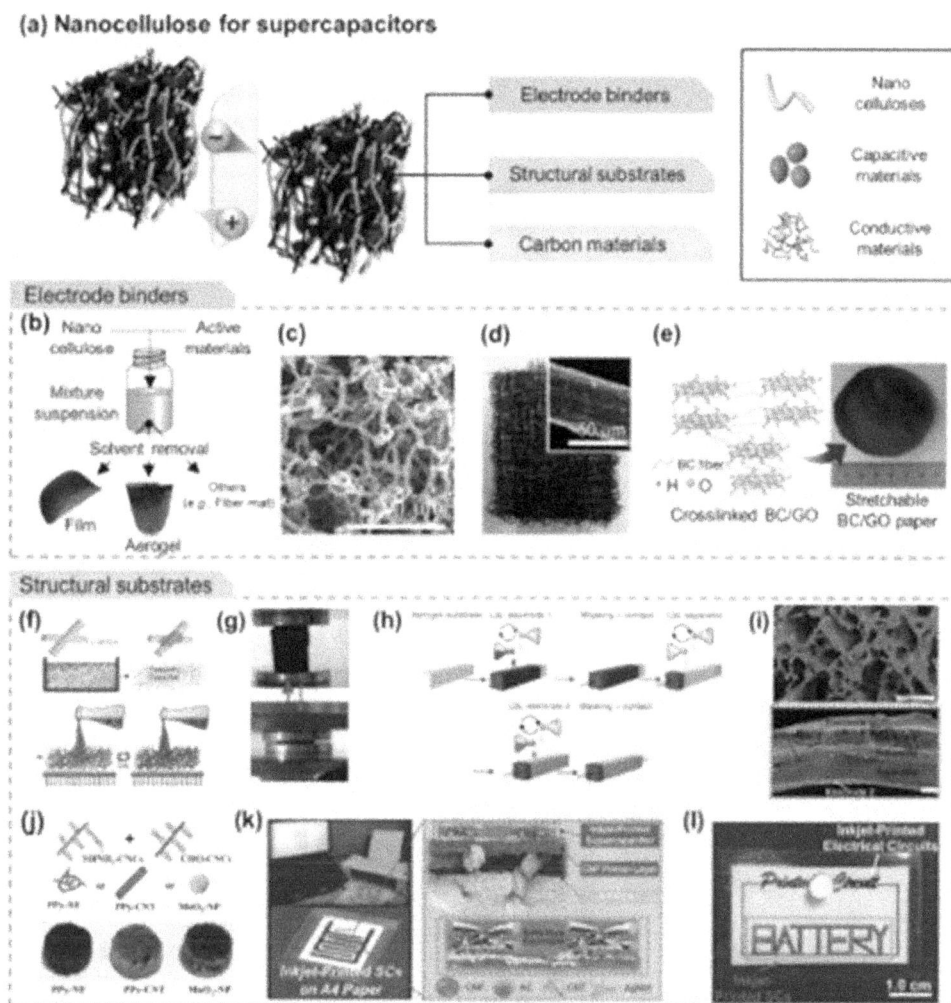

FIGURE 9.8 Electrode binders and substrates made from nanocellulose for supercapacitors. (a) Nanocellulose applications in supercapacitor electrodes. (b) Process of fabricating composite electrodes based on nanocellulose. (c) The hybrid CNF/MWCNT aerogel formed through supercritical CO_2 drying. Reproduced with permission. Copyright 2013, the Royal Society of Chemistry. (d) SEM image of CNF/SWCNT composite mats. Reproduced with permission. Copyright 2014, the Royal Society of Chemistry. (e) Crosslinked BC/GO and photograph of composite paper. Reproduced with permission. Copyright 2015, the Royal Society of Chemistry. (f) Fabricated PEI/SWCNT on a crosslinked CNF aerogel via LbL assembly. (g) Reversibly compressible (SWCNT/PEI/)/CNF aerogel. (f, g) Reproduced with permission. Copyright 2013, Wiley-VCH. (h) 3D supercapacitor from (CNT/PEI) and CNF based on aerogels produced via LbL assembly. (i) Cross-sectional SEM images of a (CNT/PEI) and CNF hybrid aerogel. (h,i) Reproduced with permission. Copyright 2015, Springer Nature. (j) Schematic illustration of aerogel components and resultant hybrid aerogels. Reproduced with permission. Copyright 2015, Wiley-VCH. (k) Illustration and photograph of inkjet-printed SCs. (l) Depiction of the letter-shaped, inkjet-printed SC (marked by a box). (k,l) Reproduced with permission. Copyright 2016, the Royal Society of Chemistry.

constructed unwoven supercapacitor exhibited exceptional electrochemical characteristics, damage resistance and extreme customizability, as well as the possibility for usage as a wearable power source. Intermolecular esterification of the mixing components resulted in the formation of a two-dimensional composite paper composed of crosslinked BC/GO (KH et al., 2014) The composite paper exhibited excellent stretchability, with a 24 percent extension and tensile strength of 18.5

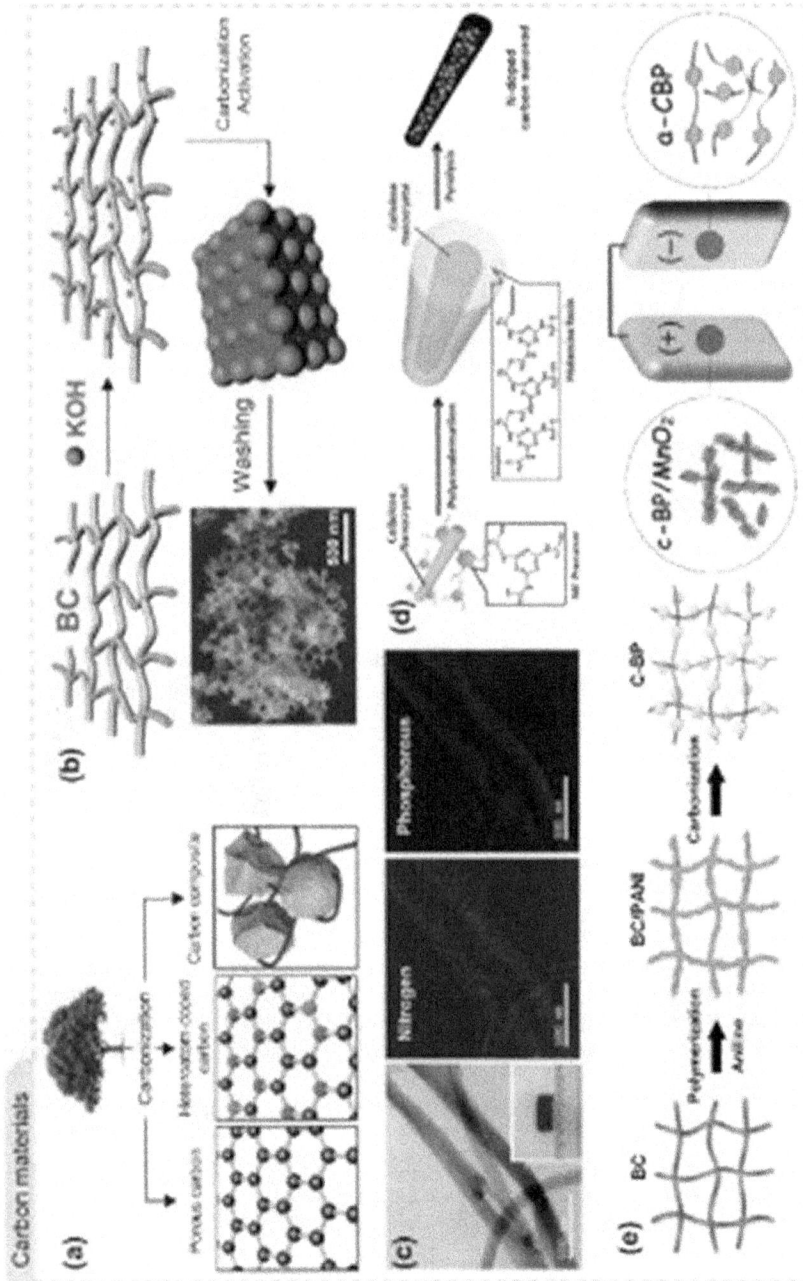

FIGURE 9.9 Carbonaceous materials derived from nanocellulose for supercapacitors. (a) Detailed illustration of cellulose-derived carbonaceous materials: hetero-atom-doped carbon, porous carbon, and carbon composites. (b) Depiction of the synthetic procedure of carbon aerogels with 3D interconnected hierarchical network honeycomb-like structure. Reproduced with permission. Copyright 2016, the Royal Society of Chemistry. (c) Energy-filtered TEM images of BC doped with N,P and elemental mapping images of N and P. Reproduced with permission. Copyright 2014, Wiley-VCH. (d) Schematic illustration of N-doped carbon nanorods from CNC coated with melamine-formaldehyde (MF) via the two-step synthesis. Reproduced with permission. Copyright 2015, the Royal Society of Chemistry. (e) An illustration representating an asymmetric supercapacitor based on nitrogen-doped c-BP (a-CBP) and MnO$_2$-coated pyrolyzed BC (c-BP) as a negative and positive electrode respectively. Reproduced with permission. Copyright 2014, Wiley VCH.

MPa. Basically, a 2D flexible Supercapacitor is an appealing and practical component for wearable and portable electronics. The two-dimensional paper supercapacitor was created by directly depositing carbon nanotube ink on cellulose paper of A4 size. The paper's stability and mechanical strength were found to be superior to those of standard paper. It performed better than when PET was employed as a substratum. Similarly, flexible supercapacitors may be made using metal oxides, conductive polymers and carbonated materials by different processes of coating, filtering or printing (W. B. et al., 2013; Z. Chen et al., 2003; Y. Li et al., 2014). Conformal coating outperforms other manufacturing methods in terms of mass loading, flexibility, and strength. On cationic NC at 300 mA/cm², the conformal-coated pyrrole exhibited a gravimetric capacity of 127 F/g and a substantial standardized volumetric capacity of 122 F/cm³. A flexible paper-based supercapacitor that incorporates CNTs and NC was developed and it was discovered to have more mechanical strength than conventional supercapacitors (Han et al., 2019; Y. Liu, Zhou, Zhu et al., 2015; D. Xu et al., 2016). Using an electrochemical deposition method, extremely porous nanocomposite electrodes were created using CNC and PPy. The electrodes had a capacitance of 336 F/g and were more stable, retaining 70% and 47% of their capacitance after 10,000 and 50,000 cycles respectively, which was much higher than the capacitance of electrodes produced with PPy doped chlorine (Electrochimica Acta, 2016). Capacitances of 69 and 488 F/g were achieved when CNC was employed in electrodes containing poly(3,4-ethylenedioxythiophen) and polyaniline, when CNC was not used (C. Chen, Zhang, Li, Kuang et al., 2017; Y. Zhang et al., 2017). To solve the conductivity issue, conductive polymer/NC ternary composites containing a third phase primarily composed of graphite is developed (Chiappone et al., 2011). By incorporating metal hydroxide or metal oxide into nanocellulose composites, the electrochemical performance of electrodes can be enhanced (Chai et al., 2017; J. Zhang et al., 2016). When PPy, cellulose paper and GO electrodes were tested, 1.2 F/cm² was the capacitance obtained with a current of 2 mA/cm², and the permittivity after 5000 cycles remained at 89 percent (F. Li et al., 2015).

An electrode composed of graphene-cellulose hybrid, aniline and silver passed the electrode cycle life test. These supercapacitors are extremely mechanical in nature, have a large specific surface area, rapid charge-discharge rate, a high power density of 1749 mW/g and excellent cyclic stability (X. J et al., 2019; Khosrozadeh, Xing et al., 2015; Y. Liu, Zhou, Zhu et al., 2015; Wan et al., 2017). With a current density of 1.6 A/g, the electrodes retained a an 84 percent capacitance, 98 percent energy density and 108 percent power density after 2400 cycles, indicating their exceptional cycle life and mechanical robustness (Khosrozadeh, Darabi et al., 2015).

9.3.2 Structural Substrates

Aerogels derived from nanocellulose and films have been utilized as structural substrates for the production of flexible electrodes and supercapacitors, which were then coated with electroactive or electrically conductive compounds (Hamedi et al., 2013; S. Li, Huang, Zhang et al., 2014; Liew et al., 2010; R. Liu et al., 2017; Nyström et al., 2009, 2015; H. Wang, Zhu et al., 2012; Z. Wang, Carlsson et al., 2015; Z. Wang, Tammela, Strømme, Nyholm et al., 2015; Yang et al., 2015). A method for combining single-walled carbon nanotubes (SWCNTs) with cross-linked CNF aerogel was proposed to create a composite aerogel with structural integrity and strong resistance to water, nearing 99 percent porosity. The supercapacitor made of CNF aerogel exhibited consistent electrode capacity and excellent compressibility. The layer-by-layer (LbL) self-assembly of multilayered films on a CNF-based open cell aerogel substrate surface led in the formation of a compressible three-dimensional supercapacitor (Nyström et al., 2015). A cross-linked, negatively charged CNF aerogel served as the substratum, which was then coated with positive and negative electrodes. As an electrode layer, an anionic (TENCOOH-functionalized) SWCNT layer was employed in conjunction with a polyethylenimine (PEI) cationic layer, and PEI/polyacrylic acid as an electrolyte and electromagnet membrane. The LbL technique generated densely linked and ultra-thin nanoporous coating layers over the whole CNF network, enabling the active materials' electrochemical activity.

After repeated compression, resultant 3D supercapacitor's cell capacitance was maintained at 25 F g^{-1} for 400 cycles based on the active mass. As electroconductors, carbon nanotube-coated cellulose fibers were developed, with the cellulose fibers having the potential to not only function as a as an internal electrolyte store but also maintain a large surface area (Gui et al., 2013). The electrolyte was swiftly absorbed by the carbon nanotube-coated cellulose fiber support, allowing it to permeate into the electrode active material. Electro-conductive polymers (ECPs), including polypyrrole (PPy), polyaniline (PANI) and poly(3,4-ethylene dioxythiophene) (PEDOT), have been discovered as potential pseudocapacitive electrode materials for supercapacitors. Due to the fact that the hydroxyl groups on the cellulose surface promote strong intra- and intermolecular contacts between ECPs and hydrogen, nanocellulose is a suitable substrate for non-metallic electrode materials (Z. Wang, Tammela, Strømme, Nyholm et al., 2015). By combining faradaic reactions with a broad active surface, the nanocellulose scaffold's thin ECP layer decreases the strain associated with volume variations and increases the electrode's specific during discharge and load cycling. (Mihranyan et al., 2008; Nyström et al., 2009). Pyrrole is chemically polymerized with iron (III) chloride to create homogeneous and thin PPy layers of coating on CNF substrates. The resulting PPy/CNF composite electrodes displayed significantly faster reduction and oxidation reactions along with a greater power output than the thick PPy sheets. Wang et al. modified the surface of the CNF with quaternary amine groups in order to create cationically charged CNF (Z. Wang, Carlsson et al., 2015). After this, its polymerization with pyrrole results in the formation of c-CNF/PPy composites with a uniformly compact shape. Additionally, CNF, CNC, and BC were coupled with ECPs to generate supercapacitor electrodes based on nanocellulose. PANI/BC composites on BC substrates were synthesized through aniline polymerization (H. Wang, Zhu et al., 2012). However, the brittle nature of these composite electrodes resulted in limited mechanical flexibility. To solve this issue, a BC substratum is coated with MWCNT and PANI is electrically deposited on top, resulting in flexible and lightweight PANI/MWCNT/BC electrodes (S. Li, Huang, Zhang et al., 2014). Due of its superior mechanical strength and dimensional properties, CNC was extensively investigated as a substratum. During electrodeposition on anionically charged CNCs (a-CNCs) (Liew et al., 2010) by thin PPy layer to produce a CNC/PPy combination, the a-CNCs function as a counterbalance to the positive charges of the backbone polymer, thereby forming a highly porous composite film. Thus, PPy/CNC composite electrodes exhibited a greater capacitance.

On the other hand, CNC composite electrodes are mechanically weak (C. W et al., 2014) and degrade rapidly when deformed externally. A cross-linked CNC aerogel method has been developed to solve these limitations (Shi et al., 2016; Yang et al., 2015). These aerogels were created by chemically linking CNCs modified through aldehydes and hydrazides.

The CNC aerogels were used as a platform for a range of capacitive materials, like PPy nanofibers, PPy-coated nanoscale carbon nanotubes and spherical manganese dioxide nanoparticles. The capacitive materials in particular interacted with the connected CNCs via non-polar interactions and hydrogen bonding to produce gel-like structures that were then freeze dried to form hybrid aerogels. The aerogels produced after 2000 cycles at 0.1 mA cm^2 demonstrated a high retention capacitance of 94 percent. On fabricating flexible solid-state supercapacitors directly on commercial A4 paper, the CNF enabled the primary layer to be an inkjet-printed nanomat, thereby allowing inkjet printing of electrolytes and electrodes. The results indicated that inkjet-printed supercapacitors have consistent electrochemical performance, high forming factors and a high degree of mechanical flexibility.

In fact, a straightforward method for coating carbon nanotube inks on ordinary photocopying paper to create leading sheets of paper was also demonstrated. Due to the intrinsic characteristics of paper, which acts as a strong nanomaterial binder and an easy solvent adsorbent, supercapacitors based on carbon nanotube-coated conductive paper may be manufactured.

Aerogels based on NC have a wide surface area, low density and high porosity which makes them suitable for usage as substrates. Due to the huge number of active sites (Gao et al., 2013). CNC-based aerogels were able to hold a considerable load when combined with active nanoparticles such as manganese dioxide and PPy coated CNTs. Aerogels consisting of carbon nanotubes, reduced graphene oxide and cellulose exhibited a capacitance of 252 F/g at a current density of 0.5 A/g. After 1000 cycles, the capacitance was determined to be 99 percent of its initial value at 1 A/g. (Q. Zheng et al., 2015). Moreover, CNF-based aerogels were developed which, due to the hydrogen bond between them, can also keep their form in water.

Porous carbon-containing chiral-nematic mesoporous CNC was also with a capacitance of 170 F/g and a current density of 230 A/g when H_2SO_4 is employed as the electrolyte. Additionally, CNF and CNC were utilized to synthesize (JA et al., 2014; G. M et al., 2015) porous carbon compounds suitable with supercapacitors. When these materials are employed as electrodes, they exhibit fast ion mobility and a capacity of 170 F/g (Z. Li et al., 2017). Additionally, supercapacitors may be produced from carbonized cellulose and cellulose derivatives.

9.3.3 CARBONACEOUS NANOCELLULOSE MATERIALS

Carbonaceous materials make up the bulk of the materials utilized in energy storage systems.

Nanocellulose has recently attracted considerable interest as a green precursor for carbonic materials. Pyrolysis initiated at high temperatures transforms nanocellulose to conducting carbon molecules in an inert environment. Three forms of carbons produced from nanocellulose exist: porous carbon, heteroatomic carbohydrates and carbon composites.

In the manufacturing of porous carbon aerogels, CNF and BC were widely employed as building ingredients (C. Chen, Zhang, Li, Dai et al., 2017; L.-F. Chen et al., 2013, 2014; LF et al., 2013; Long, Qi et al., 2014; Shan et al., 2016; X. Wu et al., 2015; X. Xu et al., 2015). It is well established that activating carbon aerogels using catalysts such as potassium citrate (*Journal of Power Sources, 2016*) and potassium hydroxide (Shan et al., 2016) efficiently increases the specific charge surface areas of the aerogels, therefore improving their specific capacity.

Potassium hydroxide was utilized, during carbonization, to aid in the formation of macropores and interconnection of BC. The resulting carbon aerogels have a three-dimensional network with a honeycomb-like structure.

Lignin (X. Xu et al., 2015) has generated considerable attention due to its three phenolic alcohol monomers that are linked to the three-dimensional polymer networks via the carbon-oxygen-carbon and carbon-carbon bonds, as well as its network structure composed of aromatic carbons. Aerogels composed of lignin, carbon, resorcinol, and formaldehyde (LRF) are mechanically fragile. The LRF solution was incorporated with BC and polycondensed to form BC/LRF hydrogels. The aerogels of BC/LRF, as well as their carbon aerogels, were synthesized through catalyst-free carbonization and supercritical CO_2 drying. Due to all-wood structure (C. Chen, Zhang, Li, Dai et al., 2017) of a low-torque and biodegradable supercapacitor consisting of a MnO_2/wood carbon cathode, a wooden membrane separator and an activated wood carbon anode, the unit has a high energy and power density. A separator must be durable and flexible, along with possessing excellent chemical, thermal and dimensional stability. It must be extremely porous in order to retain electrolytes. The nanocellulose scale has a large surface area, which assists in the regulation of the pore shape in separators. It acts as a good diffusion route for electrolytic solutions and aids in ion migration.

Additionally, the thickness and mass loading of wood-based three-dimensional electrodes may be enhanced (C. Chen, Zhang, Li, Dai et al., 2017; C. Chen, Zhang, Li, Kuang et al., 2017). It has a non-complicated structure with straight channels, which accelerates ion transport. Top-down manufacture of these three-dimensional wood electrodes preserves the unique characteristics of wood, which implies that many canals are fully aligned. Additionally, wood can be burnt into holes to gather electricity and material for active electrodes such as sodium metals, sulfur, lithium ion phosphate, sodium metals and so on. An 800-meter-long three-dimensional wood carbon cathode was developed by injecting a lithium ion slurry into the channels (C. Chen, Zhang, Li, Kuang et al., 2017). Cyclic performance and mechanical stability were demonstrated to be superior to approaches utilizing slurry coatings. Carbonized wood is significantly less durable than contemporary aluminum or copper foil collectors. Another constraint is its scalability, owing to the size of the wood slice.

TABLE 9.1
Comparison of Electrochemical Performance of Supercapacitors

Composite Material	Capacitance	Energy density	Power density	Capacitance Retention	Ref
Cellulose based aniline, graphene, and silver	-	98	108	2400 cycles 84%	(Khosrozadeh, Darabi et al., 2015)
PPy/CNC	336 F g⁻¹, 258 F g⁻¹ when doped with Cl	-	-	10000 cycles 70% 50000 cycles 47%	(Y. Liu, Zhou, Zhu et al., 2015)
Natural wood/MnO₂	3.6 F cm⁻¹ (at 1 mA cm⁻²)	1.6 mWh cm⁻²	24 W cm⁻²	10000 cycles 93%	(C. Chen, Zhang, Li, Dai et al., 2017)
Three-dimensional cellulose graphene structures	160	-	-	2000 cycles 90.3%	(Y. Liu, Zhou, Zhu et al., 2015)
Three-dimensional structures of NC/graphene/Ppy	-	-	-	2000 cycles 93.5%	(Y. Liu, Zhou, Zhu et al., 2015)
Aerogels comprising of cellulose RGO and CNT	252 F g⁻¹ (at 0.5 A/g)	7.1 mWh/g	2375 mw/g	1000 cycles 99.5%	(Q. Zheng et al., 2015)
CNF/RGO	158 mF cm⁻², 207 F g⁻¹	20 µWh cm⁻²	15.5 mW cm⁻²	5000 cycles 99.1%	(Gao et al., 2013)
CNC/PPy/polyvinylpyrrolidone (PVP)	322.6 F g⁻¹	-	-	1000 cycles 91% 2000 cycles 87%	(W. Chen et al., 2018; X. J et al., 2019)
CNF's/GO	300 F g⁻¹	-	-	3000 cycles 95.4%	(J. Zhang et al., 2017)
CNF/GO/PPy	334 F g⁻¹	18.5 mWh/g	500 mw/g	2000 cycles 100%	(J. Zhang et al., 2017)
Melamine-formaldehyde (MF) coated CNCs	352 F g⁻¹ (at 5 A/g)	-	-	2000 95.4%	(X. Wu et al., 2015)
PEDOT-PSS (poly(3,4-ethylene-dioxyiophene)-poly(styrene sulfonate)/SnO2/RGO/BC	373 F g⁻¹	-	-	2500 84.1%	(K.-K. Liu et al., 2018)
TEMPO oxidized CNF/CNT	178 F g⁻¹ (at 5 mV s⁻¹)	20 mWh cm⁻², 5.06 Wh kg⁻¹	13.6 mW cm⁻² 7.67 kW kg⁻¹	1000 cycles 99.9%	(Gao et al., 2013)
BC/GO as electrode	160 (at 0.4 A g⁻¹)	-	-	2000 cycles 90.3%	(Y. Liu, Zhou, Zhu et al., 2015)
TEMPO oxidized CNF/CNT	3.29 mF cm⁻² (at 0.02 mA cm⁻²)	0.702 µWh cm⁻²	2.435 mW cm⁻²	5000 cycles 97%	(Q. Liu et al., 2016)
TEMPO oxidized CNF/GO, CNT	252 F g⁻¹ (at 0.5 A g⁻¹)	28.4 mWh cm⁻²	9.5 mW cm⁻²	1000 cycles 99.5%	(W. Zheng et al., 2017)
MFC/-COOH introduced MWCNT	154.5 mF cm⁻² (at 20 mVs⁻¹)	-	-	-	(X. Zhang et al., 2013)

(Continued)

Material	Capacitance	Energy density	Power density	Cycles/Retention	Reference
BC/r-Bi2O3	6.675 F cm^{-2} (at 1 mA cm^{-2})	0.449 mWh cm^{-2}, 7.74 mWh cm^{-3}	40 mW cm^{-2}, 690 mW cm^{-3}	1000 cycles 90%	(R. Liu et al., 2017)
CNF/PEDOT:PSS	470 F g^{-1} (at 0.5 A g^{-1})	-	-	1000 cycles 85%	(A et al., 2015)
CNC/PPy nanofiber, PPy-coated CNT, MnO$_2$ nanoparticle as	3.32 mF cm^{-2}, 2.42 mF cm^{-2}, 2.14 mF cm^{-2} (PPy-NF, -CNT, MnO$_2$-NP) (at 2 mV s^{-1})	-	-	2000 cycles 84.19, 61.66, 92.28% (PPy-NF, -CNT, MnO$_2$-NP)	(Yang et al., 2015)
BC/PEDOT:PSS, GO	470 F g^{-1} (at 0.5 A g^{-1})	-	-	1000 cycles 85%	(Q. Jiang et al., 2017)
CNF/Ppy	127 F g^{-1}, 122 F cm^{-3} (at 33 A g^{-1})	4.0 Wh kg^{-1}	3.5 kW kg^{-1}	-	(Z. Wang, Carlsson et al., 2015)
Anionically charged CNF/SWCNT	25 F g^{-1} (at 60 C)	1 Wh kg^{-1}	-	400 cycles 75%	(Nyström et al., 2015)
CNF/Ppy	370 F g^{-1}	-	-	60 cycles 85%	(Mihranyan et al., 2008)
CNF/Ppy	38–50 mAh g^{-1}	-	-	100 cycles 94%~	(Nyström et al., 2015)
CNF/Ppy	10–60 F g^{-1}	-	-	1000 cycles	(Nyström et al., 2009)
CNF/PPy, Carbon filaments as sS	60–70 F g^{-1} (at 31 A g^{-1})	1.75 Wh kg^{-1}	2.7 kW kg^{-1}	1500 cycles ~100%	(Z. Wang, Tammela, Strømme, & Nyholm, 2015)
CNF/Ppy	38.3 F g^{-1}, 2.1 F cm^{-2}	-	-	10000 cycles 80–90%	(Nyström et al., 2012)
CNF/PAH/HA, PEI/PEDOT:PSS, PEI/ADS200P, PEI/SWCNT	419 F g^{-1}	-	-	6 cycles	(Hamedi et al., 2013)
CNF/RGO	1.73 mF cm^{-2} (at 0.012 mA cm^{-2})	-	-	5000 cycles ~81%	(Gao et al., 2013)
CNF/Ppy	36.3 F g^{-1}, 1.54 F cm^{-2} (at 1.35 mA cm^{-2})	3 Wh kg^{-1}	1.2 kW kg^{-1}	1200 cycles 95%	(Z. Wang, Tammela et al., 2017)
CNF/Ppy	354 F cm^{-3}, 5.66 F cm^{-2}	3.7 Wh L^{-1}	-	8500 cycles 84%	(Z. Wang, Tammela et al., 2017)
CNF/PANI, Ag	176 mF cm^{-2} (at 10 mVs^{-1})	10.6 Wh kg^{-1}	225 kW kg^{-1}	-	(Zhang, X.; Lin, Z.; Chen, B.; Zhang, W.; Sharma, S.; Gu, W.; Deng, Y. J. Power Sources. 2014, 246, 283 — Google Search, n.d.)

TABLE 9.1 (Continued)

Composite Material	Capacitance	Energy density	Power density	Capacitance Retention	Ref
CNF/PPy, Carbon filaments	120 F g⁻¹ (at 5 mV s⁻¹)	-	-	-	(Z. Wang, Tammela, Strømme, & Nyholm, 2015)
CNF/PANI, MWCNT	791.13 F g⁻¹ (at 0.2 A g⁻¹)	-	-	3000 cycles 82.14%	(F. Li et al., 2015)
CNF/PANI nanofiber, MWCNT	249.7 F g⁻¹ (at 10 mV s⁻¹)	-	-	1000 cycles 82.4%	(F. Li et al., 2015)
CNF/PPy, GO	198 F cm⁻³ (at 5 A g⁻¹)	3.4 Wh L⁻¹, 5.1 Wh kg⁻¹	1.1 kW L⁻¹, 1.5 W kg⁻¹	16000 cycles	(Z. Wang, Tammela, Strømme, Nyholm et al., 2015)
CNF/PEDOT	90 F g⁻¹, 920 mF cm⁻², 54 F cm⁻³	1.8 mWhcm⁻³	14.4 mW cm⁻³	15000 cycles 93%	(Z. Wang et al., 2016)
TEMPOoxidized CNF/PANI, cellulose-derived carbon sheet	3297.2 mF cm⁻², 220 Fg⁻¹ (at 1 mA cm⁻²)	-	-	3000 cycles 83%	(Q. Liu et al., 2016)
TEMPO oxidized CNF/PANI, Graphene nanosheet	421.5 F g⁻¹ (at 1 A g⁻¹)	31.3 Wh kg⁻¹	335.6–10604.6 W kg⁻¹	335.6–10604.6 W kg⁻¹	(W. Zheng et al., 2017)
BC/CNT	50.5 F g⁻¹ (20.2 mF cm⁻²)	15.5 mWh g⁻¹	1.5 W g⁻¹	5000 cycles 99.5%	(YJ et al., 2012)
BC/PANI	273 F g⁻¹ (at 0.2 A g⁻¹)	-	-	1000 cycles 94.3%	(H. Wang, Zhu et al., 2012)
BC/PPy	316 F g⁻¹ (at 0.2 A g⁻¹)	-	-	1000 cycles 88.2%	(H. Wang, Bian et al., 2012)
BC/PPy	101.9 mAh g⁻¹ (459.5 Fg⁻¹) (at 0.16 A g⁻¹)	-	-	50 cycles 70.3%	(L. Zhu et al., 2014)
BC/PANI, MWCNT	656 F g⁻¹ (at 1 A g⁻¹)	-	-	1000 cycles	(S. Li, Huang, Zhang et al., 2014)
BC/PPy, MWCNT	2.43 F cm⁻²	-	-	5000 cycles 94.5%	(S. Li, Huang, Yang et al., 2014)
BC/PPy, GO	278 F cm⁻³	77.2 Wh kg⁻¹	200.1 W kg⁻¹	5000 cycles 95.2%	(Y. Liu, Zhou, Tang et al., 2015)
BC/PPy	153 F g⁻¹ (at 0.2 A g⁻¹)	21.22 Wh kg⁻¹	6.59 kW kg⁻¹	100 cycles 93%	(F. Wang et al., 2016)
CNC/PPy	336 F g⁻¹	-	-	5000 cycles	(Liew et al., 2010)

Material	Specific capacitance	Energy density	Power density	Cycles / Retention	Reference
CNC/Ppy, MWCNT	2.1 F cm⁻²	-	-	5000 cycles 93.3%	(Shi et al., 2016)
Cotton/SWCNT, MnO_2	0.48 F cm⁻²	20 Wh kg⁻¹	10 kW kg⁻¹	130000 cycles 98%	(L et al., 2010)
Commercial paper/SWCNT	200 F g⁻¹	30–47 Wh kg⁻¹	200000 W kg⁻¹	40000 cycles	(L et al., 2009)
Conventional A4 paper, CNF Activated carbon SWCNT	100 mF cm⁻² (at 0.2 mA cm⁻²)	~ 12 Wh kg⁻¹	3024 W kg⁻¹	10000 cycles	(Choi et al., 2016)
BC/3D honeycomblike hierarchical structured carbon	422 F g⁻¹ (at 2 mV s⁻¹ Rate capability = 73.7% (at 500 mV s⁻¹)	-	-	(Asymmetric cell) 10000 cycles 113%	(Shan et al., 2016)
BC/MnO_2, N-doped carbon	113 F g⁻¹ (at 20 mV s⁻¹)	63 Wh kg⁻¹	227 kW kg⁻¹	5000 cycles 92%	(Long, Qi et al., 2014)
Cellulose acetate/CNT	241 F g⁻¹	4.1 Wh kg⁻¹	19570 W kg⁻¹	1000 cycles	(Volodymyr Kuzmenko et al., 2017)
BC/Lignin-based carbon	124 F g⁻¹ (at 0.5 A g⁻¹)	-	-	10000 cycles 98%	(X. Xu et al., 2015)
BC/N,P-doped carbon and B,P-doped carbon	204.9 F g⁻¹ (at 1.0 A g⁻¹) for N,P-CNF	7.76 Wh kg⁻¹	186.03 kW kg⁻¹	4000 cycles	(L.-F. Chen et al., 2014)
BC/N-doped p-BC	195.44 F g⁻¹ (at 1.0 A g⁻¹)	-	390.53 kW kg⁻¹	5000 cycles 95.9%	(L.-F. Chen et al., 2013)
Cellulose acetate/N-doped CNF	27.8 F g⁻¹	-	-	1000 cycles 144.9%	(Cellulose Nanoparticles: Volume 2: Synthesis and Manufacturing; Volodymyr Kuzmenko et al., 2017)
BC/N, S-doped carbon	171.2 F g⁻¹ (at 0.5 A g⁻¹)	-	-	-	(Cellulose Nanoparticles: Volume 2: Synthesis and Manufacturing; Z. Li et al., 2017)
BC/K-birnessite type MnO_2	328.2 F g⁻¹ (at 0.2 A g⁻¹)	-	-	2000 cycles 91.6%	(Applied Surface Science, 2018)
CNC/N-doped carbon nanorod	328.5 F g⁻¹ (@0.01 V s⁻¹), 352 F⁻¹ (at 5 A g⁻¹)	48.8 Wh kg⁻¹	39.85 kW kg⁻¹	2000 cycles 95.4%	(X. Wu et al., 2015)
BC/MnO_2, N-doped p-BC	-	32.91 Wh kg⁻¹	284.63 kW kg⁻¹	2000 cycles 95.4%	(L.-F. Chen et al., 2013)

Carbonic materials have been proven to successfully modify their electron donor characteristics by substituting heteroatoms such as iodine, sulfur, boron, phosphorus and nitrogen for particular atoms (L.-F. Chen et al., 2014). For example, several carbon-doped nitrogen compounds have been synthesized by carbonization of nitrogen-enriched precursors or via post-treatment with ammonia gas, both of which require hazardous working conditions or lengthy synthetic procedures. The pyrolysis of BC resulted in the formation of a network of nitrogen-doped 3D carbon nanofibers (L.-F. Chen et al., 2013). Nitrogen was introduced, via a hydrothermal reaction, into the pyrolyzed BC with an aqueous ammonia solution. A similar technique has been developed that utilizes heteroatomic molecules as a dopant (L.-F. Chen et al., 2014). Furthermore, BC slices were submerged in aqueous solutions of H_3PO_4/H_3BO_3, H_3PO_4 and $NH_4H_2PO_4$. Since functional BC groups are abundant, phosphorus, nitrogen-phosphorus and boron-phosphorus are easily doped. The 3D heteroatomic-doped carbon nanofiber networks were formed following carbonization and drying. CNC was used to control the development of a nitrogen precursor in order to generate CNC coated with melamine formaldehyde (MF) as a carbon source (X. Wu et al., 2015). Pyrolysis of this mixture resulted in the formation of materials with N-doped carbon structure consisting of varying small sized pores (LF et al., 2013). To generate a supercapacitor asymmetry, a doped c-BP negative electrode, a pyrolyzed MnO_2, positive electrode were utilized (Long, Jiang et al., 2014; Long, Qi et al., 2014) BC was used as a raw material for the fabrication of a 3D nanofibrous c-BP carbon network using a 1000 °C ring technique. The c-BP coated with MnO_2 through a hydrothermal method was then synthesized using a $KMnO_4$ or K_2SO_4 aqueous solution.

9.4 CHALLENGES

The optical transmittance and electrical conductivity trade-offs continue to be a significant challenge when it comes to fully using nanocellulose-based conductive coatings. Given the widespread usage of nanofiber-based materials in the society, it is predicted that future research will focus on ways to incorporate these materials more completely into daily life. As a result, considerable effort has been invested in the development of optical sensing and detection applications based on bright nanocellulose films under present conditions. Numerous research studies have concentrated on more complex applications, including holographic displays, mechanoluminescent sensors, and screen. TEMPO has been discovered to catalyze the synthesis of nanocellulose, which has been widely employed in the manufacturing of nanocellulose-based lighting materials. Owing to its sole employment of renewable resources only, one of the current problems is the requirement for these renewable resources in the development of high-performance materials. To create flexible and resilient systems, it is necessary to integrate multiple components with disparate properties into a single device. Luminosity and conductivity are mutually exclusive properties that must be included into next-generation green electronics. Due to intermolecular interactions, NCs are not evenly distributed in polymers. It is challenging to get great dispersion with it. Numerous researchers have concentrated their efforts on successfully conveying their findings using polymer media. NCs may be integrated into polymer matrices by post polymerization compounding and in situ polymerization (Miao & Hamad, 2013). In situ polymerization is carried out in the presence of a solvent that makes the monomers soluble while easily dispersing the NCs. This leads to the formation of percolation networks and monomers react with NC and crosslinking takes place (My Ahmed Saïd Azizi Samir, Fannie Alloin et al., 2014). When sonification is employed, it is feasible to disperse NCs in organic solvents because only water-soluble polymer matrices are permitted to be utilized in this method (Sapkota et al., 2014). Due to the fact that dispersion in aqueous fluid is adequate for these nanocomposites, they may be synthesized in liquid media. Despite this, film casting is not a widespread practice. Cellulose nanocomposites must be designed in such a way that they can be manufactured economically. Injection molding and hold great potential due to their lack of reliance on organic solvents, simplicity, affordability, and environmental friendliness. One of the two critical issues that must be resolved before these technologies can be mass produced is the aspect ratio, which is decreased as a result of the screws' shear pressures. The second problem is the aggregation of the nanofiller as the NC does not completely dry. As an undesirable outcome, it can reagglomerate and clump when molten polymer is extruded (Kalia et al., 2011). Due to the hydrophilicity of nanocellulose, it does not mix well with hydrocarbon-based polymers such as polyethylene. Nonpolar matrices will aggregate due to increased hydrogen bonds between particles as a result of interparticle interaction. Attempts to resolve these processing problems have been undertaken in the past.

9.5 CONCLUSION AND ROAD MAP FOR FUTURE WORK

Biopolymers and next-generation materials based on low-cost and sustainable feedstock have garnered considerable interest. Simpler materials for food applications (packaging etc.) have gained from nanocellulose's reinforcing action, with nanocellulose often used as a reinforcing component in these applications. Nanocellulose is economically prohibitive for a number of applications due to the energy required to make it. When we examine nanocellulose, it is acceptable to conclude that its usage is adequate to enable the development of increasingly sophisticated applications without limiting the production of large-scale high-value items. Cellulose in the form of nanocrystals or nanofibers presently serves as the foundation for next-generation energy solutions. Electrical components and gadgets made of petroleum can be replaced with more ecologically friendly and cost-effective functionalized nanocellulose. The load-bearing characteristics, flexibility and mechanical toughness of CNF are well-known, crediting/enhancing its commercialization. By grafting or integrating various characteristics onto CNF and then immobilizing nanoparticles via one of these methods, functional free-standing films have been created. CNC, on the other hand, is advantageous for developing sustainable and environmentally friendly medical, electrical, food and chemical products due to its percolation properties and variable surface chemistry. Current research indicates that the use of renewable resources in the manufacture of carbon dioxide, carbon, and graphene nanotubes is growing. Carbon footprint reduction is critical for maintaining hybrid nanocomposite characteristics. Due to its many characteristics and performance requirements, cellulose-based components stand out in the development of innovative bio-based products. Developing and maintaining a bio-based device that uses nanocellulose as a template to bind functional components and control the surface would be exceedingly challenging, though resourceful.

Recent research has established that nanocellulose is a viable material for actuators and sensors. The nanocellulose- $BaTiO_3$ composites exhibits comparable piezoelectricity to PVDF-based composites. The initiative is to determine how plasticizers impact the piezoelectricity of biobased nanocomposites as well as the orientation of CNC crystals in the film. When combined with CNC machining, high optical haze may be utilized to increase the efficiency of solar cells' power conversion. Solar cells made of nanocellulose perform better in humid temperature conditions and have a longer shelf life.

The design and fabrication processes for nanocellulose-based aerogels and three-dimensional structures are being investigated as a means of improving energy density. Additional research is necessary to develop an industrially viable approach for producing large electrodes. Efforts must be made to ensure that sustainable materials are used in energy applications while also considering environmental protection. Despite the fact that nanocellulose has a lengthy production time and high manufacturing cost, the issue is not with the material itself but with the manufacturing time and cost. New ways for producing nanocellulose on an enormous scale are likely to develop.

ACKNOWLEDGEMENT

The authors gratefully acknowledge the financial support from the Institute of Chemical Technology Mumbai to carry out this work.

ABBREVIATIONS

BC	Bacterial Cellulose
CNCs	Cellulose nanocrystals
CNFs	Cellulose nanofibers
DMSO	Dimethyl sulfoxide
ECPs	Electro-conductive polymers
EDLC	Electrical double layer supercapacitor
LBL	Layer-by-layer
MF	Melamine formaldehyde
MWCNT	Multi-walled carbon nanotube
NC	Nanocellulose
PAC	Polyacetylene
PANI	Polyaniline
PEDOT	Poly-3,4-ethylene-dioxy-thiophene
PEDOT:PSS	Polypyrrole and polystyrene sulfonate
PEI	Polyethyleneimine
PPV	Poly(p-phenylic vinylene)
PPV PEO	Polyethylene glycol
PPY	Polypyrrole
PT	Poly(p-thiophene)
PVA	Polyvinyl alcohol
rGO	Reduced graphene oxide
SCs	Supercapacitors
SWCNT	Single-walled carbon nanotube
TEMPO	2,2,6,6-Tetramethylpiperidin-1-yl)oxyl
TONC	TEMPO-oxidized Nanocellulose

REFERENCES

A, M., & D, C. (2015). Characterization of nanocellulose reinforced semi-interpenetrating polymer network of poly(vinyl alcohol) & polyacrylamide composite films. *Carbohydrate Polymers*, *134*, 240–250. https://doi.org/10.1016/J.CARBPOL.2015.07.093

A, M., J, E., H, G., ZU, K., JW, A., X, L., D, Z., H, Z., Y, Y., JW, B., I, E., M, F., L, W., X, C., & M, B. (2015). An organic mixed Ion-electron conductor for power electronics. *Advanced Science (Weinheim, Baden-Wurttemberg, Germany)*, *3*(2). https://doi.org/10.1002/ADVS.201500305

and, M. W., & Brodd., R. J. (2005). What are batteries, fuel cells, and supercapacitors? (Chem. Rev. 2003, 104, 4245−4269. Published on the Web 09/28/2004.). *Chemical Reviews*, *105*(3), 1021. https://doi.org/10.1021/CR040110E

Anirudhan, T. S., & Rejeena, S. R. (2014). Poly(acrylic acid-co-acrylamide-co-2-acrylamido-2-methyl-1-propanesulfonic acid)-grafted nanocellulose/poly(vinyl alcohol) composite for the in vitro gastrointestinal release of amoxicillin. *Journal of Applied Polymer Science*, *131*(17), 8657–8668. https://doi.org/10.1002/APP.40699

Applied Surface Science | Vol 433, Pages 1–1198 (1 March 2018) | ScienceDirect.com by Elsevier. (n.d.). Retrieved September 5, 2021, from www.sciencedirect.com/journal/applied-surface-science/vol/433/suppl/C

Azizi Samir, M.A.S., Alloin, F., Sanchez, J.Y. and Dufresne, A., 2004. Cross-linked nanocomposite polymer electrolytes reinforced with cellulose whiskers. Macromolecules, 37(13), pp. 4839–4844.

B, D., E, A., BM, C., A, B., JJ, B., LA, P., AL, L., SF, de S., & M, K. (2011). Structure, morphology and thermal characteristics of banana nanofibers obtained by steam explosion. *Bioresource Technology*, *102*(2), 1988–1997. https://doi.org/10.1016/J.BIORTECH.2010.09.030

B, W., X, L., B, L., J, Y., X, W., Q, S., S, C., & L, Z. (2013). Pyrolyzed bacterial cellulose: A versatile support for lithium ion battery anode materials. *Small (Weinheim an Der Bergstrasse, Germany)*, *9*(14), 2399–2404. https://doi.org/10.1002/SMLL.201300692

Beidaghi, M., & Gogotsi, Y. (2014). Capacitive energy storage in micro-scale devices: Recent advances in design and fabrication of micro-supercapacitors. *Energy & Environmental Science*, *7*(3), 867–884. https://doi.org/10.1039/C3EE43526A

Biofibres, biodegradable polymers and biocomposites: An overview—Mohanty—2000 — Macromolecular Materials and Engineering—Wiley Online Library. (n.d.). Retrieved September 4, 2021, from https://onlinelibrary.wiley.com/doi/abs/10.1002/%28SICI%291439-2054%2820000301%29276%3A1%3C1%3A%3AAID-MAME1%3E3.0.CO%3B2-W

Cellulose Nanoparticles: Volume 2: Synthesis and Manufacturing—Google Books. (n.d.). Retrieved September 6, 2021.

Chai, J., Liu, Z., Ma, J., Wang, J., Liu, X., Liu, H., Zhang, J., Cui, G., & Chen, L. (2017). In situ generation of poly (vinylene carbonate) based solid electrolyte with interfacial stability for LiCoO2 lithium batteries. *Advanced Science, 4*(2). https://doi.org/10.1002/ADVS.201600377

Chen, C., Zhang, Y., Li, Y., Dai, J., Song, J., Yao, Y., Gong, Y., Kierzewski, I., Xie, J., & Hu, L. (2017). All-wood, low tortuosity, aqueous, biodegradable supercapacitors with ultra-high capacitance. *Energy & Environmental Science, 10*(2), 538–545. https://doi.org/10.1039/C6EE03716J

Chen, C., Zhang, Y., Li, Y., Kuang, Y., Song, J., Luo, W., Wang, Y., Yao, Y., Pastel, G., Xie, J., & Hu, L. (2017). Highly conductive, lightweight, low-tortuosity carbon frameworks as ultrathick 3D current collectors. *Advanced Energy Materials, 7*(17), 1700595. https://doi.org/10.1002/AENM.201700595

Chen, L.-F., Huang, Z.-H., Liang, H.-W., Gao, H.-L., & Yu, S.-H. (2014). Three-dimensional heteroatom-doped carbon nanofiber networks derived from bacterial cellulose for supercapacitors. *Advanced Functional Materials, 24*(32), 5104–5111. https://doi.org/10.1002/ADFM.201400590

Chen, L.-F., Huang, Z.-H., Liang, H.-W., Yao, W.-T., Yu, Z.-Y., & Yu, S.-H. (2013). Flexible all-solid-state high-power supercapacitor fabricated with nitrogen-doped carbon nanofiber electrode material derived from bacterial cellulose. *Energy & Environmental Science, 6*(11), 3331–3338. https://doi.org/10.1039/C3EE42366B

Chen, W., Yu, H., Lee, S.-Y., Wei, T., Li, J., & Fan, Z. (2018). Nanocellulose: A promising nanomaterial for advanced electrochemical energy storage. *Chemical Society Reviews, 47*(8), 2837–2872. https://doi.org/10.1039/C7CS00790F

Chen, Z., Christensen, L., & Dahn, J. R. (2003). Comparison of PVDF and PVDF-TFE-P as binders for electrode materials showing large volume changes in lithium-ion batteries. *Journal of The Electrochemical Society, 150*(8), A1073. https://doi.org/10.1149/1.1586922

Cheng, D., Wen, Y., An, X., Zhu, X., & Ni, Y. (2016). TEMPO-oxidized cellulose nanofibers (TOCNs) as a green reinforcement for waterborne polyurethane coating (WPU) on wood. *Carbohydrate Polymers, 151*, 326–334. https://doi.org/10.1016/J.CARBPOL.2016.05.083

Chiappone, A., Nair, J. R., Gerbaldi, C., Jabbour, L., Bongiovanni, R., Zeno, E., Beneventi, D., & Penazzi, N. (2011). Microfibrillated cellulose as reinforcement for Li-ion battery polymer electrolytes with excellent mechanical stability. *Journal of Power Sources, 23*(196), 10280–10288. https://doi.org/10.1016/J.JPOWSOUR.2011.07.015

Choi, K.-H., Yoo, J., Lee, C. K., & Lee, S.-Y. (2016). All-inkjet-printed, solid-state flexible supercapacitors on paper. *Energy & Environmental Science, 9*(9), 2812–2821. https://doi.org/10.1039/C6EE00966B

Dong, L., Liang, G., Xu, C., Ren, D., Wang, J., Pan, Z.-Z., Li, B., Kang, F., & Yang, Q.-H. (2017). Stacking up layers of polyaniline/carbon nanotube networks inside papers as highly flexible electrodes with large areal capacitance and superior rate capability. *Journal of Materials Chemistry A, 5*(37), 19934–19942. https://doi.org/10.1039/C7TA06135H

Du, X., Zhang, Z., Liu, W., & Deng, Y. (2017). Nanocellulose-based conductive materials and their emerging applications in energy devices—A review. *Nano Energy, 35*, 299–320. https://doi.org/10.1016/J.NANOEN.2017.04.001

Dubal, D. P., Chodankar, N. R., Kim, D.-H., & Gomez-Romero, P. (2018). Towards flexible solid-state supercapacitors for smart and wearable electronics. *Chemical Society Reviews, 47*(6), 2065–2129. https://doi.org/10.1039/C7CS00505A

Electrochimica Acta | Vol 211, Pages 1–1092 (1 September 2016) | ScienceDirect.com by Elsevier. (n.d.). Retrieved September 4, 2021, from www.sciencedirect.com/journal/electrochimica-acta/vol/211/suppl/C

Feng, E., Peng, H., Zhang, Z., Li, J., & Lei, Z. (2017). Polyaniline-based carbon nanospheres and redox mediator doped robust gel films lead to high performance foldable solid-state supercapacitors. *New Journal of Chemistry, 41*(17), 9024–9032. https://doi.org/10.1039/C7NJ01478C

Feng, Y., Zhang, X., Shen, Y., Yoshino, K., & Feng, W. (2012). A mechanically strong, flexible and conductive film based on bacterial cellulose/graphene nanocomposite. *Carbohydrate Polymers, 1*(87), 644–649. https://doi.org/10.1016/J.CARBPOL.2011.08.039

Fukui, S., Ito, T., Saito, T., Noguchi, T., & Isogai, A. (2018). Counterion design of TEMPO-nanocellulose used as filler to improve properties of hydrogenated acrylonitrile-butadiene matrix. *Composites Science and Technology, 167*, 339–345. https://doi.org/10.1016/J.COMPSCITECH.2018.08.023

Gao, K., Shao, Z., Wang, X., Zhang, Y., Wang, W., & Wang, F. (2013). Cellulose nanofibers/multi-walled carbon nanotube nanohybrid aerogel for all-solid-state flexible supercapacitors. *RSC Advances*, *3*(35), 15058–15064. https://doi.org/10.1039/C3RA42050G

Gui, Z., Zhu, H., Gillette, E., Han, X., Rubloff, G. W., Hu, L., & Lee, S. B. (2013). Natural Cellulose Fiber as Substrate for Supercapacitor. *ACS Nano*, *7*(7), 6037–6046. https://doi.org/10.1021/NN401818T

H, Y., T, S., A, I., H, K., & M, N. (2015). Chemical modification of cellulose nanofibers for the production of highly thermal resistant and optically transparent nanopaper for paper devices. *ACS Applied Materials & Interfaces*, *7*(39), 22012–22017. https://doi.org/10.1021/ACSAMI.5B06915

Habibi, Y. (2014). Key advances in the chemical modification of nanocelluloses. *Chemical Society Reviews*, *43*(5), 1519–1542. https://doi.org/10.1039/C3CS60204D

Hamedi, M., Karabulut, E., Marais, A., Herland, A., Nyström, G., & Wågberg, L. (2013). Nanocellulose aerogels functionalized by rapid layer-by-layer assembly for high charge storage and beyond. *Angewandte Chemie International Edition*, *52*(46), 12038–12042. https://doi.org/10.1002/ANIE.201305137

Han, H., Chen, X., Qian, J., Zhong, F., Feng, X., Chen, W., Ai, X., Yang, H., & Cao, Y. (2019). Hollow carbon nanofibers as high-performance anode materials for sodium-ion batteries. *Nanoscale, 11*(45), 21999–22005. https://doi.org/10.1039/C9NR07675A

HH, H., & W, Z. (2019). Nanocellulose-based conductive membranes for free-standing supercapacitors: A review. *Membranes*, *9*(6). https://doi.org/10.3390/MEMBRANES9060074

Highly conductive and stretchable conductors fabricated from bacterial cellulose. NPG Asia Mater. 2012, 4, e19.—Google Search. (n.d.). Retrieved September 4, 2021.

Hofmann, H. E., & Reid, E. W. (1929). Cellulose acetate lacquers. *Industrial and Engineering Chemistry*, *21*(10), 955–965. https://doi.org/10.1021/IE50238A017

Hsieh, M.-C., Kim, C., Nogi, M., & Suganuma, K. (2013). Electrically conductive lines on cellulose nanopaper for flexible electrical devices. *Nanoscale*, *5*(19), 9289–9295. https://doi.org/10.1039/C3NR01951A

Hsu, H. H., Khosrozadeh, A., Li, B., Luo, G., Xing, M., & Zhong, W. (2019). An eco-friendly, nanocellulose/RGO/in situ formed polyaniline for flexible and free-standing supercapacitors. *ACS Sustainable Chemistry and Engineering*, *7*(5), 4766–4776. https://doi.org/10.1021/ACSSUSCHEMENG.8B04947

Hu, L., Zheng, G., Yao, J., Liu, N., Weil, B., Eskilsson, M., Karabulut, E., Ruan, Z., Fan, S., Bloking, J. T., McGehee, M. D., Wågberg, L., & Cui, Y. (2013). Transparent and conductive paper from nanocellulose fibers. *Energy & Environmental Science*, *6*(2), 513–518. https://doi.org/10.1039/C2EE23635D

Huang, J., Zhu, H., Chen, Y., Preston, C., Rohrbach, K., Cumings, J., & Hu, L. (2013). Highly transparent and flexible nanopaper transistors. *ACS Nano*, *7*(3), 2106–2113. https://doi.org/10.1021/NN304407R

Huang, P., Wu, M., Kuga, S., Wang, D., Wu, D., & Huang, Y. (2012). One-step dispersion of cellulose nanofibers by mechanochemical esterification in an organic solvent. *ChemSusChem*, *5*(12), 2319–2322. https://doi.org/10.1002/CSSC.201200492

J, C., S, P., J, C., M, P., & J, H. (2012). Amphiphilic comb-like polymer for harvest of conductive nanocellulose. *Colloids and Surfaces. B, Biointerfaces*, *89*(1), 161–166. https://doi.org/10.1016/J.COLSURFB.2011.09.008

J, G., SC, F., I, M., & A, T. (2012). Conductive photoswitchable vanadium oxide nanopaper based on bacterial cellulose. *ChemSusChem*, *5*(12), 2323–2327. https://doi.org/10.1002/CSSC.201200516

J, S., C, C., S, Z., M, Z., J, D., U, R., Y, L., Y, K., Y, L., N, Q., Y, Y., A, G., UH, L., HA, B., JY, Z., A, V., H, L., ML, M., Z, J., . . . L, H. (2018). Processing bulk natural wood into a high-performance structural material. *Nature*, *554*(7691), 224–228. https://doi.org/10.1038/NATURE25476

J, X., P, T., Z, W., C, X., X, L., & S, N. (2019). Nanocellulose-graphene composites: A promising nanomaterial for flexible supercapacitors. *Carbohydrate Polymers*, *207*, 447–459. https://doi.org/10.1016/J.CARBPOL.2018.12.010

JA, K., M, G., KE, S., WY, H., & MJ, M. (2014). The development of chiral nematic mesoporous materials. *Accounts of Chemical Research*, *47*(4), 1088–1096. https://doi.org/10.1021/AR400243M

Jabbour, L., Gerbaldi, C., Chaussy, D., Zeno, E., Bodoardo, S., & Beneventi, D. (2010). Microfibrillated cellulose—graphite nanocomposites for highly flexible paper-like Li-ion battery electrodes. *Journal of Materials Chemistry*, *20*(35), 7344–7347. https://doi.org/10.1039/C0JM01219J

Jiang, Q., Kacica, C., Soundappan, T., Liu, K., Tadepalli, S., Biswas, P., & Singamaneni, S. (2017). An in situ grown bacterial nanocellulose/graphene oxide composite for flexible supercapacitors. *Journal of Materials Chemistry A*, *5*(27), 13976–13982. https://doi.org/10.1039/C7TA03824K

Jiang, Y., Zhao, Y., Feng, X., Fang, J., & Shi, L. (2016). TEMPO-mediated oxidized nanocellulose incorporating with its derivatives of carbon dots for luminescent hybrid films. *RSC Advances*, *6*(8), 6504–6510. https://doi.org/10.1039/C5RA17242J

Journal of Power Sources | Vol 307, Pages 1–906 (1 March 2016) | ScienceDirect.com by Elsevier. (n.d.). Retrieved September 4, 2021, from www.sciencedirect.com/journal/journal-of-power-sources/vol/307/suppl/C

K, G. N., A, D., A, G., & MN, B. (2003). Crab shell chitin whiskers reinforced natural rubber nanocomposites. 3. Effect of chemical modification of chitin whiskers. *Biomacromolecules*, *4*(6), 1835–1842. https://doi.org/10.1021/BM030058G

K, G., Z, S., X, W., X, W., J, L., Y, Z., W, W., & F, W. (2013). Cellulose nanofibers/reduced graphene oxide flexible transparent conductive paper. *Carbohydrate Polymers*, *97*(1), 243–251. https://doi.org/10.1016/J.CARBPOL.2013.03.067

K, L., J, N., L, C., L, H., & Y, N. (2015). Preparation of CNC-dispersed Fe3O4 nanoparticles and their application in conductive paper. *Carbohydrate Polymers*, *126*, 175–178. https://doi.org/10.1016/J.CARBPOL.2015.03.009

Kalia, S., Dufresne, A., Cherian, B. M., Kaith, B. S., Avérous, L., Njuguna, J., & Nassiopoulos, E. (2011). Cellulose-based bio- and nanocomposites: A review. *International Journal of Polymer Science*, *2011*. https://doi.org/10.1155/2011/837875

KH, C., SJ, C., SJ, C., JT, Y., CK, L., W, K., Q, W., SB, P., DH, C., SY, L., & SY, L. (2014). Heterolayered, one-dimensional nanobuilding block mat batteries. *Nano Letters*, *14*(10), 5677–5686. https://doi.org/10.1021/NL5024029

Khosrozadeh, A., Darabi, M. A., Xing, M., & Wang, Q. (2015). Flexible cellulose-based films of polyaniline-graphene-silver nanowire for high-performance supercapacitors. *Journal of Nanotechnology in Engineering and Medicine*, *6*(1). https://doi.org/10.1115/1.4031385

Khosrozadeh, A., Darabi, M. A., Xing, M., & Wang, Q. (2016). *Flexible electrode design: Fabrication of freestanding polyaniline-based composite films for high-performance supercapacitors*. https://doi.org/10.1021/ACSAMI.5B11256

Khosrozadeh, A., Xing, M., & Wang, Q. (2015). A high-capacitance solid-state supercapacitor based on free-standing film of polyaniline and carbon particles. *Applied Energy*, *153*, 87–93. https://doi.org/10.1016/J.APENERGY.2014.08.046

Kumar, R.N.; Pizzi, A. Fundamentals of Adhesion. In Adhesives for Wood and Lignocellulosic Materials; John Wiley & Sons: Hoboken, NJ, USA, 2019; pp. 31–60

L, H., JW, C., Y, Y., S, J., F, L. M., LF, C., & Y, C. (2009). Highly conductive paper for energy-storage devices. *Proceedings of the National Academy of Sciences of the United States of America*, *106*(51), 21490–21494. https://doi.org/10.1073/PNAS.0908858106

L, H., M, P., FL, M., L, C., S, J., HD, D., JW, C., SM, H., & Y, C. (2010). Stretchable, porous, and conductive energy textiles. *Nano Letters*, *10*(2), 708–714. https://doi.org/10.1021/NL903949M

Leijonmarck, S., Cornell, A., Lindbergh, G., & Wågberg, L. (2013). Single-paper flexible Li-ion battery cells through a paper-making process based on nano-fibrillated cellulose. *Journal of Materials Chemistry A*, *1*(15), 4671–4677. https://doi.org/10.1039/C3TA01532G

LF, C., ZH, H., HW, L., QF, G., & SH, Y. (2013). Bacterial-cellulose-derived carbon nanofiber@MnO_2 and nitrogen-doped carbon nanofiber electrode materials: An asymmetric supercapacitor with high energy and power density. *Advanced Materials (Deerfield Beach, Fla.)*, *25*(34), 4746–4752. https://doi.org/10.1002/ADMA.201204949

Li, F., Jiang, X., Zhao, J., & Zhang, S. (2015). Graphene oxide: A promising nanomaterial for energy and environmental applications. *Nano Energy*, *16*, 488–515. https://doi.org/10.1016/J.NANOEN.2015.07.014

Li, S., Huang, D., Yang, J., Zhang, B., Zhang, X., Yang, G., Wang, M., & Shen, Y. (2014). Freestanding bacterial cellulose-polypyrrole nanofibres paper electrodes for advanced energy storage devices. *Nano Energy*, *9*, 309–317. https://doi.org/10.1016/J.NANOEN.2014.08.004

Li, S., Huang, D., Zhang, B., Xu, X., Wang, M., Yang, G., & Shen, Y. (2014). Flexible supercapacitors based on bacterial cellulose paper electrodes. *Advanced Energy Materials*, *4*(10), 1301655. https://doi.org/10.1002/AENM.201301655

Li, Y., Zhu, H., Shen, F., Wan, J., Han, X., Dai, J., Dai, H., & Hu, L. (2014). Highly conductive microfiber of graphene oxide templated carbonization of nanofibrillated cellulose. *Advanced Functional Materials*, *24*(46), 7366–7372. https://doi.org/10.1002/ADFM.201402129

Li, Z., Ahadi, K., Jiang, K., Ahvazi, B., Li, P., Anyia, A. O., Cadien, K., & Thundat, T. (2017). Freestanding hierarchical porous carbon film derived from hybrid nanocellulose for high-power supercapacitors. *Nano Research 2017 10:5*, *10*(5), 1847–1860. https://doi.org/10.1007/S12274-017-1573-8

Liang, H.-W., Guan, Q.-F., Zhu, Z.-, Song, L.-T., Yao, H.-B., Lei, X., & Yu, S.-H. (2012). Highly conductive and stretchable conductors fabricated from bacterial cellulose. *NPG Asia Materials 2012 4:6*, *4*(6), e19–e19. https://doi.org/10.1038/am.2012.34

Liew, S. Y., Thielemans, W., & Walsh, D. A. (2010). Electrochemical capacitance of nanocomposite polypyrrole/cellulose films. *Journal of Physical Chemistry C*, *114*(41), 17926–17933. https://doi.org/10.1021/JP103698P

Ling, S., Chen, W., Fan, Y., Zheng, K., Jin, K., Yu, H., Buehler, M. J., & Kaplan, D. L. (2018). Biopolymer nanofibrils: Structure, modeling, preparation, and applications. *Progress in Polymer Science*, *85*, 1. https://doi.org/10.1016/J.PROGPOLYMSCI.2018.06.004

Liu, F., Luo, S., Liu, D., Chen, W., Huang, Y., Dong, L., & Wang, L. (2017). Facile processing of free-standing polyaniline/SWCNT film as an integrated electrode for flexible supercapacitor application. *ACS Applied Materials and Interfaces*, *9*(39), 33791–33801. https://doi.org/10.1021/ACSAMI.7B08382

Liu, K.-K., Jiang, Q., Kacica, C., Derami, H. G., Biswas, P., & Singamaneni, S. (2018). Flexible solid-state supercapacitor based on tin oxide/reduced graphene oxide/bacterial nanocellulose. *RSC Advances*, *8*(55), 31296–31302. https://doi.org/10.1039/C8RA05270K

Liu, Q., Jing, S., Wang, S., Zhuo, H., Zhong, L., Peng, X., & Sun, R. (2016). Flexible nanocomposites with ultrahigh specific areal capacitance and tunable properties based on a cellulose derived nanofiber-carbon sheet framework coated with polyaniline. *Journal of Materials Chemistry A*, *4*(34), 13352–13362. https://doi.org/10.1039/C6TA05131F

Liu, R., Ma, L., Niu, G., Li, X., Li, E., Bai, Y., & Yuan, G. (2017). Oxygen-deficient bismuth oxide/graphene of ultrahigh capacitance as advanced flexible anode for asymmetric supercapacitors. *Advanced Functional Materials*, *27*(29), 1701635. https://doi.org/10.1002/ADFM.201701635

Liu, Y., Zhou, J., Tang, J., & Tang, W. (2015). Three-dimensional, chemically bonded polypyrrole/bacterial cellulose/graphene composites for high-performance supercapacitors. *Chemistry of Materials*, *27*(20), 7034–7041. https://doi.org/10.1021/ACS.CHEMMATER.5B03060

Liu, Y., Zhou, J., Zhu, E., Tang, J., Liu, X., & Tang, W. (2015). Facile synthesis of bacterial cellulose fibres covalently intercalated with graphene oxide by one-step cross-linking for robust supercapacitors. *Journal of Materials Chemistry C*, *3*(5), 1011–1017. https://doi.org/10.1039/C4TC01822B

Long, C., Jiang, L., Wei, T., Yan, J., & Fan, Z. (2014). High-performance asymmetric supercapacitors with lithium intercalation reaction using metal oxide-based composites as electrode materials. *Journal of Materials Chemistry A*, *2*(39), 16678–16686. https://doi.org/10.1039/C4TA03241A

Long, C., Qi, D., Wei, T., Yan, J., Jiang, L., & Fan, Z. (2014). Nitrogen-doped carbon networks for high energy density supercapacitors derived from polyaniline coated bacterial cellulose. *Advanced Functional Materials*, *24*(25), 3953–3961. https://doi.org/10.1002/ADFM.201304269

M, G., LK, B., MK, K., & MJ, M. (2015). Functional materials from cellulose-derived liquid-crystal templates. *Angewandte Chemie (International Ed. in English)*, *54*(10), 2888–2910. https://doi.org/10.1002/ANIE.201407141

M, S., S, T., A, R., N, N., & M, D. (2020). Preparation and characterization of hydrogel nanocomposite based on nanocellulose and acrylic acid in the presence of urea. *International Journal of Biological Macromolecules*, *147*, 187–193. https://doi.org/10.1016/J.IJBIOMAC.2020.01.038

Ma, X., Li, Y., Wen, Z., Gao, F., Liang, C., & Che, R. (2014). Ultrathin β-Ni(OH)2 nanoplates vertically grown on nickel-coated carbon nanotubes as high-performance pseudocapacitor electrode materials. *ACS Applied Materials and Interfaces*, *7*(1), 974–979. https://doi.org/10.1021/AM5077183

Manthiram, A., Fu, Y., Chung, S.-H., Zu, C., & Su, Y.-S. (2014). Rechargeable lithium—sulfur batteries. *Chemical Reviews*, *114*(23), 11751–11787. https://doi.org/10.1021/CR500062V

Miao, C., & Hamad, W. Y. (2013). Cellulose reinforced polymer composites and nanocomposites: A critical review. *Cellulose*, *20*(5), 2221–2262. https://doi.org/10.1007/S10570-013-0007-3

Mihranyan, A., Nyholm, L., Bennett, A. E. G., & Strømme, M. (2008). A novel high specific surface area conducting paper material composed of polypyrrole and cladophora cellulose. *Journal of Physical Chemistry B*, *112*(39), 12249–12255. https://doi.org/10.1021/JP805123W

Miloh, T., Spivak, B., & Yarin, A. L. (2009). Needleless electrospinning: Electrically driven instability and multiple jetting from the free surface of a spherical liquid layer. *Journal of Applied Physics*, *106*(11), 114910. https://doi.org/10.1063/1.3264884

MM, H., A, H., AB, F., K, H., M, S., F, L., L, W., & LA, B. (2014). Highly conducting, strong nanocomposites based on nanocellulose-assisted aqueous dispersions of single-wall carbon nanotubes. *ACS Nano*, *8*(3), 2467–2476. https://doi.org/10.1021/NN4060368

My Ahmed Saïd Azizi Samir, †,‡, Fannie Alloin, ‡, Jean-Yves Sanchez, ‡ & Alain Dufresne*, §. (2004). Crosslinked nanocomposite polymer electrolytes reinforced with cellulose whiskers. *Macromolecules*, *37*(13), 4839–4844. https://doi.org/10.1021/MA049504Y

My Ahmed Saïd Azizi Samir, †,‡, Fannie Alloin, ‡, Jean-Yves Sanchez, ‡, Nadia El Kissi, § & Alain Dufresne*, ⊥. (2004). Preparation of cellulose whiskers reinforced nanocomposites from an organic medium suspension. *Macromolecules*, *37*(4), 1386–1393. https://doi.org/10.1021/MA030532A

Nano Energy | Vol 13, Pages 1–836 (April 2015) | ScienceDirect.com by Elsevier. (n.d.). Retrieved September 4, 2021, from www.sciencedirect.com/journal/nano-energy/vol/13/suppl/C

Nguyen Dang, L., & Seppälä, J. (2015). Electrically conductive nanocellulose/graphene composites exhibiting improved mechanical properties in high-moisture condition. *Cellulose*, 22(3), 1799–1812. https://doi.org/10.1007/S10570-015-0622-2

Nguyen, H.-L., Jo, Y. K., Cha, M., Cha, Y. J., Yoon, D. K., Sanandiya, N. D., Prajatelistia, E., Oh, D. X., & Hwang, D. S. (2016). Mussel-inspired anisotropic nanocellulose and silver nanoparticle composite with improved mechanical properties, electrical conductivity and antibacterial activity. *Polymers*, 8(3), 102. https://doi.org/10.3390/POLYM8030102

Nie, S., Zhang, Y., Wang, L., Wu, Q., & Wang, S. (2019). Preparation and characterization of nanocomposite films containing nano-aluminum nitride and cellulose nanofibrils. *Nanomaterials*, 9(8), 1121. https://doi.org/10.3390/NANO9081121

Nyholm, L., Nyström, G., Mihranyan, A., & Strømme, M. (2011). Toward flexible polymer and paper-based energy storage devices. *Advanced Materials*, 23(33), 3751–3769. https://doi.org/10.1002/ADMA.201004134

Nyström, G., Marais, A., Karabulut, E., Wågberg, L., Cui, Y., & Hamedi, M. M. (2015). Self-assembled three-dimensional and compressible interdigitated thin-film supercapacitors and batteries. *Nature Communications*, 6(1), 1–8. https://doi.org/10.1038/ncomms8259

Nyström, G., Razaq, A., Strømme, M., Nyholm, L., & Mihranyan, A. (2009). Ultrafast all-polymer paper-based batteries. *Nano Letters*, 9(10), 3635–3639. https://doi.org/10.1021/NL901852H

Nyström, G., Strømme, M., Sjödin, M., & Nyholm, L. (2012). Rapid potential step charging of paper-based polypyrrole energy storage devices. *Electrochimica Acta*, 70, 91–97. https://doi.org/10.1016/J.ELECTACTA.2012.03.060

Pang, Q., Liang, X., Kwok, C. Y., & Nazar, L. F. (2016). Advances in lithium—sulfur batteries based on multifunctional cathodes and electrolytes. *Nature Energy*, 1(9), 1–11. https://doi.org/10.1038/nenergy.2016.132

Peng, S., Fan, L., Rao, W., Bai, Z., Xu, W., & Xu, J. (2017). Bacterial cellulose membranes coated by polypyrrole/copper oxide as flexible supercapacitor electrodes. *Journal of Materials Science*, 52(4), 1930–1942. https://doi.org/10.1007/S10853-016-0482-7

Postek, M. T., Moon, R. J., Rudie, A. W., & Bilodeau, M. A. (2013). *Production and applications of cellulose nanomaterials* (Peachtree Corners: TAPPI Press), 9–12.

Pras, O., Beneventi, D., Chaussy, D., Piette, P., & Tapin-Lingua, S. (2013). Use of microfibrillated cellulose and dendritic copper for the elaboration of conductive films from water- and ethanol-based dispersions. *Journal of Materials Science*, 48(20), 6911–6920. https://doi.org/10.1007/S10853-013-7496-1

R, T., G, C., J, T., & M, H. (2020). Highly transparent, weakly hydrophilic and biodegradable cellulose film for flexible electroluminescent devices. *Carbohydrate Polymers*, 227. https://doi.org/10.1016/J.CARBPOL.2019.115366

Raghuwanshi, V. S., & Garnier, G. (2019). Cellulose nano-films as bio-interfaces. *Frontiers in Chemistry*, 535. https://doi.org/10.3389/FCHEM.2019.00535

Sapkota, J., Jorfi, M., Weder, C., & Foster, E. J. (2014). Reinforcing poly(ethylene) with cellulose nanocrystals. *Macromolecular Rapid Communications*, 35(20), 1747–1753. https://doi.org/10.1002/MARC.201400382

Sethi, J., Farooq, M., Sain, S., Sain, M., Sirviö, J. A., Illikainen, M., & Oksman, K. (2017). Water resistant nanopapers prepared by lactic acid modified cellulose nanofibers. *Cellulose*, 25(1), 259–268. https://doi.org/10.1007/S10570-017-1540-2

SH, Y., HJ, J., MC, K., & YR, P. (2006). Electrically conductive bacterial cellulose by incorporation of carbon nanotubes. *Biomacromolecules*, 7(4), 1280–1284. https://doi.org/10.1021/BM050597G

Shan, D., Yang, J., Liu, W., Yan, J., & Fan, Z. (2016). Biomass-derived three-dimensional honeycomb-like hierarchical structured carbon for ultrahigh energy density asymmetric supercapacitors. *Journal of Materials Chemistry A*, 4(35), 13589–13602. https://doi.org/10.1039/C6TA05406D

Sharma, S., & Deng, Y. (2016). Dual mechanism of dry strength improvement of cellulose nanofibril films by polyamide-epichlorohydrin resin cross-linking. *Industrial and Engineering Chemistry Research*, 55(44), 11467–11474. https://doi.org/10.1021/ACS.IECR.6B02910

Shi, K., Yang, X., Cranston, E. D., & Zhitomirsky, I. (2016). Efficient lightweight supercapacitor with compression stability. *Advanced Functional Materials*, 26(35), 6437–6445. https://doi.org/10.1002/ADFM.201602103

Song, Y., Jiang, Y., Shi, L., Cao, S., Feng, X., Miao, M., & Fang, J. (2015). Solution-processed assembly of ultrathin transparent conductive cellulose nanopaper embedding AgNWs. *Nanoscale*, 7(32), 13694–13701. https://doi.org/10.1039/C5NR03218K

*Studies of Starch Esterification: Reactions with Alkenylsuccinates in Aqueous Slurry Systems—Jeon—1999 —
Starch—Stärke—Wiley Online Library.* (n.d.). Retrieved September 4, 2021, from https://onlinelibrary.
wiley.com/doi/abs/10.1002/%28SICI%291521-379X%28199903%2951%3A2%3C90%3A%3A
AID-STAR90%3E3.0.CO%3B2-M

Takagi, H., Nakagaito, A. N., Nishimura, K., & Matsui, T. (2016). Mechanical characterisation of nanocellulose composites after structural modification. *High Performance and Optimum Design of Structures and
Materials II*, *1*, 335–341. https://doi.org/10.2495/HPSM160311

Tan, C., Peng, J., Lin, W., Xing, Y., Xu, K., Wu, J., & Chen, M. (2015). Role of surface modification and mechanical orientation on property enhancement of cellulose nanocrystals/polymer nanocomposites. *European
Polymer Journal, Complete* (62), 186–197. https://doi.org/10.1016/J.EURPOLYMJ.2014.11.033

Tao, J., Wang, R., Yu, H., Chen, L., Fang, D., Tian, Y., Xie, J., Jia, D., Liu, H., Wang, J., Tang, F., Song, L., &
Li, H. (2020). Highly transparent, highly thermally stable nanocellulose/polymer hybrid substrates for
flexible OLED devices. *ACS Applied Materials & Interfaces*, *12*(8), 9701–9709. https://doi.org/10.1021/
ACSAMI.0C01048

Thakur, V. K. (2014). Nanocellulose polymer nanocomposites: Fundamentals and applications. *9781118871904*,
1–513. https://doi.org/10.1002/9781118872246

Trinh, B. M., & Mekonnen, T. (2018). Hydrophobic esterification of cellulose nanocrystals for epoxy reinforcement. *Polymer*, *155*, 64–74. https://doi.org/10.1016/J.POLYMER.2018.08.076

Vaca-Garcia, C., Thiebaud, S., Borredon, M. E., & Gozzelino, G. (1998). Cellulose esterification with fatty
acids and acetic anhydride in lithium chloride/N,N-dimethylacetamide medium. *Journal of the American
Oil Chemists' Society*, *75*(2), 315–319. https://doi.org/10.1007/S11746-998-0047-2

Vadakkekara, G. J., Thomas, S., & Nair, C. P. R. (2020). Sodium itaconate grafted nanocellulose for facile elimination of lead ion from water. *Cellulose*, *27*(6), 3233–3248. https://doi.org/10.1007/S10570-
020-02983-4

Volodymyr Kuzmenko, Nan Wang, Mazharul Haque, Olga Naboka, Mattias Flygare, Krister Svensson,
Paul Gatenholm, Johan Liu, & Peter Enoksson. (2017). Cellulose-derived carbon nanofibers/graphene
composite electrodes for powerful compact supercapacitors. *RSC Advances*, *7*(73), 45968–45977. https://
doi.org/10.1039/C7RA07533B

W, C., Q, L., Y, W., X, Y., J, Z., H, Y., Y, L., & J, L. (2014). Comparative study of aerogels obtained from
differently prepared nanocellulose fibers. *ChemSusChem*, *7*(1), 154–161. https://doi.org/10.1002/
CSSC.201300950

W, H., S, C., J, Y., Z, L., & H, W. (2014). Functionalized bacterial cellulose derivatives and nanocomposites.
Carbohydrate Polymers, *101*(1), 1043–1060. https://doi.org/10.1016/J.CARBPOL.2013.09.102

Wan, C., Jiao, Y., & Li, J. (2017). Flexible, highly conductive, and free-standing reduced graphene oxide/
polypyrrole/cellulose hybrid papers for supercapacitor electrodes. *Journal of Materials Chemistry A*, *5*(8),
3819–3831. https://doi.org/10.1039/C6TA04844G

Wang, F., Jeon, J.-H., Kim, S.-J., Park, J.-O., & Park, S. (2016). An eco-friendly ultra-high performance ionic
artificial muscle based on poly(2-acrylamido-2-methyl-1-propanesulfonic acid) and carboxylated bacterial cellulose. *Journal of Materials Chemistry B*, *4*(29), 5015–5024. https://doi.org/10.1039/C6TB01084A

Wang, G., Zhang, L., & Zhang, J. (2012). A review of electrode materials for electrochemical supercapacitors.
Chemical Society Reviews, *41*(2), 797–828. https://doi.org/10.1039/C1CS15060J

Wang, H., Bian, L., Zhou, P., Tang, J., & Tang, W. (2012). Core—sheath structured bacterial cellulose/polypyr-
role nanocomposites with excellent conductivity as supercapacitors. *Journal of Materials Chemistry A*,
1(3), 578–584. https://doi.org/10.1039/C2TA00040G

Wang, H., Pei, Y., Qian, X., & An, X. (2020). Eu-metal organic framework@TEMPO-oxidized cellulose nanofibrils photoluminescence film for detecting copper ions. *Carbohydrate Polymers*, *236*, 116030. https://
doi.org/10.1016/J.CARBPOL.2020.116030

Wang, H., Zhu, E., Yang, J., Zhou, P., Sun, D., & Tang, W. (2012). Bacterial cellulose nanofiber-supported polyaniline nanocomposites with flake-shaped morphology as supercapacitor electrodes. *Journal of Physical
Chemistry C*, *116*(24), 13013–13019. https://doi.org/10.1021/JP301099R

Wang, X., Chen, Y., Schmidt, O. G., & Yan, C. (2016). Engineered nanomembranes for smart energy storage
devices. *Chemical Society Reviews*, *45*(5), 1308–1330. https://doi.org/10.1039/C5CS00708A

Wang, Y., Gu, F., Ni, L., Liang, K., Marcus, K., Liu, S., Yang, F., Chen, J., & Feng, Z. (2017). Easily fabricated and lightweight PPy/PDA/AgNW composites for excellent electromagnetic interference shielding.
Nanoscale, *9*(46), 18318–18325. https://doi.org/10.1039/C7NR05951E

Wang, Z., Carlsson, D. O., Tammela, P., Hua, K., Zhang, P., Nyholm, L., & Strømme, M. (2015). Surface
modified nanocellulose fibers yield conducting polymer-based flexible supercapacitors with enhanced
capacitances. *ACS Nano*, *9*(7), 7563–7571. https://doi.org/10.1021/ACSNANO.5B02846

Wang, Z., Tammela, P., Huo, J., Zhang, P., Strømme, M., & Nyholm, L. (2016). Solution-processed poly(3,4-ethylenedioxythiophene) nanocomposite paper electrodes for high-capacitance flexible supercapacitors. *Journal of Materials Chemistry A, 4*(5), 1714–1722. https://doi.org/10.1039/C5TA10122K

Wang, Z., Tammela, P., Strømme, M., & Nyholm, L. (2015). Nanocellulose coupled flexible polypyrrole@graphene oxide composite paper electrodes with high volumetric capacitance. *Nanoscale, 7*(8), 3418–3423. https://doi.org/10.1039/C4NR07251K

Wang, Z., Tammela, P., Strømme, M., & Nyholm, L. (2017). Cellulose-based Supercapacitors: Material and Performance Considerations. *Advanced Energy Materials, 7*(18), 1700130. https://doi.org/10.1002/AENM.201700130

Wang, Z., Tammela, P., Strømme, M., Nyholm, L., Wang, Z., Nyholm, L., Tammela, P., Zhang, P., & Strømme, M. (2015). Nanocellulose coupled flexible polypyrrole@graphene oxide composite paper electrodes with high volumetric capacitance. *Nanoscale, 7*(8), 3418–3423. https://doi.org 10.1039/C4NR07251K

Wegner, T. H., & Jones, P. E. (2006). Advancing cellulose-based nanotechnology. *Cellulose, 13*(2), 115–118. https://doi.org/10.1007/S10570-006-9056-1

Wei, J., Chen, Y., Liu, H., Du, C., Yu, H., Ru, J., & Zhou, Z. (2016). Effect of surface charge content in the TEMPO-oxidized cellulose nanofibers on morphologies and properties of poly(N-isopropylacrylamide)-based composite hydrogels. *Industrial Crops and Products, 92*, 227–235. https://doi.org/10.1016/J.INDCROP.2016.08.006

Wu, B., Zhu, G., Dufresne, A., & Lin, N. (2019). Fluorescent aerogels based on chemical crosslinking between nanocellulose and carbon dots for optical sensor. *ACS Applied Materials & Interfaces, 11*(17), 16048–16058. https://doi.org/10.1021/ACSAMI.9B02754

Wu, X., Shi, Z., Tjandra, R., Cousins, A. J., Sy, S., Yu, A., Berry, R. M., & Tam, K. C. (2015). Nitrogen-enriched porous carbon nanorods templated by cellulose nanocrystals as high performance supercapacitor electrodes. *Journal of Materials Chemistry A, 3*(47), 23768–23777. https://doi.org/10.1039/C5TA07252B

Wu, Z. Y., Hu, B. C., Wu, P., Liang, H. W., Yu, Z. L., Lin, Y., Zheng, Y. R., Li, Z., & Yu, S. H. (2016). Mo2c nanoparticles embedded within bacterial cellulose-derived 3d n-doped carbon nanofiber networks for efficient hydrogen evolution. *NPG Asia Materials, 8*(7). https://doi.org/10.1038/AM.2016.87

Xu, D., Chen, C., Xie, J., Zhang, B., Miao, L., Cai, J., Huang, Y., & Zhang, L. (2016). A hierarchical N/S-codoped carbon anode fabricated facilely from cellulose/polyaniline microspheres for high-performance sodium-ion batteries. *Advanced Energy Materials, 6*(6), 1501929. https://doi.org/10.1002/AENM.201501929

Xu, H., Xie, Y., Zhu, E., Liu, Y., Shi, Z., Xiong, C., & Yang, Q. (2020). Supertough and ultrasensitive flexible electronic skin based on nanocellulose/sulfonated carbon nanotube hydrogel films. *Journal of Materials Chemistry A, 8*(13), 6311–6318. https://doi.org/10.1039/D0TA00158A

Xu, X., Zhou, J., Nagaraju, D. H., Jiang, L., Marinov, V. R., Lubineau, G., Alshareef, H. N., & Oh, M. (2015). Flexible, highly graphitized carbon aerogels based on bacterial cellulose/lignin: Catalyst-free synthesis and its application in energy storage devices. *Advanced Functional Materials, 25*(21), 3193–3202. https://doi.org/10.1002/ADFM.201500538

XW, P., JL, R., & RC, S. (2010). Homogeneous esterification of xylan-rich hemicelluloses with maleic anhydride in ionic liquid. *Biomacromolecules, 11*(12), 3519–3524. https://doi.org/10.1021/BM1010118

Y, H., M, Z., Y, H., Z, P., H, L., Z, W., Q, X., & C, Z. (2016). Multifunctional energy storage and conversion devices. *Advanced Materials (Deerfield Beach, Fla.), 28*(38), 8344–8364. https://doi.org/10.1002/ADMA.201601928

Y, Y., D, Y., H, W., & L, G. (2017). Smart electrochemical energy storage devices with self-protection and self-adaptation abilities. *Advanced Materials (Deerfield Beach, Fla.), 29*(45). https://doi.org/10.1002/ADMA.201703040

Yang, X., Shi, K., Zhitomirsky, I., & Cranston, E. D. (2015). Cellulose nanocrystal aerogels as universal 3d lightweight substrates for supercapacitor materials. *Advanced Materials, 27*(40), 6104–6109. https://doi.org/10.1002/ADMA.201502284

YJ, K., SJ, C., SS, L., BY, K., JH, K., H, C., SY, L., & W, K. (2012). All-solid-state flexible supercapacitors fabricated with bacterial nanocellulose papers, carbon nanotubes, and triblock-copolymer ion gels. *ACS Nano, 6*(7), 6400–6406. https://doi.org/10.1021/NN301971R

Yoo, Y., & Youngblood, J. P. (2016). Green one-pot synthesis of surface hydrophobized cellulose nanocrystals in aqueous medium. *ACS Sustainable Chemistry and Engineering, 4*(7), 3927–3938. https://doi.org/10.1021/ACSSUSCHEMENG.6B00781

Yu, D., Qian, Q., Wei, L., Jiang, W., Goh, K., Wei, J., Zhang, J., & Chen, Y. (2015). Emergence of fiber supercapacitors. *Chemical Society Reviews, 44*(3), 647–662. https://doi.org/10.1039/C4CS00286E

Z, W., S, Z., A, H., S, Z., & J, L. (2018). Mussel-inspired codepositing interconnected polypyrrole nanohybrids onto cellulose nanofiber networks for fabricating flexible conductive biobased composites. *Carbohydrate Polymers*, *205*, 72–82. https://doi.org/10.1016/J.CARBPOL.2018.10.016

Zeng, Z., Wu, T., Han, D., Ren, Q., Siqueira, G., & Nyström, G. (2020). Ultralight, flexible, and biomimetic nanocellulose/silver nanowire aerogels for electromagnetic interference shielding. *ACS Nano*, *14*(3), 2927–2938. https://doi.org/10.1021/ACSNANO.9B07452

Zhang, J., Jiang, G., Goledzinowski, M., Comeau, F. J. E., Li, K., Cumberland, T., Lenos, J., Xu, P., Li, M., Yu, A., & Chen, Z. (2017). Green solid electrolyte with cofunctionalized nanocellulose/graphene oxide interpenetrating network for electrochemical gas sensors. *Small Methods*, *1*(11), 1700237. https://doi.org/10.1002/SMTD.201700237

Zhang, J., Ma, C., Xia, Q., Liu, J., Ding, Z., Xu, M., Chen, L., & Wei, W. (2016). Composite electrolyte membranes incorporating viscous copolymers with cellulose for high performance lithium-ion batteries. *Journal of Membrane Science*, *C*(497), 259–269. https://doi.org/10.1016/J.MEMSCI.2015.09.056

Zhang, K., Chen, G., Li, R., Zhao, K., Shen, J., Tian, J., & He, M. (2020). Facile preparation of highly transparent conducting nanopaper with electrical robustness. *ACS Sustainable Chemistry & Engineering*, *8*(13), 5132–5139. https://doi.org/10.1021/ACSSUSCHEMENG.9B07266

Zhang, K., Ketterle, L., Järvinen, T., Hong, S., & Liimatainen, H. (2020). Conductive hybrid filaments of carbon nanotubes, chitin nanocrystals and cellulose nanofibers formed by interfacial nanoparticle complexation. *Materials & Design*, *191*, 108594. https://doi.org/10.1016/J.MATDES.2020.108594

Zhang, X., Lin, Z., Chen, B., Sharma, S., Wong, C., Zhang, W., & Deng, Y. (2013). Solid-state, flexible, high strength paper-based supercapacitors. *Journal of Materials Chemistry A*, *1*(19), 5835–5839. https://doi.org/10.1039/C3TA10827A

Zhang, X., Lin, Z., Chen, B., Zhang, W., Sharma, S., Gu, W., Deng, Y. J. Power Sources. 2014, 246, 283 — Google Search. (n.d.). Retrieved September 4, 2021, wiz&ved=0ahUKEwiquJ_n7uXyAhV68XMBHcah D9gQ4dUDCA4&uact=5

Zhang, Y., Luo, W., Wang, C., Li, Y., Chen, C., Song, J., Dai, J., Hitz, E. M., Xu, S., Yang, C., Wang, Y., & Hu, L. (2017). High-capacity, low-tortuosity, and channel-guided lithium metal anode. *Proceedings of the National Academy of Sciences of the United States of America*, *114*(14), 3584–3589. https://doi.org/10.1073/PNAS.1618871114

Zhang, Y.-Z., Wang, Y., Cheng, T., Lai, W.-Y., Pang, H., & Huang, W. (2015). Flexible supercapacitors based on paper substrates: A new paradigm for low-cost energy storage. *Chemical Society Reviews*, *44*(15), 5181–5199. https://doi.org/10.1039/C5CS00174A

Zhao, J., Wei, Z., Feng, X., Miao, M., Sun, L., Cao, S., Shi, L., & Fang, J. (2014). Luminescent and transparent nanopaper based on rare-earth up-converting nanoparticle grafted nanofibrillated cellulose derived from garlic skin. *ACS Applied Materials and Interfaces*, *6*(17), 14945–14951. https://doi.org/10.1021/AM5026352

Zheng, Q., Cai, Z., Ma, Z., & Gong, S. (2015). Cellulose nanofibril/reduced graphene oxide/carbon nanotube hybrid aerogels for highly flexible and all-solid-state supercapacitors. *ACS Applied Materials and Interfaces*, *7*(5), 3263–3271. https://doi.org/10.1021/AM507999S

Zheng, W., Lv, R., Na, B., Liu, H., Jin, T., & Yuan, D. (2017). Nanocellulose-mediated hybrid polyaniline electrodes for high performance flexible supercapacitors. *Journal of Materials Chemistry A*, *5*(25), 12969–12976. https://doi.org/10.1039/C7TA01990D

Zhou, T., Chen, D., Jiu, J., Nge, T. T., Sugahara, T., Nagao, S., Koga, H., Nogi, M., Suganuma, K., Wang, X., Liu, X., Cheng, P., Wang, T., & Xiong, D. (2013). Electrically conductive bacterial cellulose composite membranes produced by the incorporation of graphite nanoplatelets in pristine bacterial cellulose membranes. *Express Polymer Letters*, *7*(9), 756–766. https://doi.org/10.3144/EXPRESSPOLYMLETT.2013.73

Zhu, C., Monti, S., & Mathew, A. P. (2020). Evaluation of nanocellulose interaction with water pollutants using nanocellulose colloidal probes and molecular dynamic simulations. *Carbohydrate Polymers*, *229*, 115510. https://doi.org/10.1016/J.CARBPOL.2019.115510

Zhu, L., Wu, L., Sun, Y., Li, M., Xu, J., Bai, Z., Liang, G., Liu, L., Fang, D., & Xu, W. (2014). Cotton fabrics coated with lignosulfonate-doped polypyrrole for flexible supercapacitor electrodes. *RSC Advances*, *4*(12), 6261–6266. https://doi.org/10.1039/C3RA47224H

10 MOF-Based Nanocomposites for Supercapacitor Applications

*Deeksha Nagpal, Astakala Anil Kumar, Ajay Vasishth,
Shashank Priya, Ashok Kumar, and Shyam Sundar Pattnaik*

CONTENTS

10.1 INTRODUCTION

Metal organic frameworks (MOFs) are presently described as intensely crystalline, sponge-like substances comprised of organic linker molecules and metallic ions [1]. The MOF materials are characterized by their enormous surface regions, geometrical adaptability, and remarkable mechanical and thermal stability in addition to their ease of fabrication and good porosity. There has been an increase in interest in the use of MOFs in supercapacitors, lithium ion batteries, and other electrochemical devices in recent decades because of their unique features, such as their stability and long-lasting porosity, high surface area, and the ability to achieve size dependent pore patterns. This is because of the fact that the metallic clusters in MOF frameworks function like redox active sites in the electrodes [2].

MOFs' geometries can be altered during their fabrication or afterward. In the first strategy, MOFs are formed with the addition of appropriate ingredients. Due to the better bonding between the metal component and the organic binder, this approach helps in producing more persistent functional structures [3–4]. When a composite is formed, it can enhance the characteristics of particular components, which can have a beneficial influence on functionality. With this method, it is possible to achieve the necessary conductivity levels while conserving the inherent properties of MOFs [5–7].

Metal—organic frameworks, which are composed of organic molecules as well as metal ions/clusters through a coordination bond, seem to be a form of intriguing functional and porous substance with structural and compositional adjustability along with extremely high surface areas. Since the finding of graphene, several two-dimensional (2D) nanomaterials, like graphic carbon

DOI: 10.1201/9781003174646-10

171

nitride, molybdenum di-sulfide (MoS_2), hexagonal boron nitride, and MXene, have been investigated and reported. In spite of their general features, such as extremely high surface areas and larger interior layer-spaces, these sheet-like 2D materials provide a wide range of performances due to their different internal structures and composition. Recently, 2D MOFs featuring conductive MOFs and nanosheets of MOFs have also been reported as 2D materials and investigated in energy conservation applications [8]. For supercapacitors, MOFs have significant promise because of their redox movement, enormous surface regions, and appropriate porous designs. They can also act as precursors for several kinds of electrode materials, owing to their structural and compositional versatility. The applications of the various materials and their capacitive behavior are reported in the current chapter.

10.2   PRISTINE MOFs

The MOFs displayed pseudocapacitive behavior in general owing to a faradaic redox response. The redox response occurs among electrode and electrolyte, bringing about remarkable specific capacitance values (around 2000 F/g theoretically) for the synthesized electrodes [9–10]. The cyclic durability of pseudocapacitive MOFs, on the other hand, degrades with repeated charging-discharging cycles, because of electrode-electrolyte unsuitability and rare material dissolution throughout the reduction cycle [11–13].

The formulation of MOF structures is particularly important since, it has a direct impact on MOF characteristics and, as a result, their implementations in a variety of associated domains. Despite due to the larger active surface area and inherent redox characteristics, the potential of pristine MOF components as supercapacitor electrodes has received a colossal excitement. The electrochemical behavior of pure MOFs is greatly influenced by their structures and compositions. The electrochemical performance and number of active sites are determined by the composition (particularly the metal core), revealing the pseudocapacitive nature of MOFs. To alter the structure and composition of pure MOFs, the chemical synthesis methods will be described.

Metal ions as well as organic ligands often self-assemble in a particular atmosphere that comprises a compatible solvent and adequate energy supply. Surprisingly, MOFs may also be generated by vapor-phase growth in the absence of any kind of solvent [14–15]. By using a two-step method that involves metal oxide vapor coating followed by a vapor-solid reaction to deposit a thin film using microporous zeolitic imidazolate framework (ZIF)-8 [14]. It has a consistent and regulated thickness and a high aspect ratio.

Several two-dimensional pure MOFs have been prepared, for example, Ni_3(2,3,6,7,10,11-hexaiminotriphenylene)$_2$ ($Ni_3(HITP)_2$) and hexa-amino benzene (HAB)-derived MOFs exhibiting substantial porosity and surface area along with excellent electrical conductivity [16]. Electrical communication involving charge delocalization and metal centers is responsible for the high conductivity in 2D MOFs [17–18].

Zirconium (Zr) has become a preferred metal ion for the production of Zr-based MOF materials due to its affordable availability, lower toxic effects, and higher oxidation state (Zr (IV)). Most renowned Zr-based MOFs, such as MOF-808, MIL-153, UIO-66, UIO-67, and UIO-68, possess 3-D structures. However, with the introduction of DUT-84, 2D Zr-based MOFs have acquired much interest [19]. The DUT-84 was the 1st framework in the Zr-based MOF collection to have a 2D configuration. Two-dimensional MOFs based upon Zr have considerable porosity and surface area, along with convenient affordability to functional groups and metal, enhanced chemical activity, opacity, as well as mechanical elasticity (owing to their thinner layer architectures). With these characteristics, and with the last one, it is possible to create flexible materials that may be adapted in energy storage devices.

When MOFs are doped with Zn, their electrochemical efficiency can be boosted. In continuation, when analyzed under current densities 0.25 and 10 A/g, the Zn containing Ni-MOF shows a specific capacitance of about 1620 and 854 F/g respectively. It is observed that with the current electrode, the specific capacitance was maintained about 92% with the activity of 3000 cycles [11]. Researchers have created a Ni-based MOF with a multilayer structure that can reach capacitance of 1127 and 668 F g^{-1} at the current densities of 0.5 and 10 Ag^{-1}, respectively. However, after 3000 cycles, approximately the capacitivity of about 90% is retentive. The Ni-containing MOF may be responsible for these outstanding electrochemical capabilities [20]. Specific capacitance of 236.1 mA h g^{-1} and 122 mA h g^{-1} were obtained at 1 Ag^{-1} (3M KOH electrolyte) for Ni-Co MOF and Ni-Zn MOF, respectively [21].

Inadequate immersion of non-conductive and conductive phases of the native Fe metal centers was blamed for Fe-insulating MOF's properties. As a consequence, electrons cannot pass through the lattice optimally throughout the redox reaction of the iron core. As a result of the existence of such a little number of redox active iron oxides in the framework architecture, Fe-MOF would not be a durable electrode. Furthermore, several scientists are concerned about using bare Fe-MOF electrodes since some disintegration happened throughout the reduction cycle [13].

A few two-dimensional MOFs with adjustable porosity and substantial specific surface area have recently been described, together with a sufficiently significant electronic conductivity in their pure state. As a result of in-plane charge delocalization and expanded pi-conjugation, the conductivity of 2D MOFs is largely determined by electronic communication between metal nodes [17]. There has also been discussion on the use of HAB-derived 2D MOFs used in the construction of high-performance electrodes [16]. It is still beginning for the research of MOF electrodes for supercapacitors, but the area is developing rapidly. Insufficient electrical conductivity makes it difficult to attain significant specific capacitance and cyclic stability in pristine MOF materials. That is why, researchers have attempted various ways to deal with fitting the electrochemical properties of MOFs while keeping up with the core material properties like pore size distribution and explicit surface area [22–24].

10.3 MOF COMPOSITES USED IN SUPERCAPACITORS

Owing to weak conductivity of the pristine MOFs, several efforts have been made to synthesize novel composite MOFs with desired characteristics. The structure of MOFs can be changed via synthesis processes in situ or through post-processing techniques. A prevalent technique for making MOF composites using carbon involving the combination of ligands, metallic ions and the solvents used along with the carbon/carbon related components, followed by allowing the MOF to develop in situ on to the surfaces of the substrate made using carbon/carbon derivatives at a certain temperature and time. For the combination of MOFs with the other functional materials, an adhesive agent is generally advantageous. Dopamine is a well-known adhesive agent that may be utilized to direct the synthesis of CFs@UiO-66/polypyrrole [25]. Furthermore, tiny molecules may be incorporated straightforwardly into the framework of MOFs during the development phase using solvothermal or hydrothermal processes [26].

In capacitive behavior, under the influence of external potential, the diffusion of ions will take place in the electrolyte from one electrode to another. With the incorporation of the materials such as activated carbon, conductive polymers, metal and their oxides in the MOFs, the enhancement in the conductivity and stability is observed. As a result of electron transmission, and the interfacial impact between two distinct components in the hybrids has a synergistic effect.

10.4 MOF-DERIVED NANOPOROUS CARBONS AS ENERGY STORAGE DEVICE

In an inert environment, MOFs may be subjected to high-temperature pyrolysis to produce nanoporous carbons. While maintaining a high specific surface area, the resultant NPCs can exhibit enhanced electrical conductivity [27]. In contrast to the pseudocapacitive mechanism of MOFs, it has been demonstrated that NPCs, when employed as a supercapacitor electrode, may store electric charge through the electric double-layer capacitor (EDLC) process. When the electrodes are

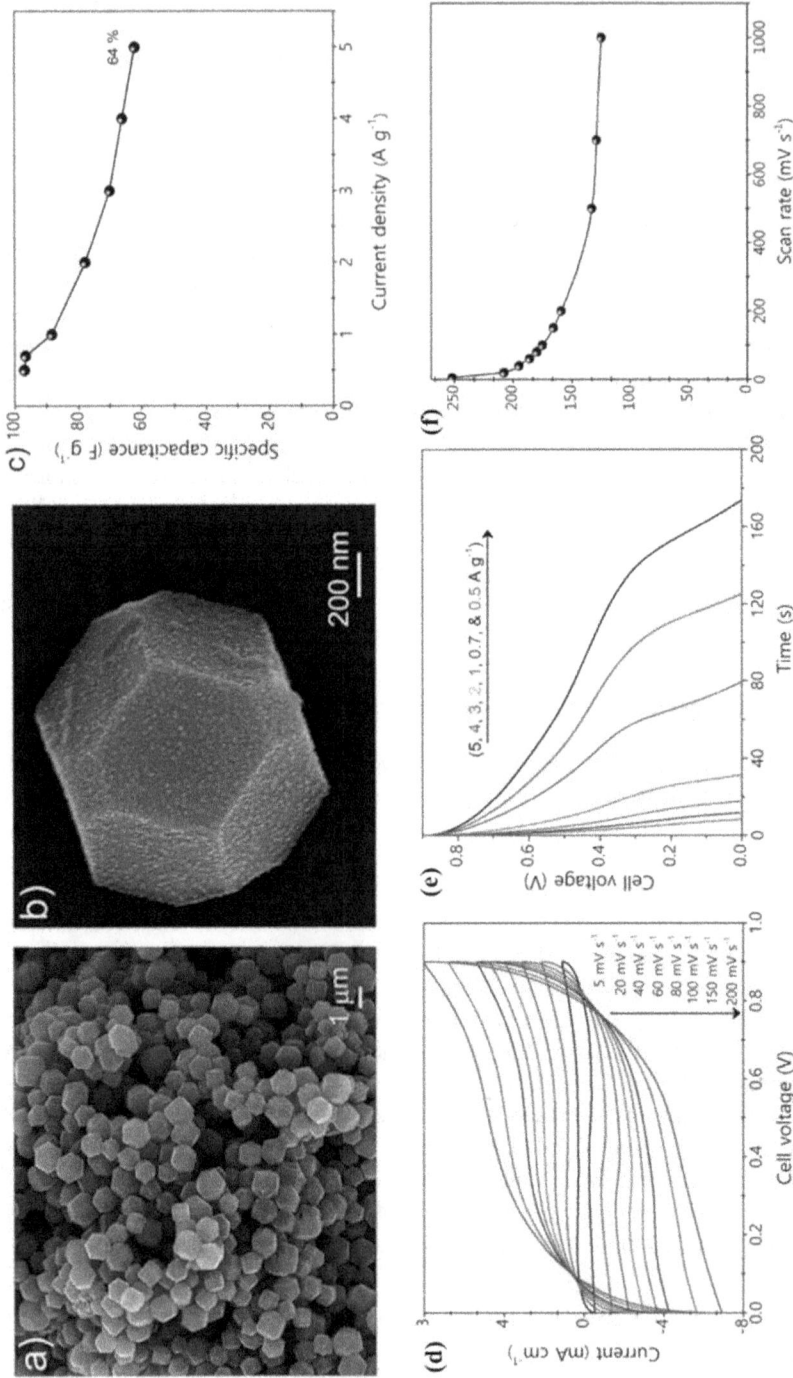

FIGURE 10.1 (a & b) Different magnifications of SEM micrographs of a large-sized NPC, (c) specific capacitance for varied current densities in the region of 0.25 to 5 A/g, (d) CV curves of the large-sized NPC at several scan speeds, (e) the curves of galvanostatic–discharge at varying current densities in the potential region of 0 to 0.9 V and (f) Specific capacitance varies with the scan frequency [35].

equipped with this characteristic, they have superior electrochemical performance and better cyclic ability. Pore size distribution in MOF-derived NPCs is more dynamic due to the formation of different morphologies during pyrolysis. Because of this, MOF-derived NPCs have a greater porosity and higher specific surface area than conventional activated carbons.

The NPCs were first synthesized using MOF-5 by early research organizations using it as a template [28]. A hierarchical pore size distribution may be used to make the NPC highly porous. Chemical activation with KOH can be used to achieve this. An NPC with 182 F/g specific capacitance at quite a 2 mV/s scanning speed was obtained when a KOH activation stage was added simultaneously to the procedure [29]. There were a number of instances when the supercapacitor electrode efficiency of pyrolyzed pure MOF NPCs was inadequate, owing to their poor electrochemical efficacy as supercapacitor electrodes [30]. For this reason, researchers sought to introduce a little amount of nitrogen to MOFs before pyrolyzing them. As a result of N_2 infusion, a more logical and uniform distribution of pores was to be created. Its specific capacitance was 239 Fg^{-1} at 5 mV scan rate when it was utilized to construct an electrode. An aqueous electrolyte containing 6 M KOH and N-doped ZIF-8 produced a specific capacitance of 285.8 Fg^{-1} when the electrode material was converted into an electrode material using 6 M KOH [31].

Another method for improving the electrochemical effectiveness was to activate N-containing ZIF-11 polyhedra in the presence of alkoxy medium (KOH) in order to further increase their electrochemical performance. An electrode made from the NPC produced 307 F/g specific capacitance at 1 A/g current density in the ambiance of 1 M H_2SO_4 electrolytic solution [32]. Supercapacitor electrodes have also been prepared from Mg and Al-derived MOF materials that have been altered to NPC materials. The NPC generated from Al- and Cu-MOFs was also used to create an electrode for the same purpose [33–34]. The fabrication of innovative symmetric supercapacitors based on NPC was made possible by carbonizing Zn-based metal-organic frameworks (MOFs) devoid of the requirement of extra precursor. An electrolyte containing 1 M H_2SO_4 and a 251 Fg^{-1} highest specific capacitance is achieved utilizing NPC materials [35]. In addition, findings of prepared material are shown in Figure 10.1.

Porous carbons are the most often utilized component in the production of electrodes in energy storage devices due to their excellent conductivity, stability, and surface area. The existence of organic linkers within MOF architectures provides an excellent source of carbon for the synthesis of MOF-derived carbons. More intriguingly, depending on the type of MOFs used as precursors, the resultant carbons might have varied textural characteristics [36].

10.5 MOF-DERIVED METAL-OXIDE-BASED SUPERCAPACITORS

Likewise, carbon, metal-oxides have gained interest owing to their pseudocapacitive characteristics in the production of components for supercapacitors. Metal oxides, on the other side, have a limited specific surface area that makes ion transportation challenging across electrodes. As a result, metal oxides with such a greater surface area that are produced utilizing MOFs as templates or even precursors to develop MOF-based metal oxides have been revealed as a viable substituent used in the construction of electrode materials. The most prevalent technique of calcination, in which the MOFs are treated in a furnace in an air environment at several hundred degrees, produces these metal oxides [37–38]. In supercapacitors devices, transition metal oxide (TMO) electrodes are extensively used. As a result of their higher porosity, MOF-derived TMOs have a larger inter-accessible surface area [24].

A supercapacitor using the produced oxide like an electrode material displayed specific capacitance of 208 F/g, when the current density is 1 A/g. After 1000 charge-discharge cycles, this prepared electrode preserved 97% of its own specific capacitance [39]. At 2 mV/s, copper oxides (Cu_2O/CuO) of various compositions were obtained by calcining Cu-MOF (MOF-199) in air. The specific capacitance reached 750 F/g and the electrode's cyclic stability was 94.5% after 3000 cycles of testing [40]. The Cr_2O_3 nanoribbon shape resulted in a faster rate of charge transport. Despite, after 3000 charging-discharging cycles, the electrode maintained remarkable cyclic stability of around 95.5% [41]. To produce Fe_3O_4 the Fe -centered MOF (MIL-88B) was used. The material was utilized to make an electrode with a specific capacitance of 139 F/g with the current density of 0.5 A/g. Additionally, the current electrode is determined to have a specific capacity, which is almost 83.3% after being analyzed for 4000 charged-discharge cycles [42].

Hierarchical ZnO/NiO composites generated from Zn/Ni MOFs were produced. It is widely established that morphology is directly connected to specific surface area as well as pore size distribution. When employed as supercapacitor electrodes, they demonstrated a significant capacitance (435.1 Fg^{-1}) at 1 Ag^{-1} current density and excellent rate capacitance [43]. The Zn/Ni-MOF spheres including a diameter of approximately of 800 nm were used as the starting material to synthesize ZnO-NiO composite material for electrode of supercapacitor. The ZnO was employed as a powerful mechanical support and electron directing route due to its outstanding chemical stability and possess better electrical conductivity. The ZnO-NiO composite electrode had a specific capacitance of 471.1 Fg^{-1} under a current density of 1 Ag^{-1}, and after 1000 cycles, 81.3% capacitance was maintained [44]. Because of the comparatively small activation energy for transfer of electrons between cations, mixed-metal oxides containing multiple metal species have stronger redox reactions and greater electrical conductivity than single-metal oxides, generally resulting to enhanced capacitance. To make a porous carbon Zn—Co MOF and $ZnCo_2O_4$, was used as a starting materials. According to the measurements, the energy density corresponding to the asymmetric device having two electrodes was 28.6 Wh kg^{-1}, while the power density was 100 W kg^{-1} [45].

Hollow structures have acquired considerable interest because of their better mass diffusion and capacity to alleviate volume growth [46–48]. For example, MOF graphene wrapped $NiGa_2O_4$ hollow spheres and yolk-shell $NiFe_2O_4$ hollow spheres, that could be used as positive and negative electrodes in supercapacitors, have a greater energy density of 118.97 Wh kg^{-1}. These have been synthesized via the use of a solvent-thermal reaction accompanied by pyrolysis [49].

MOF-derived metal oxides could be combined with various functional materials such as carbon-carbon, germanium, graphene oxide, cellulose nanofibers, carbon nanotubes, non-ferrous, and metal oxides to create a synergistic impact amongst the various components. The unique $CoFe_2O_4$ nanorods (produced from a Co-Fe MOF)/MXene nanosheets were used as the electrode material in supercapacitor. MXene coating on the electrode material not only serves like a binder and also acts like a conductive component, but it also adds to the composite's high flexibility, allowing ion movement and charge transfer.

10.6 MOF/CARBON NANOCOMPOSITE MATERIALS USED AS ELECTRODE MATERIAL IN SUPERCAPACITORS

10.6.1 MOF/Carbon

To enhance electron transport and ionic diffusion within MOF composites, carbon materials may be used to remove MOF aggregation, and can sometimes function like a current collector for MOF systems. The Co_3O_4/C composite was made by annealing in the atmosphere and the nanowire arrays of Co containing MOF were developed. Its areal capacitance at 1 mA/cm^2 was 1.32 F/cm^2 because of the porous nanowire framework and the presence of carbon. The capacitance retained was 78.3% even after 5000 cycles [50]. The Ni-ZIF8 is transformed into a nanosheet of N-doped carbon/NiO, which possess a specific capacitance of 449 Fg^{-1} at 5 Ag^{-1} due to its interconnected nanosheet shape, sufficient active sites and electric double-layer capacitors from N-doped mesoporous carbon. The 3000 cycles with the carbon/NiO nanosheet provide a specific capacitance retention of 92.2% (414 Fg^{-1} at 5 Ag^{-1}) [51].

Supercapacitor electrode material featuring wide specific surface region and considerable nitrogen doping level have been reported as g-C_3N_4 coated MOF-derived nanocarbon materials

(PMGCN). With a specific capacity of 106 F g^{-1} at a current density of one amp per square inch, the PMGCN-based supercapacitor has an excellent rate capability at 10 amps per square inch. Capacitance retention of 91% following 10,000 cycles at 1 A g^{-1} below 0.8 V shows strong air-working stability [52]. Supercapacitor fabricated from this has excellent energy storage capabilities due to its enhanced ion accommodation and electrical characteristics as shown in Figure 10.2.

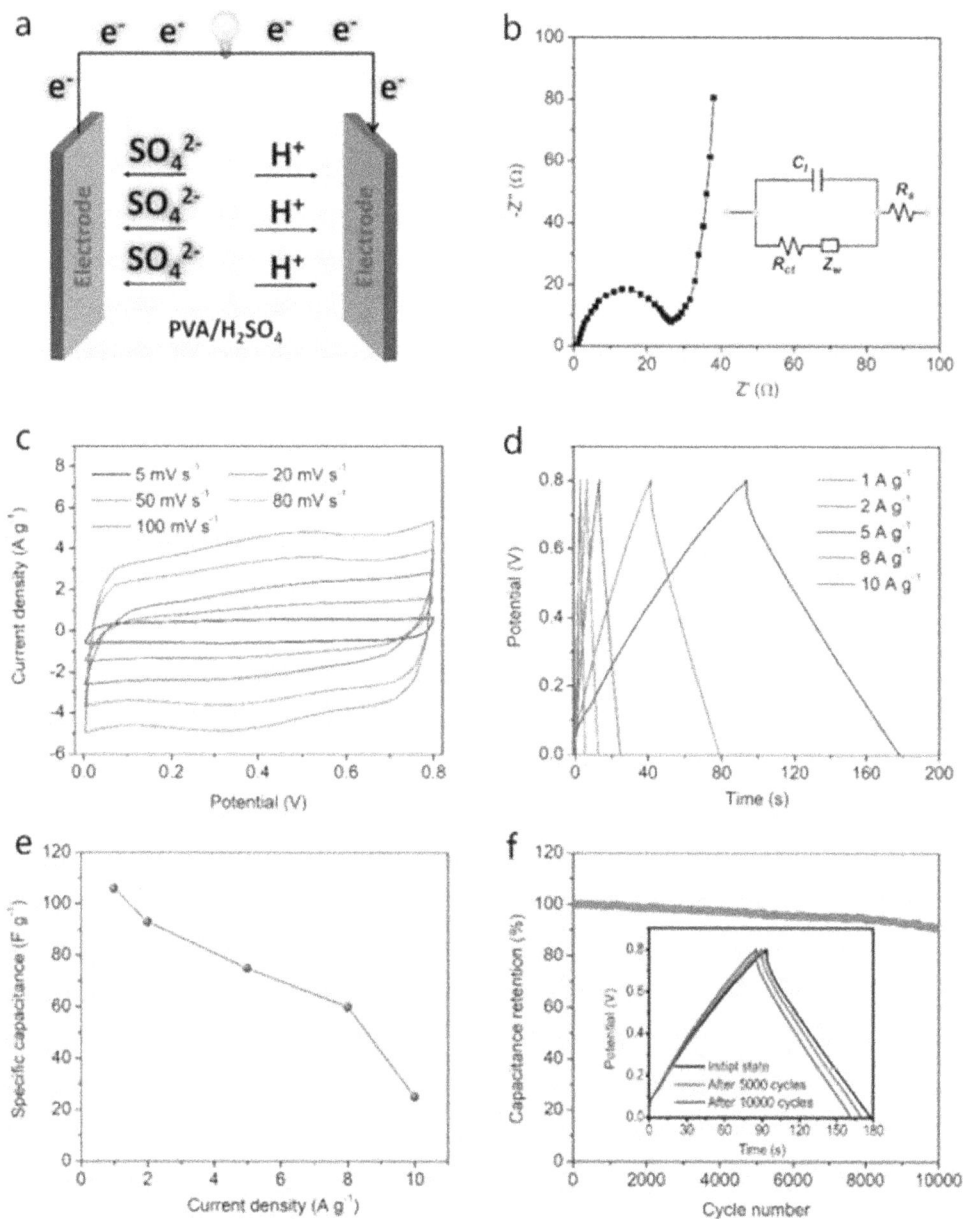

FIGURE 10.2 Supercapacitor with PMGCN-based electrochemical functionality. (a) supercapacitor archi-tecture, (b) Electrochemical impendence spectroscopy plot represent analogous circuit diagram, (c) Cyclic voltammograms curve at different scan rates, (d) Charge–discharge curves for varying current densities, (e) Specific capacitances at distinct current densities, and (f) Operating stability at 1 A g-1 current density over a period of ten thousand times and inner representation depicts CD curves after several cycles [52].

10.6.2 MOF/Graphene

Graphene materials have widely been employed in as an active material in the various energy storage devices, owing to their excellent conductivity, enormous active surface area, and remarkable flexibility. As dual metal ions in MOFs could have a synergetic impact amongst metal species, the addition of dual metal ions in MOFs might increase electrochemical behavior in a MOF/graphene composite. On the contrary, the metal-doped graphene in a MOF/GA composite system is capable to impact the electric double-layer properties of the hybrid [53–54].

A Co_3O_4/3DGN/NF hybrid without binder was produced by pyrolysis in the presence of Ar and air atmospheres. Since 3DGN/NF has excellent electron transport properties and Co_3O_4 has numerous open mesopores/macropores that facilitate transport of ions, the hybrid electrode may gain a good specific capacitance of 321 F/g at 1A. According to Figure 13.3(A), there are two distinct redox peak patterns in the usual CV curve at distinct scan speeds (5 to 500 mV/s). Even as scan rate increases from 5 to 500 mV, the form of the CV curves does not change much. The capacitance of 88% as compared to the initial value is retained after 2000 charge-discharge cycles operated at a current densities of 10 A/g. In continuation, the energy density and power densities of 7.5 Wh/kg and 794 W/kg is observed in the two-electrode arrangement [55].

It was discovered that rGO coating/sandwiching Co_3O_4 composites may improve electrical conductivity and structural stability. Researchers found that rGO/Co_3O_4 composites had greater electrical conductivity and lesser charge transfer resistance in comparison of Co_3O_4-rGO-Co$_3$. Consequently, they were recommended as the electrode components in supercapacitors based on electrochemical impedance spectroscopy measurements. Composite materials with low current densities demonstrated significant specific capacitance, and after 10,000 cycles at 5 A/g, they preserved 90% of their original capacitance at 5 A/g [56]. The synthesized rGO/MoO_3, when exposed to Ar and air in two steps, its specific capacitance reached 617 F/g at an operated current density of 1 A/g. At 6 A/g, then about 87.5% of capacitance is retained following 6000 charging/discharging cycles. In order to analyze the electrochemical behavior of the rGO/MoO_3 composite, cyclic voltammetry measurements have been conducted. Even as scan rate raised, the form of the CV curves stayed mostly unchanged; showing rGO/MoO_3 electrode's capacity to operate at high rates. Contrary to rGO/MoO_3 composite electrodes, MoO_3 electrodes produced by immediate annealing of Mo-MOFs and widely accessible MoO_3 powder had lower rate capabilities (Figure 10.3B,C) [57].

The rGO/ZIF-67 nanocomposite demonstrated the best specific capacitance worth of 210 F/g under 1 A/g current density, that is a lot more prominent than ZIF-67 for the same current density (103.6 F/g). The produced nanocomposite demonstrated a great cycling execution (80% retention after the application of 1000 cycles for current density1 A/g). The voltammograms illustrate how the redox reaction using rGO/ZIF-67 occurs at smaller potentials than in ZIF-67 (Figure 10.3D, E), demonstrating the greater electrical conductivity of rGO/ZIF-67 nanocomposite electrode [58]. When rGO/ZIF-67 was compared to ZIF-67, the currents enhanced, which was linked to elevated electrical conductivity and the availability of active sites in the nanocomposite electrode, which led to better electron movement and ionic dispersion through the electrode fabricated using the current composition [59].

10.6.3 MOF/Transition Metal Dichalcogenides

The layered two-dimensional (2D) materials attracted the researchers' attention due to their extensive electrical properties. Among 2D materials after graphene, the transition metal dichalcogenides were recently used as an anodic material in the supercapacitors. The Ni-MOF which is grown on the MoS_2, after alkaline treatment the Ni based MOF is utilized in the supercapacitor. The MOS_2/Ni(OH)$_2$ shows specific capacitance of 2192 Fg^{-1}. The capacitance of about 85% remains after 10000 cycles. The MoS_2/Ni(OH)$_2$ based device with the activated carbon as cathode has shown an energy and power density as 50.58 and 800 W Kg^{-1}, respectively. The nickel hydroxide deposited on the MoS_2 shows a greater potential with better energy storage capacity [60]. In continuation, the supercapacitor

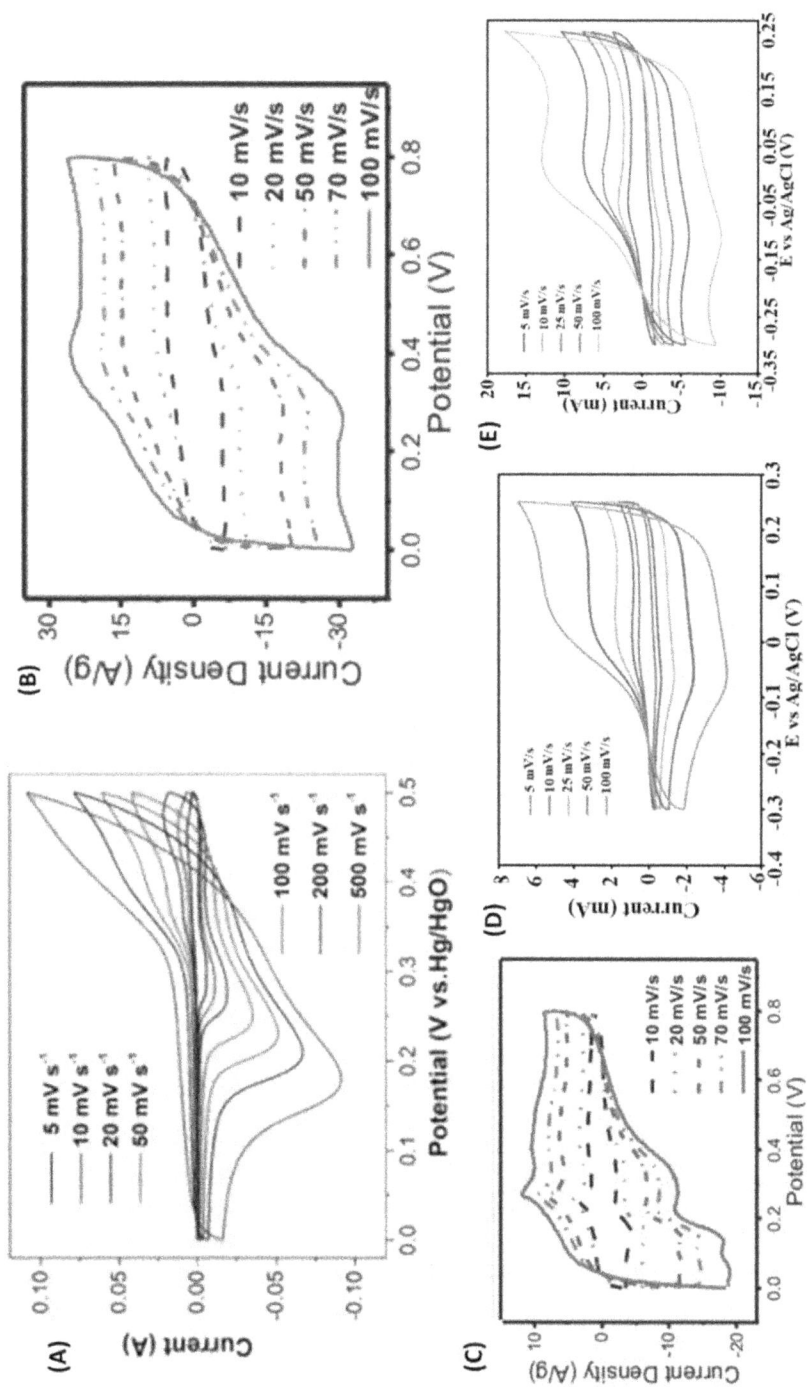

FIGURE 10.3 CV curve behavior A) Co₃O₄/3DGN/NF electrode at different scan rates [55], B) rGO/MoO₃ electrode in the voltage window of 0–0.8 V by varying scan rate values [57], C) MoO₃ electrode obtained by directly annealing Mo-MOFs [57], D) ZIF-67 at various scan levels from 5 to 100 mV/s [58] and E) The rGO/ZIF-67 electrode [58].

TABLE 10.1
Various Capacitive Properties of TMD-Based Supercapacitors

S. No	Composition	Specific capacitance (Fg⁻¹)	Energy density (Wh kg⁻¹)	Power density (W kg⁻¹)	Ref
1	$MoS_2/Ni(OH)_2/carbon$	2192	50.5	800	[60]
2	$MoS_2/Ni/carbon$	1590	72.9	375	[61]
3	$CeO_2/MoS_2/carbon$	1325	34.5	666	[62]
4	$Co_3O_4/MoS_2/carbon$	1162	31.0	388	[63]

fabricated with Ni centered MOF deposited on the surface of MOS_2 and activated charcoal as the other electrode shows the specific capacitance of 1590.24 Fg⁻¹ at 1 A g⁻¹, and shows a capacitive retention of 87.97% post to 20,000 cycles. The supercapacitor with MOS_2/Ni-MOF/activated carbon shows an energy and power densities as 72.93 Wh kg⁻¹, 375 W Kg⁻¹, respectively [61]. Later, the cerium oxide-based MOF deposited on the surface of MoS_2 is developed and used as an anodic material for the supercapacitor applications. The CeO_2/C/MoS_2 device architecture has shown better specific capacity of 1325.67 Fg⁻¹ and 92% of capacitance retention after 1000 cycles. The CeO_2/MoS_2 based device shown the energy and power density as 34.55 Wh kg⁻¹ and 666.7 Wkg⁻¹, respectively [62]. Similarly, the MnS based MOF developed on the surface of MoS_2 and the carbon flakes forms the supercapacitor and the two electrodes shows the specific capacitance of 1162 Fg⁻¹ at the 0.5 A g⁻¹ in the 2 M KOH electrolytic medium. The device shows a capacitive retention of about 81% after 5000 cycles and shows the energy and power densities as 31.0 W h kg⁻¹ and 388.3 W kg⁻¹, respectively [63]. The cobalt oxide and MoS_2 core shells were synthesized with the precursor of cobalt-based MOF and other constituent precursors. The supercapacitor is constructed from a synthetic core shell structure material, and carbon is served as the electrodes in the supercapacitor, which has a specific capacitance of 1076 Fg-1 at 10 A g-1 and also a capacitance retention of 64.5% following 5000 cycles [64]. The other 2D materials such as $MoSe_2$, WS_2, WSe_2 and other related materials also possess the similar properties comparable to graphene. The synthesis and deposition of these materials and the surface interactions in the alkoxy medium/the interaction with the other materials are to be explored. Table 10.1 shows the variation of capacitive properties of transition metal dichalcogenides in the supercapacitor applications, of which the MOS_2/Ni(OH)₂/carbon-based material shows the highest specific capacitance, energy density, and power density as of 2190 F/g, 50.5 WhKg⁻¹ and 800 Wkg⁻¹, respectively.

10.6.4 MOF/CARBON NANOTUBES

MOFs are combined with CNTs, which have excellent electrical characteristics, such as distinctive pore structure and strong thermal/mechanical stability. According to theoretical data, supercapacitor electrode materials made from MOF composites and CNTs can be very efficient [65]. When it comes to the creation of electrolytes, both single-wall and multiwall carbon nanotube materials have been studied. Due to their large specific surface area and superior pore size distribution properties (according to MOFs), MOFs/CNT hybrid composites also exhibit a high level of electrical conductivity [66].

Using hydrothermal method, Mn-MOF/CNTs necklace was synthesized exhibiting specific capacitance of 203.1 F/g at 1 A/g current density (electrolyte = 1 M Na_2SO_4) that was superior to Mn-MOF. It is possible to get a specific capacitance of 50.3 F/g⁻¹ with an optimized cell that operates at 0.25 A/g. If we increase the current density from 0.5 A/g to 1, 2, 3, 5 and 10 A/g, respectively, the specific capacitance of the cell drops from 48.9 F/g to 23.4 F/g, and their retaining rate is 97.4%, 95%, 81.6%, 73.8%, 62, and 59.6%, respectively. They were able to get such high rates because, the Mn-MOF combined with functional CNTs, makes CNTs/Mn-MOF material that exhibit good specific capacitance. It also has an amazing capacitance retention rate of 88% after 3000 cycles as illustrated in Figure 10.4a, b [67]. An electrode made of hierarchical micro-mesoporous architecture shows the specific capacitance

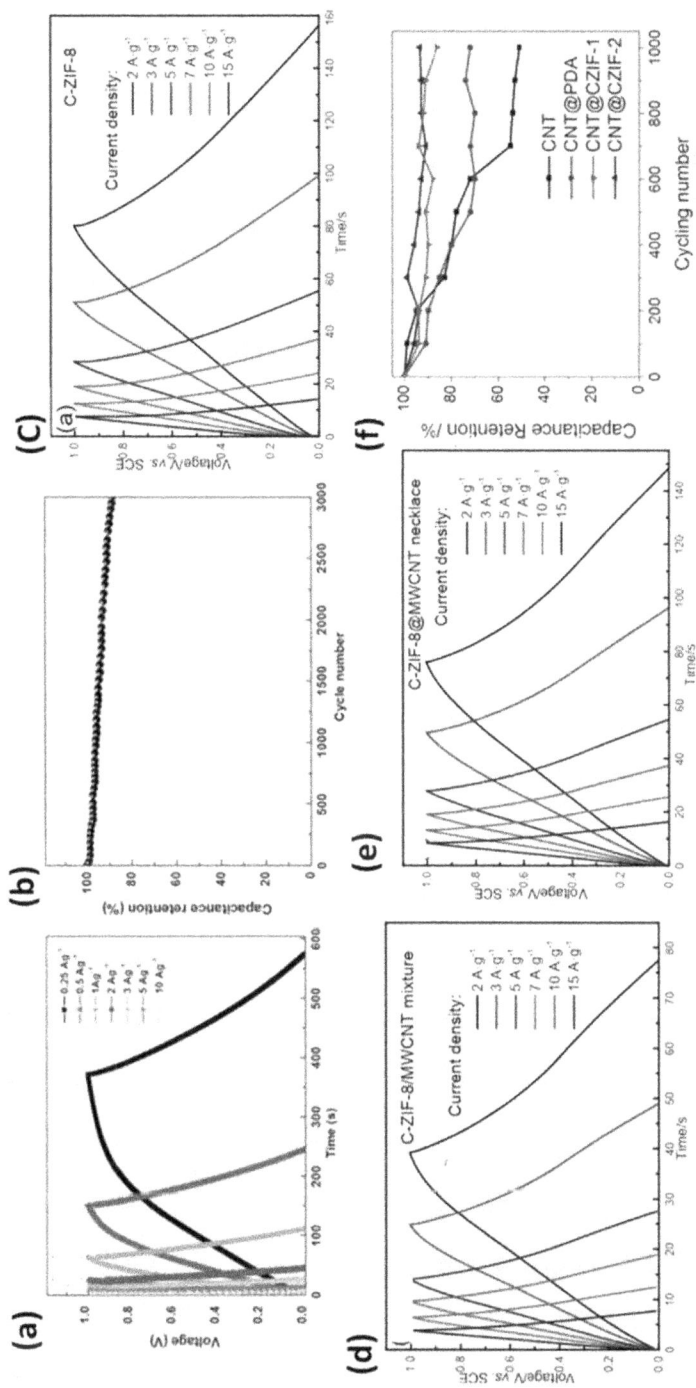

FIGURE 10.4 (a) Comparative GCD curves for symmetrical cells, (b) Cycle efficiency at 5 A/g^{-1} of a symmetrical supercapacitor [67], (c) GCD behavior of C-ZIF-8, (d) GCD characteristics of C-ZIF-8/MWCNTs mixture, (e) GCD curves of C-ZIF-8@MWCNTs necklace [68] and (f) for 1000 cycles compares the cycling stability of carbon nanotubes with polydopamine (PDA) nanotubes, CNT@CZIF-1, and CNT@CZIF-2 at discharging rate of 2 A g^{-1} [70].

of about 326 F/g at an operated current density of 1 A/g using a ZIF-8/MWCNT composite necklace architecture, where ZIF-8 nanocrystals were functionalized on MWCNTs [68]. There is a galvanostatic charge-discharge (GCD) curve for every carbon in Figure 10.4c, d, e. In comparison to the spherical Ni-MOF, the MWCNT/Ni-MOF has a specific capacity of 115 mAhg^{-1} (2 Ag^{-1}) and outstanding characteristics. Researchers have also discovered conductive Cu-MOF nanowire arrays produced on carbon fiber sheets as conductive fillers and binder-free electrodes [69].

For the composite, nanoporous carbons (NPC) generated from MOF (N-doped ZIF-8) was utilized, which possess a higher value of specific capacitance of 324 F/g at the operated current density of 0.5 A/g, according to the experiment performed. After 1000 cycles, 93.5% of the original capacitance was still there, which is impressive. From Figure 10.4f, it is apparent that CNT@CZIF-2's outstanding performance is due to its high nitrogen concentration without compromising its primary CNT frameworks. It is also possible to create extremely porous and conductive electrodes using NPCs in combination with CNT [70]. Till now, the full potential of MOF@CNT electrodes has not been exploited. That is why, there is a need for deeper research in this field. Improved supercapacitors with better energy and power densities will need the development and testing of electrodes with enhanced functional capabilities.

10.7 MOF/CONDUCTIVE-POLYMER-BASED SUPERCAPACITORS

Electrically conducting materials such as polypyrrole (PPy), polyaniline (PANI), and polyethylene dioxythiophene (PEDOT), etc. have been extensively studied and are regarded to be a promising material for supercapacitor electrodes. It has been shown that conducting polymers may be used to provide smooth charge paths between the outside circuit and the interior surface of MOFs. They can also be used to support uniformly disseminated MOF nanostructures, resulting in more active sites and greater specific surface areas [71].

Due to its ease of fabrication, higher conductivity, and excessive pseudocapacitance, PANI is among the most utilized conductive polymers in such applications [72]. For a solid-state adaptable electrode, researchers explored interweaving conductive PANI chains in MOF crystals, and then deposited this composite over carbon cloth fibers. Over two thousand cycles, around 80% of the original capacitance was still there. Using the potential of 10 mV/s, the electrode's specific capacitance was evaluated at 371 Fg^{-1} in the existence of 3 M KCl electrolyte [73].

PANI layers were interconnected utilizing ZIF-8-derived carbon. An effective electrochemical pathway to electrolyte ions was enabled by this composite. An extremely high capacitance value (300–1100 F/g) was achieved as a consequence. More than 86 percent of the original capacitance was still there even after twenty thousand cycles [74]. Electrode made from MOF-5/PANI, ZIF-8/PANI and ZIF-8 decorated N-doped carbon/PANI showed specific capacitance 477, 236 and 755 F/G, respectively at the operated current density of 1 A/g using 1 M H$_2$SO$_4$ [74–76]. An impressive 1835 Fg^{-1} specific capacitance is achieved by the NiCo-LDH@PANI@CC nanocomposite material at same value of current density [77]. Figure 10.5 illustrates that ZIF-8/PANI has extremely less value of capacitance in the comparison of other nanocomposites. In recent years, a composite electrode consisting of Cu containing MOF and the polymer with the structure Poly(o-aminophenol) (POAP) has been developed. The electrode had a specific capacitance of 241 F/g (in existence of HClO$_4$ electrolyte) and 90% cyclic stability after 1000 cycles [78].

The PPy may not always act like a substrate for the development of MOFs, yet it can even encapsulate their structures [79–80]. Unexpectedly, PPy was incorporated into a Zn/Ni-MOF for supercapacitor utilization [81]. Cu containing MOF in combination of PEDOT (depositional composite made with PEDOT/HKUST-15G-CNTF) had already been reported as a supercapacitor. The hybrid's capacitive performance was improved with the introduction of graphene oxide. After fabricating this electrode, it was utilized to build a symmetrical supercapacitor with such a remarkable areal capacitance of 37.8 mF/cm^2 and even an energy density of 0.051 mWh/cm^3 at the applied volumetric power density of 2.1 mW/cm^3 [82]. The incorporation of MOFs with such conductive

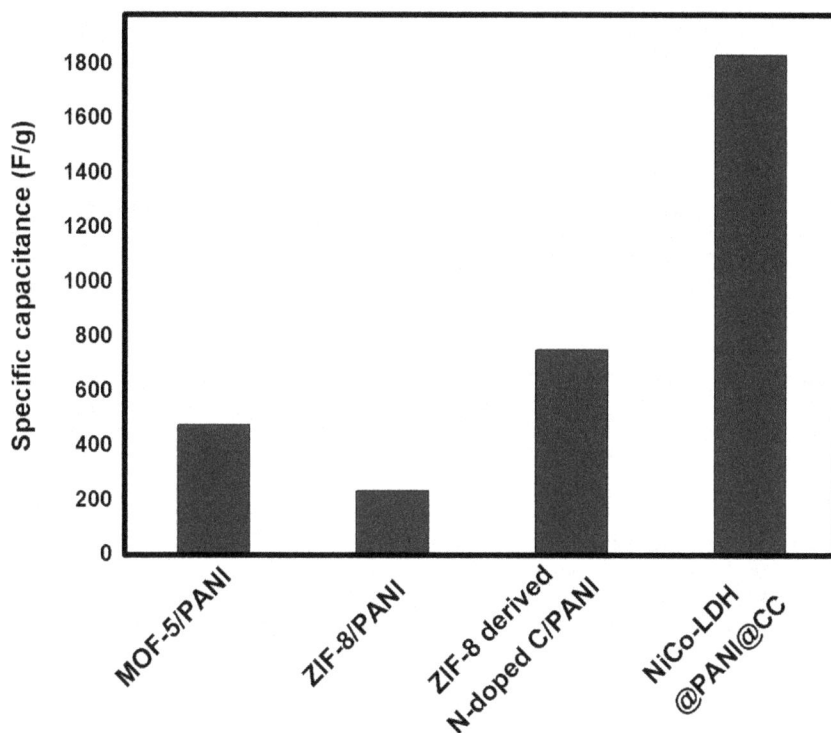

FIGURE 10.5 Comparison of specific capacitance of different MOF based nanocomposites with doping of PANI.

polymers may influence the development of some really efficient and multifunctional electrodes for supercapacitors, resulting in enhanced efficiency [82–84].

10.8 CONCLUSIONS AND FUTURE SCOPE

Since the electrochemical behavior of the existing materials used in supercapacitors has not been up to the mark, several materials with better active surface area and porosity have been investigated as an electrode. Metal organic frameworks and their functional metal/metal oxide derivatives along with their composites of other materials have been progressively used in supercapacitors in the recent years. Optimization of operating conditions can lead to MOF-inspired structures that can be applied to enhance supercapacitor electrodes that tend to outperform traditional carbon electrodes in terms of electrochemical efficacy. An important benefit of using MOF-derived materials for supercapacitors is that they may retain their distinctive properties, such as large specific surface areas and plentiful porous structures, which make MOFs unique. An increased number of active sites and a larger surface area are provided by MOFs and their composites. The capacitive behavior is enhanced with the functionalization of the MOF with conducting polymers. The MOFs typically have conductivity difficulties in their original forms, and they have been examined extensively in aspects of stability throughout the electrolyte ion insertion phase. By developing MOFs with enhanced native conductivities, these obstacles may be overcome in near future. Graphene, CNTs, conducting polymers, etc., would have to provide a plethora of new options for the creation of innovative composite electrodes. As a result of continued study in this vast sector, MOF research will be useful in the future for the establishment of next-era and sustainable energy storage technologies. Considering the rapid growth of industry, it's realistic to predict the continued development of quite efficient energy storage technologies.

REFERENCES

1. Furukawa, H., Cordova, K.E., Keeffe, M., and Yaghi, O.M., 2013. The Chemistry and applications of metal-organic frameworks, *Science*, 341(6149), p. 1230444. Doi: 10.1126/science.1230444.
2. Sundriyal, S., Kaur, H., Bhardwaj, S.K., Mishra, S., Kim, K., and Deep, A., 2018. Metalorganic frameworks and their composites as efficient electrodes for supercapacitor applications, *Coord. Chem. Rev.*, 369, p. 15–38. Doi: 10.1016/j.ccr.2018.04.018.
3. Dey, C., Kundu, T., Biswal, B.P., Mallick, A., and Banerjee, R., 2014. Crystalline metal-organic frameworks (MOFs): Synthesis, structure and function, *Acta Crystallogr., Sect. B: Struct. Sci. Crystal Eng. Mater.*, 70, pp. 3–10. Doi: 10.1107/S2052520613029557.
4. Lee, Y., Kim, J., and Ahn, W., 2013. Synthesis of metal-organic frameworks: A mini review, *Kor. J. Chem. Eng.*, 30(9), pp. 1667–1680. Doi: 10.1007/s11814-013-0140-6.
5. Khan, N.A., Haque, E., and Jhung, S.H., 2010. Rapid syntheses of a metal—organic framework material Cu₃(BTC)₂(H₂O)₃ under microwave: A quantitative analysis of accelerated syntheses, *Phys. Chem. Chem. Phys.*, 12(11), pp. 2625–2631. Doi: 10.1039/B921558A.
6. Haque, E., Khan, N.A., Park, J.H., and Jhung, S.H., 2010. Synthesis of a metal-organic framework material, iron terephthalate, by ultrasound, microwave, and conventional electric heating: A kinetic study, *Chemistry*, 16(3), pp. 1046–1052. Doi: 10.1002/chem.200902382.
7. Kim, J., Yang, S., Choi, S.B., Sim, J., Kim, J., and Ahn, W., 2011. Control of catenation in CuTATB-*n* metal—organic frameworks by sonochemical synthesis and its effect on CO₂ adsorption, *J. Mater. Chem.*, 21(9), pp. 3070–3076. Doi: 10.1039/C0JM03318A.
8. Ko, M., Mendecki, L., and Mirica, K.A., 2018. Conductive two-dimensional metal—organic frameworks as multifunctional materials, *Chem. Commun.*, 54(57), pp. 7873–7891. Doi: 10.1039/C8CC02871K.
9. Zhao, Y., Song, Z., Li, X., Sun, Q., Cheng, N., Lawes, S., and Sun, X., 2016. Metal organic frameworks for energy storage and conversion, *Energy Storage Mater.*, 2, pp. 35–62. Doi: 10.1016/j.ensm.2015.11.005.
10. Wang, L., Han, Y., Feng, X., Zhou, J., Qi, P., and Wang, B., 2016. Metal—organic frameworks for energy storage: Batteries and supercapacitors, *Coord. Chem. Rev.*, 307(2), pp. 361–381. Doi: 10.1016/j.ccr.2015.09.002.
11. Yang, J., Zheng, C., Xiong, P., Li, Y., and Wei, M., 2014. Zn-doped Ni-MOF material with a high supercapacitive performance, *J. Mater. Chem. A*, 2(44), pp. 19005–19010. Doi: 10.1039/C4TA04346D.
12. Ramachandran, R., Xuan, W., Zhao, C., Leng, X., Sun, D., Luo, D., and Wang, F., 2018. Enhanced electrochemical properties of cerium metal—organic framework based composite electrodes for high-performance supercapacitor application, *RSC Adv.*, 8, pp. 3462–3469. Doi: 10.1039/C7RA12789H.
13. Campagnol, N., Vara, R.R, Deleu, W., Stappers, L., Binnemans, K., Vos, D.E., and Fransaer, J., 2014. A hybrid supercapacitor based on porous carbon and the metal-organic framework MIL-100(Fe), *ChemElectroChem*, 1(7), pp. 1182–1188. Doi: 10.1002/celc.201402022.
14. Stassen, I., Styles, M., Grenci, G., Gorp, H.V., Vanderlinden, W., Feyter, S.D., Falcaro, P., Vos, D.D., Vereecken, P., and Ameloot, R., 2016. Chemical vapour deposition of zeolitic imidazolate framework thin films, *Nat. Mater.*, 15, pp. 304–310. Doi: 10.1038/nmat4509.
15. Liu, T., Li, P., Yao, N., Kong, T., Cheng, G., Chen, S., and Luo, W., 2019. Self-sacrificial template-directed vapor-phase growth of MOF assemblies and surface vulcanization for efficient water splitting, *Adv. Mater.*, 31(21), p. 1806672. Doi: 10.1002/adma.201806672.
16. Feng, D., Lei, T., Lukatskaya, M., Park, J., Huang, Z., Lee, M., Shaw, L., Chen, S., Yakovenko, A.A., and Kulkarni, A., 2018. Robust and conductive two-dimensional metalorganic frameworks with exceptionally high volumetric and areal capacitance, *Nat. Energy*, 3(1), pp. 30–36. Doi: 10.1038/s41560-017-0044-5.
17. Campbell, M.G., Sheberla, D., Liu, S.F., Swager, T.M., Dincă, M., 2015. Cu₃ (hexaiminotriphenylene)2: An electrically conductive 2D metal—organic framework for chemiresistive sensing, *Angew. Chemie Int. Ed.*, 54(14), pp. 4349–4352. Doi: 10.1002/anie.201411854.
18. Ajdari, F.B., Kowsari, E., Ehsani, A., Schorowski, M., and Ameri, T., 2018. New synthesized ionic liquid functionalized graphene oxide: Synthesis, characterization and its nanocomposite with conjugated polymer as effective electrode materials in an energy storage device, *Electrochim. Acta*, 292, pp. 789–804. Doi: 10.1016/j.electacta.2018.09.177.
19. Bon, V., Senkovska, I., Weiss, M.S., and Kaskel, S., 2013. Tailoring of network dimensionality and porosity adjustment in Zr-and Hf-based MOFs, *CrystEngComm.*, 15(45), pp. 9572–9577. Doi: 10.1039/C3CE41121D.
20. Yang, J., Xiong, P., Zheng, C., Qiu, H., and Wei, M., 2014. Metal—organic frameworks: A new promising class of materials for a high-performance supercapacitor electrode, *J. Mater. Chem. A*, 2(39), pp. 16640–16644. Doi: 10.1039/C4TA04140B.

21. Jiao, Y., Pei, J., Chen, D., Yan, C., Hu, Y., Zhang, Q., and Chen, G., 2017. Mixed-met allic MOF based electrode materials for high performance hybrid supercapacitors, *J. Mater. Chem. A*, 5, pp. 1094–1102. Doi: 10.1039/C6TA09805C.

22. Tan, Y., Zhang, W., Gao, Y., Wu, J., and Tang, B., 2015. Facile synthesis and supercapacitive properties of Zr-metal organic frameworks (UiO-66), *RSC Adv.*, 5(23), pp. 17601–17605. Doi: 10.1039/C4RA11896K.

23. Lee, D.Y., Shinde, D.V., Kim, E., Lee, W., Oh, I., Shrestha, N.K., Lee, J.K., and Han, S., 2013. Supercapacitive property of metal—organic-frameworks with different pore dimensions and morphology, *Microporous Mesoporous Mater.*, 171, pp. 53–57. Doi: 10.1016/j.micromeso.2012.12.039.

24. Ke, F., Wu, Y., and Deng, H., 2015. Metal-organic frameworks for lithium-ion batteries and supercapacitors, *J. Solid State Chem.*, 223, pp. 109–121. Doi: 10.1016/j.jssc.2014.07.008.

25. Qi, K., Hou, R., Zaman, S., Qiu, Y., Xia, B.Y., and Duan, H., 2018. Construction of metal-organic framework/conductive polymer hybrid for all-solid-state fabric supercapacitor, *ACS Appl. Mater. Interfaces*, 10(21), pp. 18021–18028. Doi: 10.1021/acsami.8b05802.

26. Wang, H., Zhang, M., Zhang, A., Shen, F., Wang, X., Sun, S., Chen, Y., and Lan, Y., 2018. Polyoxometalate-based metal-organic frameworks with conductive polypyrrole for supercapacitors, *ACS Appl. Mater. Interfaces*, 10(38), pp. 32265–32270. Doi: 10.1021/acsami.8b12194.

27. Liu, B., Shioyama, H., Akita, T., and Xu, Q., 2008. Metal-organic framework as a template for porous carbon synthesis, *J. Am. Chem. Soc.*, 130(16), pp. 5390–5391. Doi: 10.1021/ja7106146.

28. Liu, B., Shioyama, H., Jiang, H., Zhang, X., and Xu, Q., 2010. Metal—organic framework (MOF) as a template for syntheses of nanoporous carbons as electrode materials for supercapacitor, *Carbon*, 48(2), pp. 456–463. Doi: 10.1016/j.carbon.2009.09.061.

29. Su, P., Jiang, L., Zhao, J., Yan, J., Li, C., and Yang, Q., 2012. Mesoporous graphitic carbon nanodisks fabricated via catalytic carbonization of coordination polymers, *Chem. Commun.*, 48(70), pp. 8769–8771. Doi: 10.1039/C2CC34234K.

30. Jeon, J., Sharma, R., Meduri, P., Arey, B.W., Schaef, H.T., Lutkenhaus, J.L., Lemmon, J.P., Thallapally, P.K., Nandasiri, M.I., McGrail, B.P., Nune, S.K., 2014. In situ one-step synthesis of hierarchical nitrogen-doped porous carbon for high-performance supercapacitors, *ACS Appl. Mater. Interfaces,* 6(10), pp. 7214–7222. Doi: 10.1021/am500339x.

31. Zhong, S., Zhan, C., and Cao, D., 2015. Zeolitic imidazolate framework-derived nitrogen-doped porous carbons as high-performance supercapacitor electrode materials, *Carbon*, 85, pp. 51–59, 2015. Doi: 10.1016/j.carbon.2014.12.064.

32. Hao, F., Li, L., Zhang, X., Chen, J., 2015. Synthesis and electrochemical capacitive properties of nitrogen-doped porous carbon micropolyhedra by direct carbonization of zeolitic imidazolate framework-11, *J. Chen, Mater. Res. Bull.*, 66, pp. 88–95. Doi: 10.1016/j.materresbull.2015.02.028.

33. Fujiwara, Y., Horike, S., Kongpatpanich, K., Sugiyama, T., Tobori, N., Nishihara, H., and Susumu Kitagawa, 2015. Control of pore distribution of porous carbons derived from Mg^{2+} porous coordination polymers, *Inorg. Chem. Front.*, 2(5), pp. 473–476. Doi: 10.1039/C5QI00019J.

34. Yan, X., Li, X., Yan, Z., Komarneni, S., 2014. Porous carbons prepared by direct carbonization of MOFs for supercapacitors, *Appl. Surf. Sci.*, 308, pp. 306–310. Doi: 10.1016/j.apsusc.2014.04.160.

35. Salunkhe, R.R., Kamachi, Y., Torad, N.L., Hwang, S.M., Sun, Z., Dou, S.X., Kim, J.H., and Yamauchi, Y., 2014. Fabrication of symmetric supercapacitors based on MOF-derived nanoporous carbons, *J. Mater. Chem. A*, 2(46), pp. 19848–19854. Doi: 10.1039/c4ta04277h.

36. Wang, C., Kim, J., Tang, J., Kim, M., Lim, H., Malgras, V., You, J., Xu, Q., Li, and Yamauchi, Y., 2020. New strategies for novel MOF-derived carbon materials based on nanoarchitectures, *Chem.*, 6(1), pp. 19–40. Doi: 10.1016/j.chempr.2019.09.005.

37. Li, Y., Xu, Y., Yang, W., Shen, W., Xue, H., and Pang, H., 2018. MOF-derived metal oxide composites for advanced electrochemical energy storage, *Small*, 14(25), p. 1704435. Doi: 10.1002/smll.201704435.

38. Bigdeli, F., Lollar, C.T., Morsali, A., and Zhou, H., 2020. Switching in metal-organic frameworks, *Angew. Chemie Int. Ed.*, 59(12), pp. 4652–4669. Doi: 10.1002/anie.201900666.

39. Meng, F., Fang, Z., Li, Z., Xu, W., Mengjiao Wang, M., Liu, Y., Zhang, J., Wang, W., Zhao, D., and Guo, X., 2013. Porous Co_3O_4 materials prepared by solid-state thermolysis of a novel Co-MOF crystal and their superior energy storage performances for supercapacitors, *J. Mater. Chem. A*, 1(24), pp. 7235–7241. Doi: 10.1039/C3TA11054K.

40. Khan, I.A., Badshah, A., Nadeem, M.A., Haider, N., and Nadeem, M.A., 2014. A copper-based metal-organic framework as single source for the synthesis of electrode materials for high-performance supercapacitors and glucose sensing applications, *Int. J. Hydrogen Energy*, 39(34), pp. 19609–19620. Doi: 10.1016/j.ijhydene.2014.09.106.

41. Ullah, S., Khan, I.A., Choucair, M., Badshah, A., Khan, I., and Nadeem, M.A., 2015. A novel Cr_2O_3-carbon composite as a high performance pseudo-capacitor electrode material, *Electrochim. Acta*, 171, pp. 142–149. Doi: 10.1016/j.electacta.2015.04.179.

42. Meng, W., Chen, W., Zhao, L., Huang, Y., Zhu, M., Huang, Y., Fu, Y., Geng, F., Yu, J., Chen, X., and Zhi, C., 2014. Porous Fe_3O_4/carbon composite electrode material prepared from metal-organic framework template and effect of temperature on its capacitance, *Nano Energy*, 8, pp. 133–140. Doi: 10.1016/j.nanoen.2014.06.007.

43. Yang, P., Song, X., Jia, C., and Chen, H., 2018. Metal-organic framework derived hierarchical ZnO/NiO composites: Morphology, microstructure and electrochemical performance, *J Ind Eng Chem.*, 62, pp. 250–257. Doi:10.1016/j.jiec.2018.01.002.

44. Yu, F., Zhou, L., You, T., Zhu, L., Liu, X., and Wen, Z., 2017. Preparation of Zn0.65Ni0.35O composite from metal-organic framework as electrode material for supercapacitor, *Mater Lett.*, 194, pp. 185–188. Doi: 10.1016/j.matlet.2017.02.050.

45. He, D., Gao, Y., Yao, Y., Wu, L., Zhang, J., Huang, Z., Wang, M., 2020. Asymmetric supercapacitors based on hierarchically nanoporous carbon and $ZnCo_2O_4$ from a single biometallic metal-organic frameworks (Zn/Co-MOF), *Front. Chem.*, 8(719). Doi: 10.3389/fchem.2020.00719.

46. Zhu, Z., Han, C., Li, T., Hu, Y., Qian, J., and Huang, S., 2018. MOF-templated syntheses of porous Co_3O_4 hollow spheres and micro-flowers for enhanced performance in supercapacitors, *CrystEngComm*, 20(27), pp. 3812–3816. Doi: 10.1039/C8CE00613J.

47. Maiti, S., Pramanik, A., and Mahanty, S., 2015. Influence of imidazolium-based ionic liquid electrolytes on the performance of nano-structured MnO_2 hollow spheres as electrochemical supercapacitor, *RSC Adv.*, 5(52), pp. 41617–41626. Doi: 10.1039/C5RA05514H.

48. Jayakumar, A., Antony, R.P., Wang, R., and Lee, J., 2017. MOF-derived hollow cage $Ni_xCo_{3-x}O_4$ and their synergy with graphene for outstanding supercapacitors, *Small*, 13(11), p. 1603102. Doi: 10.1002/smll.201603102.

49. Zardkhoshoui, A.M., and Davarani, S.S.H., 2020.Boosting the energy density of supercapacitors by encapsulating a multi-shelled zinc—cobalt-selenide hollow nanosphere cathode and a yolk—double shell cobalt—iron-selenide hollow nanosphere anode in a graphene network, *Nanoscale*, 12(23), pp. 12476–12489, 2020. Doi: 10.1039/D0NR02642E.

50. Zhang, C., Xiao, J., Lihua X., Qian, Yuan, S., Wang, S., and Lei, P., 2016. Hierarchically porous Co_3O_4/C nanowire arrays derived from a metal—organic framework for high performance supercapacitors and the oxygen evolution reaction, *J Mater Chem A.*, 4(42), pp. 16516–16523. Doi: 10.1039/C6TA06314D.

51. Xia, Y., Wang, B., Wang, G., and Wang, H., 2015. Easy access to nitrogendoped mesoporous interlinked carbon/NiO nanosheet for application in lithium-ion batteries and supercapacitors. *RSC Adv.*, 5(120), pp. 98740–98746. Doi: 10.1039/C5RA19155F.

52. Lu, C., and Chen, X., 2019. Porous $g-C_3N_4$ covered MOF-derived nanocarbon materials for high-performance supercapacitors, *RSC Adv.*, 9(67), pp. 39076–39081. Doi: 10.1039/C9RA09254D.

53. Liu, L., Yan, Y., Cai, Z., Lin, S., Hu, X., 2018. Growth-oriented Fe-based MOFs synergized with graphene aerogels for high-performance supercapacitors, *Adv. Mater. Interfaces*, 5(8), p. 1701548. Doi:10.1002/admi.201701548.

54. Rahmanifar, M.S., Hesari, H., Noori, A., Masoomi, M.Y., Morsali, A., and Mousavi, M.F., 2018. A dual Ni/Co-MOF-reduced graphene oxide nanocomposite as a high-performance supercapacitor electrode material, *Electrochim. Acta*, 275, pp. 76–86. Doi: 10.1016/j.electacta.2018.04.130.

55. Deng, X., Li, J., Zhu, S., Fang He, F., He., Liu, E., Shi, C., Li, Q., and Zhao, N., 2017. Metal—organic frameworks-derived honeycomb-like Co_3O_4/three-dimensional graphene networks/Ni foam hybrid as a binder-free electrode for supercapacitors, *J Alloys Compd.*, 693, pp. 16–24. Doi: 10.1016/j.jallcom.2016.09.096.

56. Yin, D., Huang, G., Sun, Q., Li, Q., Wang, X., Yuan, D., Wang, C., and Wang, L., 2016. RGO/Co_3O_4 composites prepared using GO-MOFs as precursor for advanced lithium-ion batteries and supercapacitors electrodes, *Electrochim Acta*, 215, pp. 410–419. Doi: 10.1016/j.electacta.2016.08.110.

57. Cao, X., Zheng, B., Shi, W., Yang J., Fan, Z., Luo, Z., Rui, X., Chen, B., Yan, Q., and Zhang, H., 2015. Reduced graphene oxidewrapped MoO_3 composites prepared by using metal-organic frameworks as precursor for all-solid-state flexible supercapacitors, *Adv Mater.*, 27(32), pp. 4695–4701. Doi: 10.1002/adma.201501310.

58. Hosseinian, A., Amjad, A., Hosseinzadeh-Khanmiri, R., Ghorbani-Kalhor, E., Babazadeh, M., and Vessally, E., 2017. Nanocomposite of ZIF-67 metal—organic framework with reduced graphene oxide nanosheets for high-performance supercapacitor applications, *J Mater Sci: Mater Electron*, 28(23), pp. 18040–18048. Doi:10.1007/s10854–017–7747-z.

59. Zhang, W., Tan, Y., Gao, Y., Wu, J., Hu, J., Stein, A., and Tang, B., 2016. Nanocomposites of zeolitic imidazolate frameworks on graphene oxide for pseudocapacitor applications, *J. Appl. Electrochem.*, 46, pp. 441–450. Doi: 10.1007/s10800-016-0921-9.
60. Yang, W., Guo, H., Fan, T., Zhao, X., Zhang, L., Guan, Q., Wu, N., Cao, Y., and Yang, W., 2021. MoS2/Ni (OH)$_2$ composites derived from in situ grown Ni-MOF coating MoS2 as electrode materials for super-capacitor and electrochemical sensor, *Colloids Surf. A: Physicochem. Eng. Asp.*, 615, p. 126178. Doi: 10.1016/j.colsurfa.2021.126178.
61. Yue, L., Wang, X., Sun, T., Liu, H., Li, Q., Wu, N., Guo, H., and Yang, W., 2019. Ni-MOF coating MoS2 structures by hydrothermal intercalation as high-performance electrodes for asymmetric supercapacitors, *Chem. Eng. J.*, 375, p. 121959. Doi: 10.1016/j.cej.2019.121959.
62. Govindan, R., Hong, X., Sathishkumar, P., Cai, Y., and Gu, F.L., 2020. Construction of metal-organic framework-derived CeO2/C integrated MoS2 hybrid for high-performance asymmetric supercapacitor, *Electrochimica Acta*, 353, p. 136502. Doi: 10.1016/j.electacta.2020.136502.
63. Yan, Z.S., Long, J.Y., Zhou, Q.F., Gong, Y., and Lin, J.H., 2018. One-step synthesis of MnS/MoS 2/C through the calcination and sulfurization of a bi-metal—organic framework for a high-performance supercapacitor and its photocurrent investigation, *Dalton Trans.*, 47(15), pp. 5390–5405. Doi: 10.1039/c7dt04895e.
64. Wang, B., Tan, W., Fu, R., Mao, H., Kong, Y., Qin, Y., and Tao, Y., 2017. Hierarchical mesoporous Co3O4/C@ MoS2 core—shell structured materials for electrochemical energy storage with high supercapacitive performance, *Synth. Met.*, 233, pp. 101–110. Doi: 10.1016/j.synthmet.2017.09.011.
65. Korenblit, Y., Rose, M., Kockrick, E., Borchardt, L., Kvit, A., Kaskel, S., Yushin, G., 2010. High-rate electrochemical capacitors based on ordered mesoporous silicon carbide-derived carbon, *ACS Nano*, 4(3), pp. 1337–1344, 2010. Doi: 10.1021/nn901825y.
66. Nishihara, H., Itoi, H., Kogure, T., Hou, P., Touhara, H., Okino, F., Kyotani, T., 2009. Investigation of the Ion storage/transfer behavior in an electrical double-layer capacitor by using ordered microporous carbons as model materials, *Chem. Eur. J.*, 15(21), pp. 5355–5363. Doi: 10.1002/chem.200802406.
67. Zhang, Y., Lin, B., Sun, Y., Zhang, X., Yanga, H., and Wang, J., 2015. Carbon nanotubes@metal—organic frameworks as Mn-based symmetrical supercapacitor electrodes for enhanced charge storage, *RSC Adv.*, 5(72), pp. 58100–58106, 2015. Doi: 10.1039/C5RA11597C.
68. Wang, Y., Chen, B., Zhang, Y., Fu, L., Zhu, Y., Zhang, L., and Wu, Y., 2016. ZIF-8@MWCNT-derived carbon composite as electrode of high performance for supercapacitor, *Electrochim. Acta*, 213, pp. 260–269. Doi: 10.1016/j.electacta.2016.07.019.
69. Li, W., Ding, K., Tian, H., Yao, M., Nath, B., Deng, W., Wang, Y., and Xu, G., 2017. Conductive metal—organic framework nanowire array electrodes for high-performance solid-state supercapacitors, *Adv. Funct. Mater.*, 27(27), p. 1702067. Doi: 10.1002/adfm.201702067.
70. Wan, L., Shamsaei, E., Easton, C.D., Yu, D., Liang, Y., Chen, X., Abbasi, Z., Akbari, A., Zhang, X., and Wang, H., 2017. ZIF-8 derived nitrogen-doped porous carbon/carbon nanotube composite for high-performance supercapacitor, *Carbon*, 121, pp. 330–336. Doi: 10.1016/j.carbon.2017.06.017.
71. Zhang, K., Kirlikovali, K.O., Le, O.V., Jin, Z., Varma, R.S., Jang, H.W., Farha, O.K., and Shokouhimehr, M, 2020. Extended metal—organic frameworks on diverse supports as electrode nanomaterials for electrochemical energy storage, *ACS Appl. Nano Mater.*, 3(5), pp. 3964–3990. Doi: 10.1021/acsanm.0c00702.
72. Wang, Y., Zhang, W., Wu, X., Luo, C., Wang, Q., Li, J., and Hu, L., 2017. Conducting polymer coated metal-organic framework nanoparticles: Facile synthesis and enhanced electromagnetic absorption properties, *Synth. Met.*, 228, pp. 18–24. Doi: 10.1016/j.synthmet.2017.04.009.
73. Wang, L., Feng, X., Ren, L., Piao, Q., Zhong, J., Wang, Y., Li, H., Chen, Y., and Wang, B., 2015. Flexible solid-state supercapacitor based on a metal—organic framework interwoven by electrochemically-deposited PANI, *J. Am. Chem. Soc.*, 137(15), pp. 4920–4923. Doi: 10.1021/jacs.5b01613.
74. Salunkhe, R.R., Tang, J., Kobayashi, N., Kim, J., Ide, Y., Tominaka, S., Kim, J.H., and Yamauchi, Y., 2016. Ultrahigh performance supercapacitors utilizing core—shell nanoarchitectures from a metal—organic framework-derived nanoporous carbon and a conducting polymer, *Chem. Sci.*, 7(9), pp. 5704–5713. Doi: 10.1039/c6sc01429a.
75. Guo, S.N., Zhu, Y., Yan, Y.Y., Min, Y.L., Fan, J.C., Xu, Q.J., Yun, H., 2016. (Metal-Organic Framework)-Polyaniline sandwich structure composites as novel hybrid electrode materials for high-performance supercapacitor, *J. Power Sources*, 316, pp. 176–182. Doi: 10.1016/j.jpowsour.2016.03.040.
76. Guo, S.N., Shen, H.K., Tie, Z.F., Zhu, S., Shi, P.H., Fan, J.C., Xu, Q.J., and Min, Y.L., 2017. Three-dimensional cross-linked Polyaniline fiber/N-doped porous carbon with enhanced electrochemical performance for high-performance supercapacitor, *J. Power Sources*, 359, pp. 285–294. Doi: 10.1016/j.jpowsour.2017.04.100.

77. Hu, W., Chen, L., Du, M., Song, Y., Wu, Z., and Zheng, Q., 2020. Hierarchical NiCo-layered double hydroxide nanoscroll@PANI nanocomposite for high performance battery-type supercapacitor,
Electrochimica Acta, 338, p. 135869. Doi: 10.1016/j.electacta.2020.135869.

78. Naseri, M., Fotouhi, L., Ehsani, A., and Dehghanpour, S., 2016. Facile electrosynthesis of nano flower
like metal-organic framework and its nanocomposite with conjugated polymer as a novel and hybrid
electrode material for highly capacitive pseudocapacitors, *J. Colloid Interface Sci.*, 484, pp. 314–319.
Doi: 10.1016/j.jcis.2016.09.001.

79. Xu, X., Tang, J., Qian, H., Hou, S., Bando, Y., Hossain, S.A., Pan, L., and Yamauchi, Y., 2017. Three-
dimensional networked metal-organic frameworks with conductive polypyrrole tubes for flexible supercapacitors, *ACS Appl. Mater. Interfaces*, 9(44), pp. 38737–38744. Doi: 10.1021/acsami.7b09944.

80. Patterson, N., Xiaob, B., and Ignaszak, A., 2020. Polypyrrole decorated metal—organic frameworks for
supercapacitor devices, *RSC Adv.*, 10(34), pp. 20162–20172. Doi: 10.1039/D0RA02154G.

81. Jiao, Y., Chen, G., Chen, D., Pei J., and Hu, Y., 2017. Bimetal—organic framework assisted polymerization of pyrrole involving air oxidant to prepare composite electrodes for portable energy storage, *J.
Mater. Chem. A*, 5(45), pp. 23744–23752. Doi: 10.1039/C7TA07464F.

82. Fu, D., Li, H., Zhang, X.M., Han, G., Zhou, H., Chang, Y., 2016. Flexible solid-state supercapacitor fabricated by metal-organic framework/graphene oxide hybrid interconnected with PEDOT, *Mater. Chem.
Phys.*, 179, pp. 166–173. Doi: 10.1016/j.matchemphys.2016.05.024.

83. Mulzer, C.R., Shen, L., Bisbey, R.P., McKone, J.R., Zhang, N., Abruña, H.D., and Dichtel, W.R., 2016.
Superior charge storage and power density of a conducting polymer-modified covalent organic framework, *ACS Centr. Sci.*, 2(9), pp. 667–673. Doi: 10.1021/acscentsci.6b00220.

84. Ehsani, A., Khodayari, J., Hadi, M., Shiri, H.M., and Mostaanzadeh, H., 2017. Nanocomposite of p-type
conductive polymer/Cu (II)-based metal-organic frameworks as a novel and hybrid electrode material for
highly capacitive pseudocapacitors, *Ionics*, 23(1), pp. 131–138. Doi: 10.1007/s11581-016-1811-1.

11 Polymeric Blend Nano-Systems for Supercapacitor Applications

Ravindra U. Mene, Ramakant P. Joshi, Vijaykiran N. Narwade, K. Hareesh, Pandit N. Shelke, and Sanjay D. Dhole

CONTENTS

11.1 INTRODUCTION

Environmental pollution and energy shortage are serious concerns for sustainable development. The consumption of fossil fuels and other non-renewable energy resources such as coal, petrol, natural gas, etc. at an alarming rate triggered researchers to develop environment friendly high-power energy storage technologies (Lian et al. 2019). Therefore, attempts have been made to develop low-cost and efficient renewable energy storage devices through synthesis of new materials. The preparation processes of such materials are also being investigated by researchers. (Wang et al. 2012). With advancements in technology, different types of useful and efficient energy storage devices have been explored, such as fuel cells, batteries, electrochemical supercapacitors, etc. Figure 11.1 illustrates the relation between specific power and specific energy density for various electrical energy storage devices (Forouzandeh et al. 2020).

Figure 11.1 illustrates that, as compared to a battery, a capacitor delivers high power while the battery stores comparatively more energy than the capacitor. It is observed that batteries can also store a large amount of energy in the range of 120–200 Wh/kg. However, its low power delivery or uptake (0.4–3 kW/kg) and low efficiency limit its applications as a fast energy storage device. On the other hand, conventional capacitors have large specific power (10–100 kW/kg) but small specific

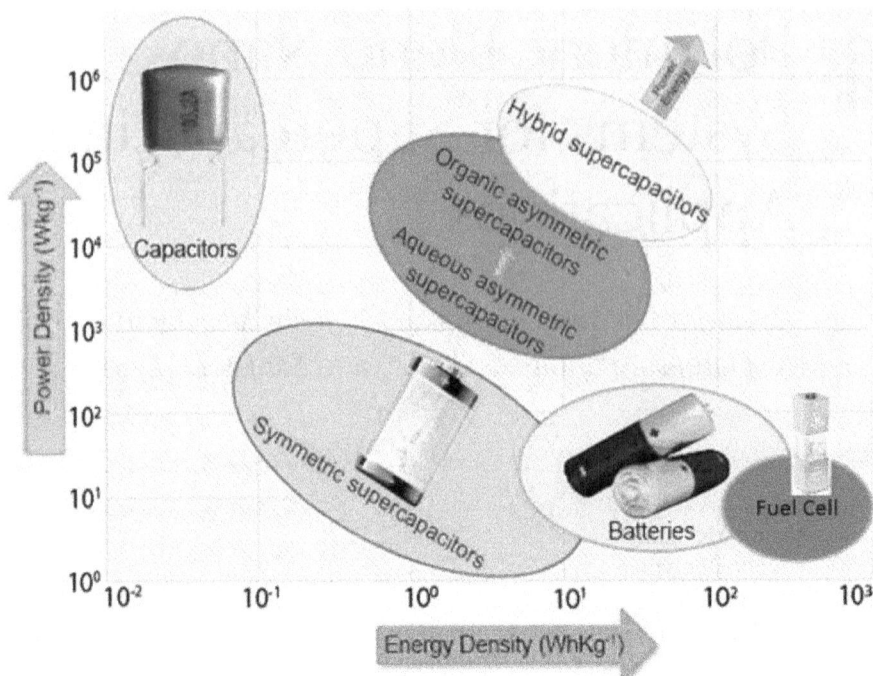

FIGURE 11.1 Ragone plot of the power-energy density range for different electrochemical energy storage devices (Forouzandeh et al. 2020).

energy (0–1 Wh/kg). The new and upcoming energy storage devices known as supercapacitors have the potential to bridge the gap between batteries and conventional capacitors. In comparison to batteries and conventional capacitors, supercapacitors have higher specific power (5 to 55 kW/kg) and specific energy (4 to 8 Wh/kg) (Yu, Tetard et al. 2015). Supercapacitors have attracted significant attention by researchers due to their excellent electrochemical performance; that is, they store a large amount of energy, they have high specific capacitance, high specific power, a wide range of operation temperatures, long life cycles (> 105 times), and rapid charge-discharge rates (Mohd Abdah et al. 2020;Shi et al. 2018; Wu et al. 2019; Atchudan et al. 2019).

11.2 SUPERCAPACITOR MECHANISM

Energy storage in supercapacitors is in the form of charge, which is stored in the surface or sub-surface of the electrode. Consequently, supercapacitors can supply a lot of power by easily liberating energy from surface or sub-surface layer compared to bulk material. The surface area of an electrode is used for charging-discharging processes, which does not bring about any drastic structural change in electroactive materials; hence supercapacitor electrodes have outstanding cycling ability. Due of these unique features, a supercapacitor is among the most efficient energy storage devices. A supercapacitor cell is made up of two electrodes, electrolyte solution, and a separator. Activated carbon is most widely used as an electrode for supercapacitors. The reason being, it has micro pores of different sizes which give a larger surface area, thus increasing the capacitance. Activated carbon electrodes are dipped in the electrolyte solution which plays significant role in enhancing the properties of the supercapacitor. When electrolyte solution is dissolved in appropriate ionizing solvents it will get ionized. A separator (such as filter paper, glassy paper, cellulose or polyacrylonitrile membranes) separates the two electrodes. The thickness of the separator is of order of a few Angstroms (0.3–0.8 nm) and has good ion permeability which facilitates ion transportation (Low et al. 2019).

Supercapacitors are categorized into three groups, based on their charge storage mechanism, Electrical Double Layer Capacitor (EDLC), Electrochemical Double Layer Capacitor (pseudocapacitor (PC)), and Hybrid Capacitor (HC) as illustrated in Figure 11.2 (Shen et al. 2017). Carbon based substances, metal oxides, metal hydroxides and conducting polymers are the most widely used electrode materials for supercapacitor applications. In case of EDLCs, the charge storage takes palace electrostatically (non-faradaically) at the electrode-electrolyte interface in the double layer and there is no charge transfer among electrode and electrolyte. In EDLCs, generally, carbon-based materials are used as electrodes. These include activated carbon, graphene, nano-architectured carbon and carbon aerogels and the charge accumulation takes place via reversible adsorption/desorption of ions at the electrode-electrolyte interface (Kulandaivalu et al. 2019; Wong et al. 2018; Huang et al. 2016; Afif et al. 2019). To store maximum charges, surface area of carbon-based materials should be large so that EDLCs will exhibit maximum output power as well as outstanding cycling ability. Whereas, EDLCs have lower energy density and specific capacitance as compared to pseudocapacitors.

The second category of supercapacitors is known as pseudocapacitors (PCs) (redox capacitor) in which charge is stored through the mechanism of faradaic action or electrochemically, i.e., charge transfer among electrode and electrolyte at surface of the active materials. The charge transfer process will be completedvia reduction-oxidation or redox reactions. On account of redox reaction process, PCs can store large number of charges which accounts to larger capacitance and higher energy density than EDLCs. The electrode materials used for PCs are mainly transition metal oxides (TMO) or conducting polymers (CPs). TMO offers higher specific capacitance and specific energy, whereas CPs possesses good intrinsic conductivity, which makes them an integral part of high-performance supercapacitors (Gan et al. 2015). For PCs application, various conducting polymers are used such as polyaniline, polypyrrole, and PEDOT and for TMOs, RuO_2, V_2O_5, MnO_2, WO_3 etc., are used (Wang et al. 2016).

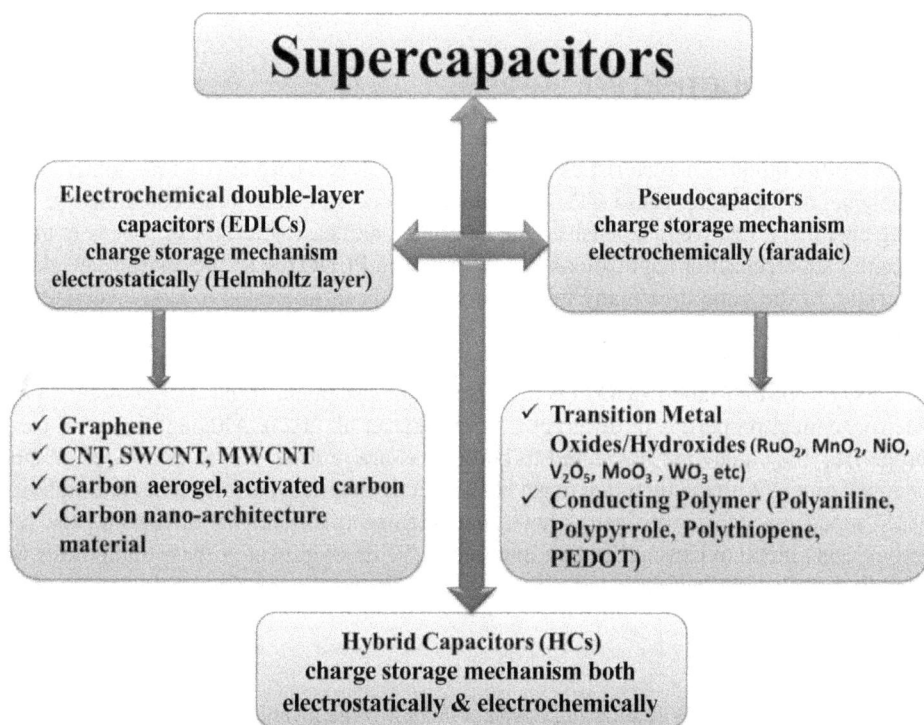

FIGURE 11.2 Classification of supercapacitors (Shen et al. 2017).

The various supercapacitor electrode materials such as metal oxides, metal hydroxides and CPs have their own merits and demerits such as; carbon material offers large density of powers and a long life cycle, but it has still low specific capacitance value mostly for double layer capacitance. So it is restricted for the applications that require large energy density devices. However, metal oxides or metal hydroxides maintain pseudocapacitance as compare to EDLC and have large potential range for charging and discharging curve. But still, this type of electrode has relatively low surface area and poor life cycle. For supercapacitor applications, CPs has various advantages for example good conductivity, high capacitance, low cost, and ease to synthesis but their mechanical stability and life cycles are relatively low. To enhance the performance of supercapacitor, it is a crucial task to organize, optimize and extent the properties and structures of electrode material. Hence, many attempt taken by the researchers to fabricate Hybrid Capacitors (HCs) which have the both advantages such as EDLCs and pseudocapacitors (Abdah, M.A.A.M. et al. 2018, 2020; Azman et al. 2016; Sulaiman et al. 2017). Properties of the nanocomposite electrode depend on individual components used as well as interfacial distinctiveness and morphology of the nanocomposite. From last few years, researcher took significant emphasis to expand various kinds of nanocomposite capacitive materials, for example, CPs with metal oxides, mixed metal oxides, carbon nano-tubes and/or metal oxide/graphene/CPs. During the fabrication and design of nanocomposite electroactive materials for supercapacitor applications, need to consider various factors for example synthesis methods, parameters of synthesis process, selection of material, electrical conductivity, nanocrystallite size, surface area, and interfacial distinctiveness, etc. Even though researchers have made significant progress in the development of nanocomposite electroactive materials used as an electrode for supercapacitor applications, a lot of challenges still remain to be conquered. This chapter presents a review of some commonly used CPs such as polyaniline (PANI), polypyrrole (PPY), and poly(3,4-ethylenedioxythiophene)(PEDOT). It also gives a concise summary of pure CP electroactive materials, CP-based carbon and metal-oxide/hydroxide binary nanocomposites, and CP-based ternary nanocomposite material in supercapacitor applications. Moreover, in the concluding remarks, future challenges and research directions are highlighted.

11.3   PURE CONDUCTING POLYMERS

To date, pure conducting polymer (CP) shows considerable potential as a pseudocapacitive material because of its unique characteristics. Electronically, CP is a derivative of monomers like aniline thiophene, and pyrrole. These CPs have unique properties, like good conductivity, doping/de-doping chemistry, relative inexpensiveness, flexibility, and ease of synthesis (Meng et al. 2017). Therefore, for supercapacitor applications PANI, PPY, and PEDOT, CPs are appropriate electroactive materials. At the same time many investigators have been interested to study electrochemical attainment of CPs electrodesand have tried to createa synthesis process to develop properties of such CP electrodes. A literature review on research development of supercapacitor electrodes based on pure CPs (PANI, PPY, and PEDOT) is discussed in this segment. Table 11.1 shows the physical and electrochemical properties of different CPs (Naskar et al. 2021). Although pure CP has their own properties, they alone are not suited to be used as active material for electrodes in supercapacitor applications. With the intention to get better electrochemical performances and stabilities of CPs based supercapacitors, investigators have been attempting to fabricate binary and also ternary composites using metal oxides and carbon materials. The development of these composites will be evaluated in thefollowing section.

11.3.1   POLYANILINE

In 1934, Polyaniline (PANI) a conducting polymer (CP) was discovered as aniline black. In the case of PANI, a CP, conversion from nonconducting to metallic conductivity depends on both induced protonation and oxidation state. Polymerization of PANI with aniline monomer using various

TABLE 11.1

Physical and Electrochemical Properties of Different Conducting Polymers

Conducting Polymers	MW (g/mol)*	Dopant Level	Voltage range (V)	Conductivity S/cm	Theoretical specific Capacitance F/g	Measured specific Capacitance F/g
Polyaniline (PANI)	93	0.5	0.7	0.1–5	750	240
Polypyrrole (PPy)	67	0.33	0.8	10–15	620	530
Poly(3,4-ethylenedioxythiophene) (PEDOT)	142	0.33	1.2	300–500	210	92

*(MW: molecular weight per unit monomer (g mol))

Source: Naskar et al. 2021

techniques has many advantages like easy synthesis, simple acid/base doping/de-doping chemistry and environmental stability (Li, Huang et al. 2009). Such synthesized material is used as an electrode in pseudocapacitors. From a literature survey, it has been observed that PANI has much higher capacitance (>600 F/g) than other CPs. The maximum capacitance of PANI is due to reversible redox reactions, wherein one electron is removed from each two-monomer unit. Furthermore, the nano-structural morphology of PANI has a vibrant impact on its electrochemical properties. Therefore, it is essential to look for suitable and highly effective synthesis techniques to fabricate PANI with appropriate nanostructures. In general, PANI can be synthesized by chemical or electrochemical routes as well as by oxidation of aniline monomers (Hatchett et al. 1999). Synthesis of PANI by chemical oxidation routes gives various morphologies in the form of nanorods, nanospheres, and nanotubes; and nanoflowers can be accomplished by using the precise control of oxidants or/and addition of dopants (Tran et al. 2011). Conversely, the electrochemical method is fast as compare to other methods and it is free from additives or oxidants and also gives morphologies, like nanofibers, nanogranules, and thin films (Kumar et al. 2015).

The interfacial polymerization method is successfully used to synthesize PANI nanofibers and examine their electrochemical (Sivakkumar et al. 2007). The synthesized supercapacitor shows maximum capacitance i.e. 554 F/g at 1.0 A/g constant current. However, the supercapacitor shows poor retention stability due to rapidly decrease in initial capacitance. Experimental as well as theoretical study is carried out to assess the electrochemical performance of PANI (Li, Wang et al. 2009). It is observed that, experimental values of specific capacitance of PANI evaluated by different ways are found to be very small as that of theoretical value (i.e. 2000 F/g). Lower specific capacitance is owing to small quantity of PANI make influence to capacitance capability. The effectiveness of PANI is depending on conductivity of PANI and also counter anions diffusion. From literature survey it is conclude that the electrochemical performance of PANI, particularly the cycling stability is still not meet requirements of practical application. Specific capacitance of PANI is decreasing rapidly due to poor cyclic stability of the supercapacitor which resulting in short cycle life. To improve electrochemical properties of supercapacitor, researchers make an effort to fabricate various PANI based nanocomposites with carbon materials and/or metal oxides (Xu et al. 2010; Meng et al. 2010).

11.3.1.1   PANI-Carbon Nanocomposites

As per literature survey, electrochemical properties of PANI based supercapacitor electrode are not excellent due to the low specific capacity and instability. Consequently, to improve the specific capacitance and stability of supercapacitors, the properties of electrode materials have great concern. (Masikhwa et al. 2017; Deng et al. 2017). Basically, PANI offers appropriate matrix for synthesis of nanocomposites, which widen their application further than original ones.

In this view, researcher employ various routes for the development of PANI-based nanocomposites like carbon, metals, metal oxides, etc. (Vellakkat et al. 2017; Qu et al. 2017). In next subsection, we will study research advancement of PANI based binary nanocomposites. From last few years, researcher uses carbon materials as supercapacitor electrodes but because of its small value of capacitance limits its use in supercapacitor application. Conversely, outstanding benefits of carbon materials are good electrical conductivity, long cyclic stability, large surface areas and very good mechanical properties. As per the literature survey it is found that instead of pristine CPs, CP with composites of carbon material shows very good electrochemical properties because carbon-based materials are superlative filler for CP based supercapacitor application. Researcher studied various type of PANI-carbon nanocomposites, for example PANI/GN (Gómez et al. 2011; Wang, Carlsson et al. 2015; Cong et al. 2013), PANI/carbon nanotubes (CNTs) in the form of single-walled (SWCNT) or multi-walled (MWCNT) (Imani et al. 2015; Niu et al. 2012; Mi et al. 2007), PANI/carbon nanofibers (Tran, C. et al. 2015), PANI/GN oxide (GO) or reduced GO (rGO) (Sun, She et al. 2015; Wang et al. 2009), PANI/carbon spheres (Shen et al. 2015) and PANI/carbon particles (Khosrozadeh et al. 2015).

The PANI/CNT nanocomposites are prepared by growing PANI on the surface of carbon nanotubes wherein surface functional groups provide active sites for polymer nucleation. However, the presence of carboxyl groups in carbon nanotubes increase 60% specific capacitance of PANI/CNT composites (He et al. 2016). The improved electrochemical performance is observed for SWCNT/PANI nanocomposite, which is strongly dependent on PANI content. When 73 wt.% PANI is deposited on SWCNT surface shows largest specific capacitance value of around 463 F/g. In addition, SWCNT/PANI nanocomposites show excellent stability because capacitance decreased only 5% after 500 cycles and only 1% after the next 1,000 cycles (Gupta and Miura 2006). The free-standing SWCNT with PANI core-shell composites are fabricated through an electrochemical polymerization cyclic process in 1 M H_2SO_4 medium. In electrochemical polymerization cyclic process, PANI is deposited on the surface of SWCNT which forms 50–100 nm tubular structure in diameter. The synthesized composite electrode confirmed the maximum capacitance of 501.8 F/g at potential scan rate of 5 mV/s for 90 polymerization cycles of synthesis and it enhances up to 706.7 F/g by five cycles of controlled electro-degradation (Liu, Sun et al. 2010). In one more study the free-standing PANI/SWCNT composite synthesized via electrochemical polymerization method (Niu et al. 2012). PANI/SWCNT nanocomposites ensure maximum electrical conductivity and improved specific capacitance 236 F/g with deposition time of 30 s, which is ten times greater than pristine SWCNTs (23.5 F/g). The simple microwave-assisted polymerization method is utilized for synthesize PANI/MWCNT composite. The result shows that capacitance of 322 F/g and energy density of 22 Wh/kg is about 12 times more than the pristine MWCNTs (Mi et al. 2007). In one more study, PANI/MWCNT composite with nanotubular morphology is synthesized in situ low-temperature procedure (Imani and Farzi 2015). When 10% MWCNT is used in the composite, it shows higher capacitance value of 552.11 F/g at 4 mA/cm² current density than the pristine PANI capacitance value (411.52 F/g). In another work PANI/MWCNT composites shows the maximum specific capacitance of 560 F/g at 1 mV/s potential scan rate. However, the capacitance decreases to 177 F/g by increasing the scan rate to 5 mV/s (Zhou et al. 2010).

Fabrication of flexible PANI/Graphene (GN) composite is carried out by two processes: in first case, free-standing GN paper is fabricated by suitable method; secondly, PANI nano-rods are electro-polymerized on GN paper. The synthesized GN paper reveals high electrical conductivity, good flexibility, and low weight. When the paper is used as a working electrode in supercapacitor, it exhibits outstanding specific capacitance of 763 F/g and good retention stability (82% of the initial value after 1000 cycles) of combined effect GN and PANI (Cong et al. 2013). In one more study, PANI/GN composites synthesized in situ polymerization route with GO in the acid (Zhang et al. 2010). The composite shows high capacitance of 480 F/g at current density of 0.1 A/g. This ensures that when bulky PANI is doped with GN/GO it reveals high capacitance and outstanding cycling stabilities.

A novel 3D self-supported PANI/reduced GN (rGN) film was prepared via a simple chemical process (Wang, Carlsson et al. 2015). The synthesized nanocomposite achieves the maximum value of specific capacitance around 740 F/g at the current density 0.5 A/g. Also, the retention of the initial capacitance is found to be 87% after 1000 cycles. From this study, it is understood that this type novel three-dimensional hierarchical structure offers large interfacial surface, low ion diffusivity channel and hence by taking full advantages of such material used as active materials in supercapacitor applications. Furthermore, fibrillar PANI doped with GO sheet is fabricated based on by a soft chemical route (Wang et al. 2009). The prepared nanocomposite confirms the improved capacitance value of 531 F/g at a current density of 0.2 A/g with high conductivity of 10 S/cm at 22 °C as compared to the individual PANI (216 F/g). It is observed that, due to GO sheets being added together with PANI, it has shown a considerable improvement in electrochemical performance. A template-directed in situ polymerization method has been successfully used to fabricate self-standing 3D PANI/rGO nanocomposite foam (Sun, She et al. 2015). The prepared composite displays a large capacitance value of 701 F/g at 1 A/g current density. In addition, prepared composite can maintain retention stability of 92% of its initial value after 1000 cycles of charge discharge. The carbon nanofiber is a promising carbon material that is applicable as an electrode in supercapacitor application. The electrochemical properties of nano-PANI with hollow carbon spheres are synthesized by a polymerization method. Result reveals that synthesized composites possessed maximum specific capacitance of 435 F/g at 0.5 A/g and 60% of retention initial value after 2000 cycles (Shen et al. 2015). The three-dimensional free-standing supercapacitor electrodes, which consist of PANI with porous carbon nanofibers, are fabricated in situ of polymerization method. As compare to the pristine electrode of carbon nanofiber, hybrid electrodes demonstrate outstanding capacitance of 366 F/g at 100 mV/s scan rate. Moreover, some report shows the electrochemical properties of PANI with carbon spheres and carbon particles electrode material (Tran et al. 2015). According to this free-standing PANI/carbon particle are formed in free-standing composite film. It shows the maximum capacitance is around 272.6 F/g at current density of 0.63 A/g. Moreover, it possesses unique advantages like stable cycling characteristics, suitable thickness, and flexibility (Khosrozadeh et al. 2015). In concluding remarks, synthesis of PANI/carbon material composites is one of the most important ways to get better performance of the electrochemical supercapacitor, particularly in cyclic and retention stabilities. Basically, freestanding three-dimensional structures have its own distinctive advantages as compared to other structures so it is essential to fabricate suitable 3-dimensional structures by variety of novel techniques to get better performance of supercapacitor. On the other hand, metal oxide has its own advantage in aspect of improving specific capacity. Consequently, study on PANI with metal oxides nanocomposites are significant aspects in supercapacitor field.

11.3.1.2 PANI-Metal Oxide Nanocomposites

Metal oxides are promising candidates due to achievable pseudocapacitance over a large range of potential. However, commonly it suffers from low electrical conductivity as well as unstability in acidic electrolyte (Wang et al. 2014). The reason behind the poor conductivity is due to wide band gap which gives low electron and hole concentrations (Jiang, Ma et al. 2012). Basically, metal oxides are transition elements and studied from long back in the field of pseudocapacitor materials and rechargeable batteries because of their excellent cycle stability, high surface area and excellent charge storage characteristics, low cost, eco-friendliness etc. For the groundwork of metal oxide/PANI supercapacitors a variety of metal oxides have been used such as MnO_2 (Zhang et al. 2012; Sun, Wang et al. 2015; Zhou et al. 2015b; Liang et al. 2016), RuO_2 (Deshmukh et al. 2014) SnO_2 (Li et al. 2012; Jin et al. 2015; Hu et al. 2009), TiO_2 (Chen, Xia et al. 2015; Gottam and Srinivasan 2015; Su et al. 2012), MnO (Han et al. 2012; Li et al. 2015), MoO_3 (Peng, Ma, Mu et al. 2014), Fe_2O_3 (Radhakrishnan et al. 2011), CuO (Ates et al. 2015; Zhu, Wu et al. 2016), ZnO (Pandiselvi and Thambidurai 2014), V_2O_5 (Shao et al. 2012; Bai et al. 2014), WO_3 (Sun, Peng et al. 2015).

Synthesis of electrodeposited MnO_2/PANI electrode shows maximum capacitance of 715 F/g and it only 3.5% capacitance loss is observed after 5000 cycles of charging and discharging (Prasad and Miura 2004). MnO_2/PANI nanocomposite prepared with the help of soaked PANI nanofibers in a $KMnO_4$ solution (Jiang et al. 2012a). The synthesized MnO_2/PANI nanocomposite among 72.4% MnO_2 shows shell and core nanostructure and reveal very high value capacitance around 383 F/g also shows superior cyclic and rate stability in 1M Na_2SO_4 aqueous solution. A novel material for supercapacitor is synthesized as PANI/MnO_2 intercalated layered composite with n octadecyl trimethyl ammonium (Zhang et al. 2007). The synthesized nanocomposite demonstrates capacitance value of 330 F/g at 1 A/g constant current density, which is twofold larger than specific capacitance of pristine PANI i.e. 187 F/g. Furthermore, the synthesized nanocomposites maintain initial value of 94% after 1000 cycles, demonstrate exceptional stability. The PANI/MnO_2 honeycomb structure of composite is fabricated, wherein PANI decorated on MnO_2 nanospheres. It is found that mass ratio of aniline and MnO_2 is 1:1, the synthesized nanocomposite demonstrates large capacitance of 565 F/g at discharge current density of 0.8 A/g. The composite retained 77% of the initial value after 1000 cycles at 8 A/g (Sun, Gan et al. 2015). In another work, MnO_2/PANI nanocomposite film is synthesized using silane coupling regent to alter MnO_2 nano-particles surface which enhance reciprocity of PANI and MnO_2. The higher capacitance of 415 F/g at 1.67 mA/cm^2 obtained compare to PANI/MnO_2 synthesized in a similar condition (Chen et al. 2010).

A novel PANI/$NiCoO_4$ nanocomposite is synthesized in situ chemical oxidation polymerization process (Xu, Wu et al. 2015). The nanocomposite demonstrated high capacitance value of 439.4 F/g and sustain initial value of ~66.11% after 1000 cycles at 5 mA/cm^2 charge-discharge current density. These results show that $NiCoO_4$/PANI nanocomposites possess excellent electrochemical performance. Microsphere composites PANI having cabbage-like structure with hydroquinone showed good electrochemical properties. An electronic conductivity of PANI offers pathway for hydroquinone which can be used as pseudocapacitance component. The composite reveal capacitance is 126.0 F/g at 5 mV/s and retention capacity of 85.1% after 500 cycles scanning at 1 A/g constant current density (Chen, Fan et al. 2015). In another study, CuO doped with PANI, PPY and PEDOT films are prepared by electrochemical deposition and studied the comparative electrochemical performance. PANI/CuO showed superior results as compared to PPY-CuO and PEDOT-CuO nanocomposites. Results exhibited maximum capacitance of PANI/CuO is 286.35 F/g at 20 mV/s, whereas PEDOT/CuO, PPY/CuO is 198.89, 20.78 F/g at 5 mV/s, respectively (Ates et al. 2015). The multi-component architecture of MoS_2 nanosheets and PANI nanoneedle is fabricated by ice reaction process (Zhu, Sun et al. 2015). Binary composite PANI/MoS_2 offers excellent electrical conductivity, a large surface area and rapid ionic diffusion. These properties of PANI/MoS_2 led to large power densities and energy densities. The PANI/MoS_2 along with electrolyte solution of 0.5 M H_2SO_4 demonstrates capacitances 669 F/g, 821 F/g and 853 F/g at 1 A/g in the voltage range of ±0.6 V, ±0.8 V and ±1.0 V, respectively. Specific capacitance of binary PANI/MoS_2 composite is higher than pure PANI polymer and exfoliated MoS_2. From literature survey it reveals capacitance of pure PANI polymers are 397 F/g, 457 F/g and 433 F/g and for exfoliated MoS_2 250 F/g, 322 F/g and 341 F/g respectively. Two-step hydrothermal method is successfully utilized to synthesis PANI/$Ni(OH)_2$ composite, wherein one-dimensional fiber-like PANI polymer is decorated among three-dimensional flowerlike $Ni(OH)_2$. The flower-like composite showed large capacity (55.50 C/g at 0.5 mA/cm^2) and extended life cycle with 79.49% at 1.5 mA/cm^2 after 2500 charging discharging cycles (Zhang et al. 2015).

11.3.1.3 PANI-Ternary Nanocomposites

In order to achieve superior electrochemical performance, researcher emphasis on next-generation of chemical supercapacitors based on ternary composite which include PANI, metallic compounds and carbonaceous materials. The most important issue is to develop the process to modify the microstructure of composites and each component interaction so as attempt maximize synergistic end product of various type of materials. The most capable active materials for supercapacitor electrodes are CPs, carbon materials and metal oxide, most of the researchers (Chen, Liu et al. 2015; Sk et al. 2015) effort to produce CP/metal oxide/carbon ternary composites using variety of techniques. The ternary composite PANI/MoO_3/GN (PMG) is through an in situ polymerization of aniline along with fabricated MoO_3 and GN nanoplatelets. The morphological analysis demonstrated that a fibrillar PANI uniformly coated on the MoO_3/GN composite. The capacitance value of PMG, PANI-GN and PANI are 593, 442 and 295 F/g, respectively, at a current density of 1 A/g. In addition to this, the PMG electrode exhibits excellent cycling stability that is 92.4% after 1000 cycles at 1 A/g as compared to the values of pristine PANI (85.84%) and PANI/GN composite (89.37%) (Das et al. 2015). In one more study, ternary composite MnO_2/carbon/PANI are prepared via polymerization process in the solution of 1M H_2SO_4. The protecting PANI nano-layer allows composites of MnO_2/carbon to be activated in the solution of acidic electrolyte. The value of maximum specific capacitance 695 F/g and the cycling stability 88% after 1000 cycles with 12% of MnO_2 loading (Yan et al. 2012). The TiO_2/GO/PANI networks, which facilitate the faradaic reactions of PANI (Renault et al. 2013). The result reveals that excellent capacitance of 1020 F/g achieved from novel PANI/TiO_2/grapheme composites. In another work, nanorod arrays of GN/MnO_2/PANI are produced (Yu et al. 2014). The specific capacitance for this composite is 755 F/g at 0.5 A/g along with 87% cycling stability after

1000 cycles. The electrodes based on freestanding nanowires array is fabricated in ternary composites NiMoO$_4$/PANI/Conductive carbon cloth (CC) (Chen, Du et al. 2015). The ternary nanocomposite exhibits excellent capacitance of 1340 F/g at 1 mA/cm^2 and 96.7% cycling stability holds after 2000 cycles. These results of ternary nanocomposites are superior to pristine NiMoO$_4$ that is, specific capacitance is 1142 F/g at 1 mA/cm^2 and retention stability of 81% after 2000 cycles. In another work, pseudocapacitance nanostructured PANI/MnO$_2$/MWCNT composites are prepared through polymerization technique. The highest capacitance, energy density and power density of the composite are 517.13 ± 15.25 F/g, 71.88 ±2.12 W h/kg and 10.08 ± 0.26 kW/kg respectively (Sk et al. 2015). For modification of binary system, introduction of metal nanoparticles (Ag, Au Cu^{2+}, Ni^{2+} and other metal particles) is crucial technique which improves the electrical conductivity, cycle life of polymer, impact strength, and thermal conductivity. For example, PANI/GN nanocomposites are decorated by Ag nanoparticles by using polymerization process (Dhibar and Das 2015). The nanocomposite reveals a high capacitance 591 F/g at 5 mV/s scan rate and also superior energy having value 20.24 Wh/kg with 749.30 W/kg power density at 0.5 A/g and 3 A/g respectively. Furthermore, supercapacitor electrodes maintain 96% retention stability after 1500 cycles. The PANI/MWCNTs modified with transition metal ions like Cu2+ and Ni2+ during polymerization reaction have shown better performance (1337 F/g at 5 mV/s). This is due to good coordination between these ions and the nitrogen atoms of PANI that facilitates effective delocalization throughout the PANI Chain (K. Sharma et al. 2017). In another work CV method was used to synthesize ternary composite that is nanoparticles of PANI/rGO/Au are deposited on the electrode of glassy carbon. The ternary composite illustrate very good specific capacitance value and capacitance stability are more than that of electrode of pristine PANI (Shayeh et al. 2015). Table 11.2 summarizes the preparation method and electrochemical performance of PANI-based electrode materials.

11.3.2 POLYPYRROLE

Polypyrrole (PPY) is an important CP that possesses a lot of advantages including easy synthesis, maximum specific capacitance, and good retention stability. The Π-conjugated backbone of an original PPY is similar to that of cis-polyacetylene, apart from that PPY has an extra nitrogen heteroatom on every four carbon atoms, connecting between the first and fourth carbon atoms (Heeger 2010). The polymer's molecular structure is stabilized due to nitrogen heteroatoms and the repeating ring structure unit. PPY is used in supercapacitor applications because of its simple synthesis process and water solubility of its pyrrole monomer, and it takes excellent environment stability and conductivity. Generally, the range of capacitance values of PPY materials are 200–500 F/g. Conversely, PPY in supercapacitors has two deficiencies: its theoretical and practical values differ, and cycling stability is very poor (Lu et al. 2014). To resolve these limitations, several efficient approaches are utilized: (i) synthesis of PPY with premeditated structure/morphology; (ii) synthesis of a variety of PPY composites with metal oxides/hydroxides and carbon materials; and (iii) novel designed flexible and/or 3D design of PPY-based electrode capacitor. Therefore, the specific capacity and cycling stability of PPY could be increased significantly.

The PPY flexible films are synthesized with methylorange and FeCl$_3$ as a reactive self-degradable template by chemical oxidation technique (Li and Yang 2015). The synthesized intrinsic PPY flexible films shows nanotubes having 5–6 µm length and 50–60 nm in diameter due to FeCl$_3$ to monomer molar ratio of 0.5. PPY film shows excellent electrochemical performance, namely, 576 F/g specific capacitance at 0.2 A/g and cycling stability of 82% after 1000 cycle. Free-standing PPY film is synthesized with or without surfactant through oil or water interfacial polymerization (Yang, Hou et al. 2015). It was observed that the prepared PPY films with surface active agent possess small size with more vesicles or pores. This revealsa capacitance of 261 F/g at 25 mV/s and maintains 75% of retention stability for 1000 cycles with the same scan rate. PPY film is prepared by an electropolymerization method in which phytic acid is used as a acting dopant (Rajesh et al. 2016). The result of prepared PPY film reveals excellent capacitance of 343 F/g at 5 mV/s. Besides this, the

TABLE 11.2

Preparation Method and Electrochemical Performance of PANI-Based Carbon, Metal Oxide, and Ternary Electrode Materials

Material	Fabrication Technique	Specific capacitance	Cyclestability (%)	No of Cycles	Reference
PANI	Interfacial polymerization	554 F/g at 1 A/g	10%	1000	(Sivakkumar et al. 2007)
PANI/GN	Oxidative polymerization	500 F/g at 0.1 A/g	-	-	(Wu et al. 2013)
	Electro- polymerization	763 F/g at 1 A/g	82%	1000	(Cong et al. 2013)
PANI/rGO	In situ polymerization	701 F/g at 1 A/g	92%	1000	(Sun et al. 2015)
PANI/rGN	Dilute polymerization,	740 F/g at 0.5 A/g	87%	1000	(Wang et al. 2015)
PANI/GO,	In situ polymerization	531 F/g at 0.2 A/g	-	-	(Wang et al. 2009)
PANI/CNT	Microwave-assisted polymerization	322 F/g at 1 mA/cm^2	-	-	(Mi et al. 2007)
	Electrochemical polymerization	236 F/g at 10 A/g	~85%	1000	(Niu et al. 2012)
	Low-temperature polymerization	552.11 F/g at 4 mA/cm^2	-	-	(Imani and Farzi 2015)
PANI/carbon particle	In situ polymerization	272.6 F/g at 0.63 A/g	95.7%	501	(Khosrozadeh et al. 2015)
PANI/carbon nanofiber	Electrochemical polymerization	366 F/g at 100 mV/s	80%	1000	(Chau et al.2015)
PANI/Carbon sphere	In situ polymerization	435 F/g at 0.5 A/g	~60%	2000	(Shen et al. 2015)
PANI/Ni(OH)2	Hydrothermal synthesis	5.50 C/g at 0.5 mA/cm^2	79.49%	2500	(Zhang et al. 2015)
PANI/NiCoO4	chemical oxidation polymerization	439.4 F/g at 5 mA/cm^2	66.11%	1000	(Xu et al. 2015a)
PANI/hydroquinone	In sit 1 polymerization	126.0 F/g at 5 mV/s	85.1%	500	(Chen et al. 2015a)
PANI/MoS2	Hydrothermal redox reaction	450 F/g at 0.5 A/g	80%	2000	(Zhu et al. 2015a)
PANI/MnO$_2$	Exchange reaction	330 F/g at 1 A/g	94%	1000	(Zhang et al. 2007)
	In situ polymerization	565 F/g at 0.8 A/g,	77%	1000	(Sun et al. 2015d)
	Oxidative polymerization	383 F/g at 0.5 A/g	75.5%	2000	(Jiang et al. 2012)
	Electrochemical polymerization	415 F/g at 1.67 mA/cm^2	> 85%	1000	(Chen et al. 2010)
PANI/rGO/Au	Electrochemical polymerization	303 F/g at 25 mV/s	80%	20,000	(Shayeh et al. 2015)
PANI/GN/Ag	In situ polymerization	591 F/g at 5 mV/s	96%	1500	(Dhibar and Das 2015)
PPY/CNT/MnO$_2$	Deposition method	529.3 F/g at 0.1 A/g	98.5%	1000	(Zhou et al. 2015a)
PPY/GN/MnO$_2$	Ultrasonic irradiation	258 F/g at 1 A/g	-	-	(Sun et al. 2016)
PANI/MnO$_2$/MWCNT	In situ polymerization	517.13 F/g at 1 mA	90%	1000	(Sk et al. 2015)
PANI/MoO$_3$/GNP	In situ polymerization	734 F/g at 10 mV/s	92.4%	1000	(Das et al. 2015)
PANI/NiMoO$_4$/CC	Chemical bath deposition	1340 F/g at 1 mA/cm^2	96.7%	2000	(Chen et al. 2015f)

PPY-based electrode maintains 91% at 10 A/g after 4000 cycles. PPY nanorodsare coated with conductive cotton fabrics by polymerization method (Xu, Wang et al. 2015). The fabric supercapacitor electrodes achieve 325 F/g value of specific capacitance and 24.7 Wh/kg energy density at 0.6 mA/cm^2 current density. It is observed that cycling stability is very poor and needs to be improved –the value of initial capacitance is 63% after 500 cycles.

From this discussion, it is observed that properties and microstructure of PPY-based electrodes affected by many aspects that are synthesis method, dopant, substrate, template, etc. By using proper proportions of these factors, electrochemical performance of PPY base electrodes can be enhanced significantly. But it is quite difficult to achieve practical application requirements by PPY-based electrodes. Therefore, to improve electrochemical performance study on PPY along with carbon composites, metal oxide composites are required.

11.3.2.1 PPY-Carbon Nanocomposites

As compare to PANI, PPY has very low conductivity and thermal stability; overall the most serious problem is that itscycle stability is poor. With the intention to get better cyclic stability of PPY, researchers have paid their attention on the exploration of PPY based carbon composites for application in supercapacitors. Alternately, to improve storage capacity of PPY based supercapacitor electrodes, researcher's emphasis on hybrid composites consists of PPY and metal oxides electroactive materials. In this segment, the main focus is on the research development of PPY/metal oxide composites and PPY/carbon nanocomposites. Wherein, CNTs and GN have various assets such as large surface area, good chemical and thermal stability, excellent electron-transfer through material and low resistivity. Electrochemical characteristics can be enhancing by using CNTs as templates or additives to produce PPY-based nanocomposite (Yang, Hou et al. 2015). From last few years, researchers have been synthesis various type PPY/CNTs composites for supercapacitor applications.

For example, CNT/MnO_2//KCl-CH_2=CH-SiO_2/polyacrylamide//CNT/PPY synthesized for supercapacitor application wherein CNT/PPY is prepared by electro-deposition method and used as an anode (Tang, Chen et al. 2015). CNT/PPY film exhibits specific capacitance of 637 mF/cm^2 at 1 mA/cm^2 current density. Moreover, it reveals high energy and power density 40 Wh/kg, 519 kW/kg respectively. A PPY/CNTs composite is synthesized in presence of cetyl-trimethyl ammonium bromide (CTAB) which used as soft template (Fu et al. 2013). The synthesized PPY/CNTs composites demonstrated 183.2 F/g specific capacitance at 8 A/g and cycling stability of 85% at 1 A/g after 1000 cycles. Concurrently, asymmetric supercapacitors were synthesized via PPY/CNTs composites. The synthesized electrode retains a capacitance of 72% at 1 A/g after 3000 cycles. The electrical conductivity of PPY/MWCNT composite synthesized through polymerization process was enhanced when MWCNT content was lower than 15%. However, it reached 72 S/cm when the MWCNT content was 15% (Wang et al. 2014). The capacitance (265 F/g) of prepared PPY/SWCNTs composite electrode shows better performance compare to pristine PPY due to improved active sites on PPY chains by addition of SWCNTs (An et al. 2002). The supercapacitor electrode is synthesized by a robust as well as flexible CNT-based material used wherein PPY is electrodeposited on freestanding vacuum-filtered CNT film (Chen, Sun et al. 2015). The synthesized supercapacitor electrode has superior mechanical properties and hence it shows outstanding electrochemical performance. From literature study, it is found that supercapacitors electrodes synthesis by solid state route showed long cyclic life and excellent flexibility and also 95% retention capacity after following 10,000 cycles. In another work, PPY/MWCNT compositesbetween core or shell and inhomogeneous structures synthesized via interfacial polymerization method. It is found that the prepared samples through in situ polymerization route reveals molecular conformation and ordered chain packing. (Song et al. 2016). The influence of short and long CNTs in composite of PPY/PSS-CNT is carried out for electrochemical application. It is observed that the capacitive property of PPY based electrodes is significantly enhanced with the addition of both types of CNT (Zhou, Zhao et al. 2015a). PPY/PSS electrodes integrated with long CNT showed higher capacitive behavior and cycling stability because of porous surface morphology, core-shell nanostructure, and nano-network interconnects conducive of very long CNT.

A chemical oxidative polymerization route is successfully utilized to prepare novel PPY/bonded CNT composite (Yang, Shi et al. 2015). Synthesized PPY/bonded CNT composite shows good conductivity and thermal stability. PPY/CNT composite-based hybrid supercapacitor electrode is fabricated and its electrochemical performance is investigated. The experimental result reveals that supercapacitor electrode retain 92% maximum capacitance more than 3000 cycles, exhibiting superb cycle stability (Warren et al. 2015). Basically, GO and rGO are derived from GN, with the intention to get better capability rate and cycling stability of PPY. GN content can be adjusted in PPY/GN composites to get "cauliflower" morphology of PPY (Zhu, Xu et al. 2015). Due to porous morphology of the composites an excellent electrochemical performance can be obtained that is large electrical conductivity provides free pathways for exchange of ions/electrodes and speedy diffusion. A chemical oxidation polymerization method is used to prepare hierarchical GN/PPYnanosheet composites (Xu et al. 2011). The prepared hierarchical GN/PPYnanosheet composites revealed 318.6 F/g electrochemical specific capacitance. The specific capacitance retained 132.9 F/g at 100 mV/s scan rate after 1000 cycles. Surface initiated polymerization method is used to synthesize a novel PPY/GO core-shell nanocomposite (Wu et al. 2015). From result analysis, it is observed that 70 nm PPYnano-spheresare homogeneously developed on graphene oxide sheets, which produced a composite structure with continuous core shell. The PPY/GO composites reveal a 370 F/g specific capacitance which is excellent than pure PPY (216 F/g) at current density 0.5 A/g. Also, the PPY/GO and PPY electrodes shows better cycling stabilities of 91.2% and 57.8% maintain of specific capacitance is large than 4000 CV with scan rate 100 mV/s, respectively. PPY and GN quantum dots are effectively synthesized (Wu et al. 2013). The highest capacitance that is 485 Fg^{-1} at 5 mVs^{-1} is obtained, for PPY to GQD mass ratio of 50:1. The GO/PPY composites are fabricated via in situ chemical oxidation polymerization method (Fan et al. 2014). In GO/PPY composites, when GO to pyrrole mass ratios is 1:10, it shows great electrochemical performance. The prepared electrode exhibits 98.6 m^2/g specific surface area and capacitance of 332.6 F/g at 0.25 A/g. A hierarchical plush PPY layers intercalated GN sheet is synthesized through in situ intercalative chemical polymerization method (Liu et al. 2013). The GN/PPY based electrode exhibits specific capacitance is 650 F/g with highest energy density of 54 Wh/kg and highest power density of 778.1 W/kg. In addition to this, GN/PPY based supercapacitor electrode demonstrated the highest electrical conductivity of 1980 S/cm. The polymerization method is successfully utilized to prepare rGO doped with PANI, PPY and PEDOT (Zhang and Zhao 2012). The synthesized composite i.e. PPY/rGO showed specific capacitance of 248 F/g at 0.3 A/g current density. An electrochemical performance of synthesized PPY/rGO composite is superior to rGO/PEDOT composite of capacitance 108 F/g whereas its performance is poorer than rGO/PANI composite of specific capacitance 361 F/g with similar conditions, and maintained 81% of the initial value over 1000 charging and discharging cycles which is poor than rGO/PEDOT (88%) and rGO/PANI (82%) with the similar condition. The surface-initialed polymerization technique used to prepare a novel GO/PPY with 3D core/shell structure in which PPY nanospheres are uniformly coated on GO sheets (Wu et al. 2015). The synthesized supercapacitor electrode displays outstanding capacitance and very good cycle stability. The specific capacitance achieved by the supercapacitorelectrode is 370 F/g at 0.5 A/g among 8.0 mg/cm^2 mass loading and retention capacity of 91.2% over 4000 cycles due to synergistic effect of GO and PPY. Besides, CNTs, GNs, and their derivatives, other carbon materials are used by researchers to prepare composites with PPY, including PPY/active carbon (Keskinen et al. 2015), PPY/carbon cloth (Gao et al. 2015), PPY/graphite sheets (Raj et al. 2015; Tao et al. 2015), PPY/carbon nanofibers (Cai et al. 2015). It is concluded that as carbon nanomaterials have a lot of appropriate advantages, they are suitable for PPYsupercapacitor electrodes with the intention of greater cycling conductivity and stability. MixingPPY/carbon material with PPY/pseudocapacitance composites producing PPY/carbon material composites gives fast charge storage ability also improves power density and specific energy of supercapacitors through synergistic effects (Cao et al. 2004). The next subsection will introducethe research in the field of supercapacitor electrodes using PPY/metal oxide composites.

11.3.2.2 PPY-Metal Oxide Nanocomposites

PPY-based electrode material is very important for supercapacitor electrodes because it owing high conductivity, low cost, high charge storage ability and easy to synthesis but the capacitance of pure PPY is in the range of 200–400 F/g (Sun et al. 2009). Till date, preparation of high capacitance supercapacitor electrode using PPY-based electrode is still a challenge. Metal oxides, for example CoO (Zhou et al. 2013), MnO_2 (Ji, Zhang et al. 2015), VOx (Yu, Zeng et al. 2015), NiO (Ji, Ji et al. 2015), CuO (Qian et al. 2015), WO_3 (Wang, Zhan et al. 2015), Co_3O_4 (Cao et al. 2004), and V_2O_5 (Sun, Peng et al. 2015;Sun, Li et al. 2015) are ideal pseudocapacitive materials due to their high theoretical capacitance.

For example, PPY/CoO is a three-dimensional hybrid nanowire array developed on nickel foam template that shows excellent pseudocapacitive performance (Zhou et al. 2013). The capacitance of three dimensional PPY-CoO hybrid nano-wire array developed on nickel foam composite attain specific capacitance 2223 F/g at current density of 1 mA/cm^2, and cyclic stability maintain 99.8% over 2000 cycles. Also, positive electrode of aqueous solution of asymmetric supercapacitor among hybrid array displayed high energy density of43.5 W h/kg and 5500 W/kg power density at 11.8 W h/kg. In addition, the cyclic ability exhibited bythis supercapacitoris exceptional that is 20000 times. It concludes that synergetic effect of composites of CoO nano-wires and conductive PPY, enhanced capacitance. CuS microspheres are synthesized successfully in which first PPY is uniformly inserted into the subunit of intertwined sheet and then decorated on CuS surface. This structure showed an outstanding retention cyclic stability with initial specific capacitance of 227 F/g (Peng, Ma, Sun et al. 2014). An asymmetric supercapacitor core/shell nanowire of WO_3/PPY was successfully utilized as negative electrode (Wang, Zhan et al. 2015). The synthesized supercapacitor electrode achieves specific capacitance of 253 mF/cm^2, in the negative potentials range of -1.0 V to 0.0 V. In another work, Co $(OH)_2$ nanowires developed on carbon fiber substrate used as positive electrode and PPY/CF-WO_3 used as negative electrode. It exhibits energy density and volumetric capacitance achieve up to 1.02 mW h/cm^3 and 2.865 F/cm^3, and retained ~90.5% specific capacitance over 4000 cycles. PPY nanotube/MnO_2 a composite through a facile approach is synthesized which revealed significant value of specific capacitance of 403 F/g at 1 A/g and 88.6% cyclic stability over 800 cycles (Ji, Zhang et al. 2015). The electrochemical co-deposition rout is used to fabricate core/shell CuO/PPY nanosheet arrays on inter digital electrode (Qian et al. 2015). All solid-state devices demonstrated outstanding specific capacitance value i.e., 1275.5 F/cm^3 and 28.35 mW h/cm^3 energy density. In addition, solid- state device retain 100% specific capacitance at 2.5 A/cm^3 current density over 3000 cycles. All exceptional performances replicate its prospective in practical applications as a supercapacitor electrode. PPY/Carbon Aerogel (CA) composite is fabricated through chemical oxidation polymerization process (An et al. 2010). The PPY/CA electrode showed specific capacitance of 433 F/g at 1 mV/s, this performance is better than pure CA. In long life cycling test, loss of specific capacitance of PPY/CA composite is large for the duration of first 500 cycles maybe due to instability of PPY, but its specific capacitance value steady after following 500 cycles. Novel composites PPY/NFs, PPY/CNTs and MnO_2/NP are prepared during the aerogel assembly (Wang et al. 2017; Yang, Shi et al. 2015). Lightweight assembly of an aerogel can stay on the top of the feather which demonstrates same internal morphology. The synthesized composites such as PPY/NFs, PPY/CNTs and MnO_2/NP shows the value of specific capacitance 3.3 mF/cm^2, 2.4 mF/cm^2, and 2.1 mF/cm^2 respectively at 2 mV/s CV scan rate. Symmetric supercapacitor cells demonstrate retain the value of specific capacitance 84.2%, 61.7%, and 92.3% at 0.1 mA/cm^2 current density more than 2000 cycles for PPY/NFs, PPY/CNTs and MnO_2/NP respectively.

From a literature survey it has been found that activated carbon (AC) shows large cyclic life and also low cost, but its specific capacitance value and electrical conductivity are very low as compared to PPY (Choudhary et al. 2020). Activated carbon (AC) with PPY nanocomposites were prepared in situ electrochemical oxidation polymerization method. Prepared PPY/AC composites revealed high value specific capacitance i.e., 354 F/g at 1 mV/s (Muthulakshmi et al. 2006). A novel PPY/ nanocellulose fiber (NCF) composite illustrate maximum capacitance of 127 F/g at 33 A/g current density (Wang, Carlsson et al. 2015). In another work the influence of temperature on TiC nano-cube and PPY-PVA based electrodes are studied with the temperature range from $-18\ ^\circ$C to $60\ ^\circ$C (Weng et al. 2015). Through a collaborative effect of the composites PVA, PPY, and TiC showed high specific capacitance and retention stability in the temperature range studied. Both metal sulfide (MoS_2) and GN are analogue and can also be easily synthesized (Ma et al. 2013). Metal sulfide forms three layers of atom i.e. S-Mo-S, these layers are stacked jointly via Vander Waals interactions, and these multi-layers strip into single or multi-layer structures through chemical or physical process that display excellent mechanical and electrical properties. The nanocomposite PPY-MoS_2 shows excellent capacitance of 553.7 F/g and the cyclic capacity is maintained up to 90% at 1 A/g after 500 charge discharge cycles. In another work, PPY/MoS_2 nanocomposite synthesized by via oxide polymerization process. The synthesized PPY/MoS_2 nanocomposite reveals excellent specific capacitance of value 695 F/g at 0.5 A/g and outstanding retention stabilityof 85% over 4000 successive cycles of charge and discharge. Moreover, they also prepared PPY/MoS_2 electrode in the form of nanowire through oxidative polymerization process which shows very good electrochemical performance. Here, PPY/MoS_2 based nanowire composite used as negative electrode. The fabricated electrode demonstrates the capacitance of 462 F/g at 1 A/g with retention stability 82% after 2000 cycles. (Tang, Wang et al. 2015). In the last few years, researchers have made efforts to improve electrochemical performance by utilizing different kinds of unique structured materials such as nickel cobalt hexacyanoferrate (Ensafi et al. 2015), Ti_3C_2 (Zhu, Huang et al. 2016), and so on. In brief, instead of using traditional carbon nanomaterials and metal oxide materials, various types of mixed composite materials have been utilized and such a composition holds superb electrochemical properties.

11.3.2.3 PPY-Ternary Nanocomposites

As stated earlier, synthesis of the binary composite PPY among metal oxide and/or carbon material forms an electrode for supercapacitors and shows good electrochemical performance. However, the limitations of each component, the electrochemical properties, cycling stability and other properties of the binary composite electrode are not as good as the requirements of the supercapacitor electrode. A synthesis of ternary composite is the effective approach toget better specific capacitance and retention stability of supercapacitor electrodes through a collective effect. Therefore, researchers have grown interested in PPY-based ternary composites.

A MnO_2-PPY/TSA ternary nanocomposite electrode is synthesized for supercapacitor application (Dong et al. 2011). In this synthesis process, to form a homogeneous solution, MnO_2-PPY/TSA nanocomposite p-Toluenesulfonic acid (p-TSA) and pyrrole are dispersed ultrasonically in deionized water. Oxidant, redox reactions occurred by addition of $KMnO_4$ or $FeCl_3\ 6H_2O$ and MnO_2-PPY/TSA nanocomposite will be formed. A nanocomposite consists of MnO_2-PPY with TSA supercapacitor electrode revealed larger capacitance that is 376 F/g at 3 mA/cm^2 and enhanced retention stability in solution of 0.5M Na_2SO_4 than composite MnO_2-PPY. A three-Dimensional hierarchical CNT with MnO_2 core/shell nanostructure with PPY is prepared (Zhou, Han et al. 2015b). The ternary composite electrode demonstrated excellent capacitance of 529.3 F/g at 0.1 A/g and 98.5% initial capacitance can be maintain over 1000 continuous cycle. Prepared asymmetric supercapacitor that is CNT/PPY/MnO_2 showed superior energy density that is 38.42 W h/kg at 100 W/kg power densities and the initial capacity can retain 59.52% at 10,000 W/kg power densities. A novel ternary composite PPY/MnO_2 deposited on CNT textile is prepared with the intention of to prevent capacitance degradation (Yun et al. 2015). Furthermore, the ternary nanocomposite supercapacitors electrode demonstrates outstanding bending capacity. Therefore, such good performance electrode reveals its potential in supercapacitor application. The electro-polymerization route is successfully utilized to prepare novel ternary composite PPY/TiN/PANI coaxial nanotube array (Xie and Wang 2016). For PPY/PANI/TiN, PANI/TiN, and PPY/TiN the specific capacitances are 1471.9 F/g, 846.1 F/g and 744.8 F/g respectively, at 0.5 A/g current density. As compare to binary systems, the ternary nanotube hybrid shows higher capacitance due to the combined effect of PPY and PANI. On the other hand, it is observed that cycling stability of PPY/TiN/PANI is poor that is 78.0%, 46.30% and 38.80% retention of specific capacitance more than 200, 500 and 1000 cycles at 10 A/g current density, respectively. In the field of EDLC, GN and their derivatives i.e. carbon materials are most significant materials, there are numerous ternary system examined associated to them. The functionalized graphene and CNTs are synthesized via negatively charged poly (sodium 4-styrenesulfonate), and then PPY/GN/CNT composites prepared through in situ chemical oxidation polymerization method (Lu et al. 2012). The synthesized PPY/GN/CNT composites show a meso- and macro-porosity with large surface area that is 112 m^2/g. PPY/GN/CNT composites achieve the value of specific capacitance of 361 F/g at 0.2 A/g current density. Specific capacitance of PPY/GN/ CNT composites is much greater than pristine PPY (176 F/g) and PPY/CNT binary composite (253 F/g) and PPY/GN (265 F/g). The composite PPY/GN/CNT maintains initial capacitance of 96% more than 2000 charging and discharging cycles at 6 A/g. In situ potentiostatic electrochemical polymerization technique is used to prepare PPY/GP/CNTs composites (Aphale et al. 2015). PPY/ GP/CNTs electrodes achieved the value specific capacitance, energy, and power density 453 F/g, 62.96, and 566.66 W/kg respectively.

One-pot redox relay strategy is used to produce novel PPY/polyoxometalate/rGO ternary nano-hybrids (TNHs) (Chen, Yuan et al. 2015). The ternary nanohybrid composites reveal large specific capacitance and also the TNHs demonstrated excellent stabilityrate, mechanical stability and outstanding flexibility. Ternary composite PPY/GN/MnO_2 is synthesized by ultrasonic irradiation technique in which dopant is used as a p-toluenesulfonic acid. The composite showed a specific capacitance of 258 F/g at 1 A/g, (Sun et al. 2016). In similar work, a flexible ternary composite PPY/ GN/MnO_2 displayed a capacitance of 258 F/g at 1 A/g and demonstrates cyclic stability 96.58% over 1000 cycles (Ng et al. 2015). Furthermore, ZnO (Jiang et al. 2015) and TiO_2 (Chee et al. 2015) have been studied with PPY and GN and its derivatives. In the field of supercapacitor, all ternary material shows excellent properties and is capable to be used as electrode material. Many research-ers work on other ternary systems for example, an asymmetric supercapacitor is synthesized with Ni/PPY/MnO_2 and Ni/MnO_2/PPY as anode and cathode which showed specific capability of 191 F/g in the voltage potential range of 1.3–1.5 V (Chen, Liu, Lin et al. 2015). In another work, Ag metal nanoparticles are used to prepare composite with PPY and GN (Kalambate et al. 2015). Ag nanoparticles play vital part to improve electrochemical storage capacity and electrical conductivity of composite electrodes. Supercapacitor integrated through Ag nanoparticles illustrate capacitance 450 F/g at 0.9 mA/g current density and 92% retention stability over 1000 cycles. Table 11.3 sum-marizes the prepration method and electrochemical performance PPY based electrode materials.

11.3.3 POLY (3,4-ETHYLENE DIOXY THIOPHENE)

In 1980, Poly (3,4-ethylene dioxy thiophene) (PEDOT) was first established by scientists in Bayer AG research lab, Germany. PEDOT is from polythiophene family and has great prospective for electrodes in pseudocapacitor application due to speedy electrochemical reaction and excellent intrinsic conductivity then supercapacitors using CPs (Groenendaal et al. 2000). In supercapacitor applications, mostly two types of PEDOT conducting polymers are used. In first type, PEDOT pre-pare by using monomers of 3,4-Ethylenedioxythiophene (EDOT) by chemically oxidization method or electrochemically polymerization method (Chu et al. 2012; Jiang et al. 2012b; Choi et al. 2010). The other type is water solution of PEDOT:PSS which contain surfactants of PEDOT suspended in water. Previously it was found that PEDOT is insoluble, but later on insolubility problem was resolved by using water soluble poly (styrene sulfonic acid) (PSS) and polyelectrolyte (Groenendaal et al. 2000). Supercapacitor community researchers are attracted their attention towards PEDOT because it has excellent conductivity in the range from small value to 500 S/cm in doped state. Moreover, it has a large potential window, superior chemical, and thermal stability. Moreover, PEDOT has very good cycling stability, with capacitance retention of 80% above 70,000 cycles as compared to other conducting polymers (Pettersson et al. 1998).

TABLE 11.3

The Preparation Method and Electrochemical Performance of PPY-Based Carbon, Metal Oxide and Ternary Electrode Materials

Material	Fabrication Technique	Specific capacitance	Cyclestability (%)	No of Cycles	Reference
PPY	Interfacial polymerization	261 F/g at 25 mV/s	75%	1000	(Yang et al. 2015)
	Electro-polymerization	343 F/g at 5 mV/s	91%	4000	(Rajesh et al. 2016)
	In situ polymerization	325 F/g at 0.6 mA/cm^2	63%	500	(Xu et al. 2015)
	Chemical oxidation	576 F/g at 0.2 A/g	82%	1000	(Li and Yang 2015)
PPY/nanocellulose fibers, Chemical polymerization		127 F/g at 300 mA/cm^2	93%	5000	(Wang, Zhan et al. 2015)
PPY/MoS$_2$	Intercalative polymerization	553.7 F/g at 1 A/g	90%	500	(Ma et al. 2013)
In situ oxide polymerization		695 F/g at 0.5 A/g	85%	4000	(Tang et al. 2013)
PPY/MnO$_2$	In situ oxidative polymerization	403 F/g at 1 A/g	88.6%	800	(Ji, Zhanget al. 2015)
PPY/WO3	Electrochemical polymerization	253 mF/cm^2 at 0.67 mA/cm^2	85%	5000	(Wang, Zhan et al. 2015)
PPY/CoO	Chemical polymerization	2223 F/g at 1 mA/cm^2	99.8%	2000	(Zhou et al. 2013)
PPY/CuO	Electrochemical co-deposition	1275.5 F/cm^3 at 2.5 A/cm^3	~100%	3000	(Qian et al. 2015)
PPY/CNT/MnO$_2$	Deposition method	529.3 F/g at 0.1 A/g	98.5%	1000	(Zhou, Han et al. 2015)
PPY/GN/MnO$_2$	Ultrasonic irradiation	258 F/g at 1 A/g	-	-	(Sun et al. 2016)
PPY/polyoxometalate/ rGO, One-pot redox relay strategy		360 F/g at 0.5 A/g	-	-	(, Yuan et al. 2015)
Ni/PPY/MnO$_2$	Electrochemical deposition	350 F/g at 2 A/g	91.3%	5000	(Chen, Liu et al. 2015)

In the supercapacitor application, PEDOT electrode material revealed 103 F/g value of specific capacitance in 1 M Et4NBF4/acetonitrile (Villers et al. 2003). Many researchers have been used PEDOT in asymmetric type supercapacitors. Asymmetric supercapacitors mechanism consists of activated carbon and PEDOTshows specific capacitance 27 F/g in LiPF6 and dimethyl carbonate (EC/DMC) with ethylene carbonate and 22 F/g in 1 propylene carbonate (PC)/M Et4NBF4. More than 1000 cycles, the cells have a capability of 50 F/g in EC/DMC and 19 F/g in PC (Ryu et al. 2004). PEDOT Electropolymerized on platinum shows ultimate capacitive properties and has130 F/g value of specific capacitance (Liu et al. 2008). Likewise, vapor phase polymerization process conducted in vacuum oven and is used forPEDOT polymerization through vapor phase PEDOT monomer. PEDOT is deposited by VPP on the foil of carbon coated aluminum which shows 134 F/g specific capacitance [(Tong et al. 2015)103]. PEDOT particles, blocks and nanorods are synthesized by electrochemical deposition method with current density controlling which revealed specific capacitance 37 F/g, 72 F/g and 109 F/g respectively (Li et al. 2010). The specific capacitance value is decreased due to area of surface decreased. In brief, PEDOT can be modified physically and chemically for the supercapacitor's applications.

Poor mechanical stability during cycling is the most important weakness of PEDOT supercapacitor electrode. Decrease in conductivity of PEDOT supercapacitor electrode materials is due to volumetric modification at the doping and de-doping progression which causes shrinkage, cracking, breaking, and swelling. Also, during the electrochemical activity, PEDOT polymer depredated due to the oxidation process (White et al. 2004). Furthermore, nearly all of conducting polymers, PEDOT also has poorly conductive in reduced state. The conductivity problem at reduced potential can be enhanced by mix with carbon materials. For example, PEDOT integrating with carbon nanotubes may increase the cycling life of supercapacitors. (Lota et al. 2004). From literature survey, it is observed that composite materials might be adapted the PEDOT volume change upon insertion/extraction process of ions. Synthesis of PEDOT with CNTs by using electrochemical deposition method shows that the value of specific capacity of 150 F/g (Peng et al. 2006).

11.3.3.1   PEDOT-Carbon Nanocomposites

From literature review, it is found that, depending on the various polymerization methods, pristine PEDOT supercapacitor exhibits specific capacitance between 70 to 130 F/g (Zhao et al. 2015). On the other hand, PEDOT/Carbon based composite improve cyclic stability and specific capacitance performance of supercapacitors. The main roles of carbon-based materials like GN/CNT in the composites as supercapacitor electrodes are

i. Carbon-based compositeshave heterogeneous structure along with PEDOT, which successfully decreases structural damage, namely, peeling off, cracking, and collapse, but it causes shrinkage and swelling (i.e. volumetric change) of PEDOT at charging and discharging cycles.

ii. The electrical conductivity of GN is larger than the conductivity of PEDOT:PSS. This makes GN/PEDOT:PSS composite more conductive.

iii. Addition of GN in nanocomposite makes three-dimensional morphology that may lead to considerable enhancement in the value of specific capacitance by giving a wide surface area for redox reactions and electrolyte penetrations (Zhao et al. 2015).

For example, Microwave-assisted synthesis route is used to fabricate PEDOT/GN composites. In this method, PEDOT is uniformly grown on rGO resulting large surface area also large electrical conductivity. The nearly rectangle shape of CV curves of PEDOT/GN composites signifies excellent specific capacitance. The PEDOT/GN composites revealed 270 F/g specific capacitance with retention of 93% above 10000 cycles (Sun et al. 2013). To create multilayers, PEDOT and GN are deposited in sequence on substrate of gold foil; this multilayer provides large specific capacitance because of strong interactions among huge charge mobility. During charge—discharge process, GN sheet prohibited the shrinking and swelling of sheets of PEDOT which gives large cycling stability. As compared to conventional GN-PEDOT film, performance of multilayer's showed specific capacitance of 154 F/g and 21% enhancement in cyclic stability (Chu et al. 2012). PEDOT/GN composite hydrogel is madeup in situ polymerization method through sulfonic acid functionalization process. The prepared composites display a porous formation having ~ 0.5 µm pore size.

GN sheets with covalent functionalization support for dispersion in matrix, as a result, electrical conductivity will be reduced to 0.17 S/cm; although the nanocomposite illustrated specific capacitance of 220 F/g (Han et al. 2013). The Langmuir—Blodgett method is successfully used to prepare GO layers followed by thermal reduction of oxygen function group coupled with vapor phase polymerization of EDOT (Wen et al. 2014). PEDOT layer of 40 nm thicknesses is deposited on top of a GN layer by varying deposition time, showing 377.2 S/cm electrical conductivity. Also, the nanocomposite demonstrated 213 F/g value of specific capacitance holding 87% over 2000 cycles. In another work, PEDOT/GN nanocomposite is prepared via in situ polymerization of EDOT confirm 108 F/g value of specific capacitance and 88% maintain over 1000 cycles. Accumulation of RGO, causes poor dispersion and also low contact among PEDOT causes poor performance of PEDOT/GN nanocomposite (Zhang and Zhao 2012). RGO/PEDOT nanocomposite is synthesized via chemical oxidation polymerization method. The electrochemical performance of RGO/PEDOT nanocomposite enhanced as compare to pristine PEDOT supercapacitors. Also, the RGO/PEDOT nanocomposite provides a more rapidly electrochemical reaction having 350 F/g specific capacity. (Alvi et al. 2011). Under hydrothermal conditions, an in situ polymerization technique has been applied for the synthesis of PEDOT-CNT composite with one-dimensional core-sheath nanostructure supercapacitor electrode (Chen et al. 2009). PEDOT-CNT composites achieve 198.2 F/g specific capacitance which is the highest value at 0.5 A/g current density when the content of PEDOT makes 50%. However, the specific capacitance of supercapacitor electrode decreases to 26.9% after 2,000 cycles. The capacitance and stability of PEDOT-CNT nanocomposite with core-sheath nanostructures supercapacitor electrode are excellent than pure PEDOT and PEDOT-CNT composite without core-sheath structures (Lu et al. 2011).

11.3.3.2 PEDOT-Metal Oxide Nanocomposites

RuO_x-deposited PEDOT supercapacitor electrode is fabricated by a dipping-hydrolysis and electrolysis method (Hong et al. 2001). RuO_x-deposited PEDOT supercapacitor electrode showed optimize composite structure with specific capacitance value of 420 F/g at current density of 50 mV/s. PEDOT/RuO_x and PEDOT having symmetrical cells stored 12.40 mA h/g and 27.50 mA h/g specific capacities respectively, during the charging of cells from 0 V to 1 V at 100 to 400 mA/cm^2 current. Surprisingly, the result shows that current density increases from 100 mA/cm^2 to 400 mA/cm^2 it could not decrease the stored capacity of supercapacitor electrode. Furthermore, PEDOT/RuO_x and PEDOT based cells result showed 12.40 Wh/kg and 27.50 Wh/kg stored energy densities respectively. Aerogels of PEDOT:PSS and MnO_2 are prepared via mechanical integration through MnO_2 particles following by freeze drying techniques. PEDOT:PSS/MnO_2 aerogel displayed porous arrangement, with 1068 F/g specific capacitance at 1 mV/s current density and 95% retention capacity over 2000 cycles. While three-dimensional morphology leads to large specific capacitance, deficient contact between MnO_2 particles and PEDOT:PSS causes a considerable reduction to 206 F/g at current density 100 mV/s (Ranjusha et al. 2014).

Various types of manganese sources, for example manganese (II) acetate (Su et al. 2013), $Mn(CH_3COO)_2$ (Tang et al. 2013), $MnSO_4$ (Hu et al. 2003), $LMnCl_2$ (Jiang and Kucernak 2002), and so on are used to formed MnO_2 by electrochemical deposition. Co-electro-deposition method with manganese (II) acetate used to synthesis PEDOT:PSS/MnO_2 composites (Su et al. 2013). MnO_2 displayed porous nano-spheres morphology of 100 nm to 500 nm having 400 F/g capacitance at current density 1 mV/s and specific capacitance of 160 F/g at 100 mV/s. As compare to directly mixing process, Co-electro-deposition route showed enhanced stability under various scan rates and maintain 99.5% retention stability above 4000 cycles. Similarly, MnO_2 nanorods are produced by redox reaction among $MnSO_4$ and $KMnO_4$ following in situ polymerized route of EDOT, revealed a value of specific capacitance 315 F/g. As compare to nanoparticles, nanorod has higher surface area. But specific capacitance value of nanorod is still lower than nanoparticles. The deprived performance is because of poor spreading of nanorod in PEDOT (Sen et al. 2013). Furthermore, urchin like MnO_2 particles from $Mn(CH_3COO)_2$ deposited electrochemically displayed particles size with average diameters of 500 nm, and nanofiber lengths of 50–250 nm were found and reveal 487 F/g specific capacitance (Tang et al. 2013). From the literature analysis, it is seen that various sources and different deposition techniques can produce different morphologies that creates a noteworthy outcome on capacitive performance. Redox reaction among reducing agents and potassium permanganate can be producing MnO_2.Forinstance, through a doping process, PEDOT can donate electrons to $KMnO_4$. Hence, MnO_2 nanoparticles are formed by soaking PEDOT films into $KMnO_4$ solution. Simultaneous deposition and formation displayed homogeneous allocation of particles, possessing 410 F/g specific capacitance (Liu, Duay et al. 2010). PEDOT with MoO_3 is prepared by EDOT monomer chemically polymerization among $FeCl_3$ as an oxidizing agent in MoO_3 suspension (Murugan et al. 2006). The PEDOT/MoO_3 nanocomposite has excellent electrochemical performance that is highest 300 F/g specific capacitance as compared to pure MoO_3,which exhibits 40 mF/g specific capacitance. An electrochemical performance improved because of increase in surface area as well as intercalation of electrically conducting between layers of MoO_3 and PEDOT. In another work, PEDOT-$NiFe_2O_4$ nanocomposite is synthesized by chemicalpolymerization of monomer of EDOT in a solution consisting of nickel ferrite nanoparticles ($NiFe_2O_4$) (Sen et al. 2010). They also prepared pristine polymer, PEDOT in n-hexane, and aqueous medium with the same method without $NiFe_2O_4$ nanoparticles. The synthesized nanocomposite of PEDOT-$NiFe_2O_4$ illustrated large capacitance value (251 F/g) whereas $NiFe_2O_4$ (127 F/g) and PEDOT (156 F/g) shows low specific capacitance. In this case, pore structure morphology plays an important role over total surface area.

11.3.3.3   PEDOT-Ternary Nanocomposites

Ternary composite is one of the most important candidates for upcoming active material for researchers in the field of supercapacitor applications. The reason behind the use of ternary composite is its excellent properties, including large specific capacitances, conductivities, and cyclic stabilities. These levels cannot be attained using binary materials alone (Li et al. 2013). Nowadays, researchers use carbon, metal oxide, and conducting polymer as a ternary composite. In a three-phase ternary composite, conducting polymer (PEDOT) is used for supercapacitor applications due to collaborative effects between these three phases. In supercapacitors, phase morphology is most important for its electrochemical performance. For example, ternary composite graphite/PEDOT/MnO_2 electrode, in which, graphite is used through pencil drawing, unlike via EDOT and the MnO_2 electrochemical deposition method. PEDOT deposited on graphite substrate displaysfiber-like morphology. The synthesized ternary nanocomposite electrodes show a maximum capacitance value of 264 F/g. PEDOT layers deposition considerably decreased electrodes internal resistance in the range of 1133 Ω to 4.60 Ω, assisting deposition of MnO_2. It is found that, for redo reactions, only very thin layer of MnO_2 is utilized on the other hand as the thickness of MnO_2 layer increases and reach at optimum point, the specific capacitance of the composites decreases gradually (Tang et al. 2014). In another work, PEDOT:PSS/RGO/MnO_2 ternary nanocomposite synthesized in two step process; first MnO_2/RGO nanocomposites are fabricated by hydrothermal method from GO and $KMnO_4$ and in second step PEDOT:PSS is added in RGO/MnO_2 composites. Synthesized ternary composite revealed low value capacitance of 169 F/g is due to poor contact between three phases (Yan et al. 2014). PEDOT/MnO_2/MWCNT supercapacitor electrodes are fabricated with the same approach. PEDOT/MnO_2/MWCNTs three-phase supercapacitor electrode composite exhibited large capacitance of 147 F/g, as compare to MnO_2/MWCNTs 141 F/g (Yoon and Kim 2013). It is found that the capacitance of three-phase supercapacitor electrode nanocomposite increases due to PEDOT layer. In one more study, the co-electrodeposition method is used to fabricate MnO_2-PEDOT composite on Co_3O_4@graphite foam, which revealed 350 F/g specific capacity up to 20,000 cycles with current rate of 5 A/g (Xia et al. 2014). Ternary solid-state supercapacitor electrodes is prepared success-fully using carbon black/carbon nanotube/MnO_2/PEDOT:PSS (Garcia-Torres and Crean 2018). As synthesized ternary electrode nanocomposite revealed a large specific capacitance value of 351 F/g. This may be due to the contributory collaboration in ternary composite of each material. To enhance electrochemical performance of ternary composite, 3D morphology provides high surface area for electrochemical reactions and electrolyte penetrations. PEDOT/GN/carbon cloth ternary composite synthesized for supercapacitor electrode applications (Jiang, Yao et al. 2012). In this ternary composite, carbon cloth plays a role of conducting 3D scaffold. In the synthesis process, filtration method is used to deposit functionalized GN sheets on carbon cloth, followed via electro-chemical polymerization of EDOT. After deposition of GN, wrinkles are observed on the smooth surface of carbon cloth that increases surface area of PEDOT for redox reaction. The outcome of this scenario is the nanocomposite revealed a maximum specific capacitance value of 714.93 F/g. Hence, 3D scaffold large surface area morphology plays a vital role for supercapacitor applications. PEDOT on flexible 3D carbon fiber cloth (CFC) is fabricated via hydrothermal process (Rajesh et al. 2017). In this research work, CFC is used as a substrate because CFC has availability of big-ger surface area, excellent conductivity, large porosity, advanced chemical stability, low weight, low-priced and flexibility. To increase, electrode and electrolyte contact area and improvement in ion diffusion is achieved due to uniform distribution and growth of PEDOT on CFC surface. As a result, synthesized PEDOT/CFC electrode delivered large value of 203 F/g specific capacitance at 5 mV/s. The capacitance of PEDOT/CFC electrode retained about ~86% over 12000 cycles. This shows good electrochemical stability of PEDOT/CFC electrode. Electro-polymerization method is preferable for growth of PEDOT directly on current collector without binder addition. Electro spin-ning and electro-polymerization method is used to synthesis PEDOT coated on PVA-GO nanofibres (Mohd Abdah et al. 2017). Images using FESEM microscopy displayed web structure of GO/PVA nanofibersare completely covered by porous cauliflower like structure of PEDOT. As result, coating of PEDOT on PVA/GO nanofiber increase electroactive surface area of nanocomposites and also improved its charge storage capacity.

Electrochemical performance of hybrid PEDOT/PVA-GO Ternary electrode showed that the value of specific capacitance is around 224.27 F/g and 9.58 Wh/kg specific energy and the specific capacitance value decreased by around 28.9% following 5000 cycles. Authors declare that cyclic stability of PEDOT/PVA-Graphene Oxide hybrid electrode was low might be because of shrinkage and swelling in structure of PEDOT and succeeding electrode material deterioration. The electro spinning and electro polymerization is successfully utilized to prepare PVA-GQD-Co3O4/PEDOT ternary composite material (Abidin et al. 2018). It is observed that average diameter size of fibers decreases from 44 nm to 13 nm after the addition of Co_3O_4 nano-particles. The decrease in diam-eter reduces the pathway of electron transfer even as providing large active sites for storage of charge. Fiber nanocomposite illustrates high value that is 361.97 F/g specific capacitance at scan rate 100 mV/s.

A similar approach has been utilized to prepare PVA/graphene quantum dot (GQD)/PEDOT (PVA/GQD/PEDOT) composites (Syed ZainolAbidin et al. 2018). Uniform coating of PEDOT on PVA/GQD nanofibres demonstrated that the presence of GQDs provides extra nucleation for the homogeneous deposition of PEDOT. Synthesized PVA-GQD/PEDOT) fiber composite revealed 291.86 F/g that is high specific capacitance value with scan rate 100 mV/s and demonstrates out-standing cyclic stability at 98% retention of specific capacitance more than 1000 cycles. Because of low value specific energy i.e.16.95 Wh/kg, use of PVA-GQD/PEDOT electrode in supercapacitor applications is still limited. This problem may be resolved by introducing TMOs and CPs into the nanofiber composites, which improves the capacitive properties of electrodes. Ternary composite CNT/MnO_2/PEDOT-PSS is prepared for electrochemical properties. The ternary nanocomposite showed a high specific capacitance value of 200 F/g. Moreover, the electrode exhibited an excellent charge/discharge rate and good cycling stability; retaining capacity is over 99% of its initial charge after 1000 cycles. The author recommended that the mechanical stability of ternary composite can

TABLE 11.4

The Preparation Method and Electrochemical Performance of PEDOT Based Carbon, Metal Oxide, and Ternary Electrode Materials

Material	Fabrication Technique	Specific capacitance	Cyclestability (%)	No of Cycles	Reference
PEDOT	In situ polymerization	103 F/g	-	-	(Villers et al. 2003)
	Electrochemical deposition	109 F/g (Nanorods),	-	-	(Li et al. 2010)
	Vapor phase Polymerization	134 F/g	-	-	(Tong et al. 2015)
	Electro-polymerization	130 F/g	-	-	(Liu et al. 2008)
PEDOT/GN	Microwave-assisted	270 F/g at 1 A/g	93%	10000	(Sun et al. 2013)
	Electrodeposition Method	154 F/g at 1 A/g	71%	1000	(Chu et al. 2012)
	In situ polymerization	220 F/G at 1 A/g	-	-	(Han et al. 2013)
	Langmuir-Blodgett method	213 F/g at 1 A/g	87%	2000	(Wen et al. 2014)
	In situ polymerization	108 F/g and	88%	1000	(Zhang and Zhao 2012)
PEDOT:PSS/MnO$_2$	Mechanical integration	1068 F/g at 1 mV/s	95%	2000	(Ranjusha et al. 2014)
	Co-electro–deposition method	400 F/g at 1 mV/s	99.5%	4000	(Su et al. 2013)
PEDOT/MoO$_3$	Chemical Polymerization	300 F/g at 1 mV/s	85%	1000	(Murugan et al. 2006)
PEDOT-NiFe$_2$O$_4$	Chemical Polymerization	251 F/g at 1 mV/s	78%	1000	(Sen and De 2010)
Graphite/PEDOT/MnO$_2$	electrochemical deposition	264 F/g at 5 mV/s	-	-	(Tang et al. 2014)
PVA-GO/PEDOT	Electrospinning	224.27 F/g	71%	5000	(Mohd Abdah et al. 2017)
PEDOT/GN/CFC	Hydrothermal	203 F/g at 5 mV/s	~86%	12000	(Rajesh et al. 2017)
MnO$_2$/RGO/PEDOT:PSS	Hydrothermal Method	169 F/g	-	-	(Yan et al. 2014)
CB/CNT/MnO$_2$/PEDOT:PSS	Wet spinning method	351 F/g at 5 mV/s	-	-	(Garcia and Crean 2018).
PVA/GQD/Co$_3$O$_4$/PEDOT	Electrospinning	361.97 F/g at 100 mV/s	92%	1000	(Abidin et al. 2018)
PVA/GQD/PEDOT	Electrospinning	291.86 F/g at 100 mV/s	98%	1000	(Syed Z. Abidin et al. 2018)
MnO$_2$/CNT/PEDOT-PSS	In situ Polymerization	200 F/g at 100 mV/s	99%	1000	(Hou et al. 2010).
CNT/PEDOT:PSS/MnO$_2$	In situ Polymerization	478.6 F/cm^3 at 0.05 A/cm^3	91%	10, 000	(Cheng et al. 2016)
PEDOT:PSS/NiFe$_2$O$_4$/rGO	One Step method	1090 F/g at 0.5 A/g	94%	750	(Hareesh et al. 2016)
MnO$_2$/rGO/PEDOT:PSS	Hydrothermal method	633 F/gat 0.5 A/g	100%	5000	(Hareesh et al. 2017)

be increased by conductive CNTs that provide a high surface area for depositing porous MnO_2 nanospheres (Hou et al. 2010). Ternary CNT/PEDOT:PSS/MnO_2 fiber electrode is fabricated to study the effect of MnO_2 and CNT in PEDOT for electrochemical properties (Cheng et al. 2016). In fabrication process, nanosheets of MnO_2 incorporated along with PEDOT:PSS layer coated on CNT fiber. The layer coating of PEDOT:PSS is formed by repetitively dipping CNT fiber into the solution of PEDOT:PSS than the composite annealing at 120°C. The MnO_2 nanosheets are grown electrochemically on ternary electrode consist of PEDOT:PSS/CNT/MnO_2 fiber. Middle layer of PEDOT:PSS provides pseudocapacitance. Also, the middle layer of PEDOT:PSS work as binder to join MnO_2 outer layer moreover it join to inner layer of CNT fiber. The electrochemical performance of ternary CNT/PEDOT:PSS/MnO_2 fiber electrode is better than its binary counterpart. A ternary CNT/PEDOT:PSS/MnO_2 fiber electrode possess high specific capacitance value of 478.6 F/cm^{-3} or 411.6 F/g and is achieve 0.05 A/cm^{-3} which is greater than CNT/MnO_2 nanocomposite i.e. 386.9 F/cm^{-3} and large specific capacitance retention rate of 91% after 10,000 cycles. In our lab, A ternary nanocomposite that is PEDOT:PSS/$NiFe_2O_4$/rGO (GNP) is fabricated by using one-step method for the supercapacitor application. The GNP nanocomposite result reveals that specific capacitance of 1090 F/g and energy density 660 Wh/kg at current density 0.5 A/g with 94% cycling stability of retention of capacitance after 750 cycles. (Hareesh et al. 2016). In one more study ternary nanocomposite MnO_2/rGO/PEDOT:PSS (MGP) which consist of rGO sheets and MnO_2 nanorods supported on PEDOT:PSS polymer is developed by hydrothermal method for supercapacitor electrodes. Specific capacitance of MGP nanocomposite is 633 F/g at 0.5 A/g with 100% specific capacitance retention over 5000 cycles. (Hareesh et al. 2017). Table 11.4 summarizes the synthesis method and electrochemical performance PEDOT based electrode materials. In outline, ternary composites based on PANI, PPY and PEDOT can take compensations of each component through collaborative effects, thus demonstrating superior cyclic stability and overall improved electrochemical performance. Hence, an investigator has been focusing their devotion on conducting polymer based ternary composites and this is the proper direction for research work future.

11.4   SUMMARY AND CONCLUSIONS

This chapter gives an overall review of CP nanocomposites for advanced electrode materials that have been studied for supercapacitor applications. The chapter began by discussing the requirement of energy storage devices for continuous development and how CPs play a crucial role for advances in supercapacitors. Furthermore, this chapter presents a number of key findings to researchers for future investigation on CPs and CP-nanocomposite supercapacitors. CPs, mostly PANI, PPY, and PEDOT, have a lot of exceptional advantages, for instance, high pseudocapacitance, flexibility, and ease of synthesis. These properties are used to resolve challenges to improve electrodes of existing supercapacitors. However, pristine CP electrodes reveal some limitations, mainly poor cycling stabilities, low energy, and power density. To overcome this issue, effective combinations of CPs and dopant materials are used to fabricate composite material electrodes. As a result, combinations of such composite nanostructures exhibit outstanding cyclic stability and improvement in energy density. At present, the specific capacitance of supercapacitors along with CP composites do not achieve significant results experimentally as compared to theoretical values. With the intention of improvement in electrochemical performance of conducting polymer-based nanocomposite, the following points should be considered:

i. In primary stages it is essential to improve crystallinity, optimization of surface morphology, and microstructure by using various polymerization methods, concentration of dopants, surfactant's type, oxidation level, etc.

ii. The preparation process is one of the most important and effective ways to get better thermal stability, mechanical properties, and processing abilities of CP nanocomposites to achieve practical applications.

iii. From recent publications it is found that fabrication of CP-based nanocomposites with various active materials, for example, CPs with metal oxides/hydroxides/sulfides, havea significant process to develop electrochemical activities of supercapacitors through synergistic effects.

iv. For energy storage mechanisms, both EDLC and pseudocapacitance are used, while CPs with carbon materials like CNTs, carbon aerogels, GN, and other carbon materials are used for improvement of capacitance, power, and energy density.

v. Electrochemical performance of binary composites can be improved by design and fabrication of ternary nanostructure composites, that is, by using three types of materials like metal oxides, CPs, and carbon materials – or else other types of pseudo material may be more consistent for effective performance of supercapacitors.

vi. The use of theoretical or computational methods is also supportive for mixtures of suitable polymers and inorganic nanostructures with metal oxides and carbon nanomaterials. Therefore, more research work could be expected on the modeling field.

vii. In the future, the electrochromic uniqueness of CPs may bring a wonderful advantage to applications in stretchable and flexible as well as cost-effective and good electrochemically performing supercapacitor electrodes.

ACKNOWLEDGMENT

Financial Support from University Grants Commission (UGC) New Delhi, India (F.No.F-47-1058/14 (GENERAL/64/WRO/XIIth Plan 2017) is gratefully acknowledged.

REFERENCES

Abdah, M.A.A.M., Azman, N.H.N., Kulandaivalu, S. and Sulaiman, Y., 2020. Review of the use of transition-metal-oxide and conducting polymer-based fibres for high-performance supercapacitors. *Materials & Design*, *186*, p. 108199.

Abdah, M.A.A.M., Edris, N.M.M.A., Kulandaivalu, S., Rahman, N.A. and Sulaiman, Y., 2018. Supercapacitor with superior electrochemical properties derived from symmetrical manganese oxide-carbon fiber coated with polypyrrole. *International Journal of Hydrogen Energy*, *43*(36), pp. 17328–17337.

Abdah, M.A.A.M., Zubair, N.A., Azman, N.H.N. and Sulaiman, Y., 2017. Fabrication of PEDOT coated PVA-GO nanofiber for supercapacitor. *Materials Chemistry and Physics*, *192*, pp. 161–169.

Abidin, S.N.J.S.Z., Mamat, M.S., Rasyid, S.A., Zainal, Z. and Sulaiman, Y., 2018. Electropolymerization of poly (3, 4-ethylenedioxythiophene) onto polyvinyl alcohol-graphene quantum dot-cobalt oxide nanofiber composite for high-performance supercapacitor. *Electrochimica Acta*, *261*, pp. 548–556.

Afif, A., Rahman, S.M., Azad, A.T., Zaini, J., Islan, M.A. and Azad, A.K., 2019. Advanced materials and technologies for hybrid supercapacitors for energy storage—A review. *Journal of Energy Storage*, *25*, p. 100852.

Alvi, F., Ram, M.K., Basnayaka, P.A., Stefanakos, E., Goswami, Y. and Kumar, A., 2011. Graphene—polyethylenedioxythiophene conducting polymer nanocomposite based supercapacitor. *Electrochimica Acta*, *56*(25), pp. 9406–9412.

An, H., Wang, Y., Wang, X., Zheng, L., Wang, X., Yi, L., Bai, L. and Zhang, X., 2010. Polypyrrole/carbon aerogel composite materials for supercapacitor. *Journal of Power Sources*, *195*(19), pp. 6964–6969.

An, K.H., Jeon, K.K., Heo, J.K., Lim, S.C., Bae, D.J. and Lee, Y.H., 2002. High-capacitance supercapacitor using a nanocomposite electrode of single-walled carbon nanotube and polypyrrole. *Journal of the Electrochemical Society*, *149*(8), p.A1058.

Aphale, A., Maisuria, K., Mahapatra, M.K., Santiago, A., Singh, P. and Patra, P., 2015. Hybrid electrodes by in situ integration of graphene and carbon-nanotubes in polypyrrole for supercapacitors. *Scientific Reports*, *5*(1), pp. 1–8.

Atchudan, R., Edison, T.N.J.I., Perumal, S., Thirukumaran, P., Vinodh, R. and Lee, Y.R., 2019. Green synthesis of nitrogen-doped carbon nanograss for supercapacitors. *Journal of the Taiwan Institute of Chemical Engineers*, *102*, pp. 475–486.

Ates, M., Serin, M.A., Ekmen, I. and Ertas, Y.N., 2015. Supercapacitor behaviors of polyaniline/CuO, polypyrrole/CuO and PEDOT/CuO nanocomposites. *Polymer Bulletin*, *72*(10), pp. 2573–2589.

Azman, N.H.N., Lim, H.N. and Sulaiman, Y., 2016. Effect of electropolymerization potential on the preparation of PEDOT/graphene oxide hybrid material for supercapacitor application. *Electrochimica Acta, 188*, pp. 785–792.

Bai, M.H., Liu, T.Y., Luan, F., Li, Y. and Liu, X.X., 2014. Electrodeposition of vanadium oxide—polyaniline composite nanowire electrodes for high energy density supercapacitors. *Journal of Materials Chemistry A, 2*(28), pp. 10882–10888.

Cai, J., Niu, H., Li, Z., Du, Y., Cizek, P., Xie, Z., Xiong, H. and Lin, T., 2015. High-performance supercapacitor electrode materials from cellulose-derived carbon nanofibers. *ACS Applied Materials & Interfaces, 7*(27), pp. 14946–14953.

Cao, L., Xu, F., Liang, Y.Y. and Li, H.L., 2004. Preparation of the novel nanocomposite Co (OH) 2/ultra-stable Y zeolite and its application as a supercapacitor with high energy density. *Advanced Materials, 16*(20), pp. 1853–1857.

Chee, W.K., Lim, H.N., Harrison, I., Chong, K.F., Zainal, Z., Ng, C.H. and Huang, N.M., 2015. Performance of flexible and binderless polypyrrole/graphene oxide/zinc oxide supercapacitor electrode in a symmetrical two-electrode configuration. *Electrochimica Acta, 157*, pp. 88–94.

Chen, C., Fan, W., Zhang, Q., Ma, T., Fu, X. and Wang, Z., 2015. In situ synthesis of cabbage like polyaniline@ hydroquinone nanocomposites and electrochemical capacitance investigations. *Journal of Applied Polymer Science, 132*(29).

Chen, G.F., Liu, Z.Q., Lin, J.M., Li, N. and Su, Y.Z., 2015. Hierarchical polypyrrole based composites for high performance asymmetric supercapacitors. *Journal of Power Sources, 283*, pp. 484–493.

Chen, J., Xia, Z., Li, H., Li, Q. and Zhang, Y., 2015. Preparation of highly capacitive polyaniline/black TiO2 nanotubes as supercapacitor electrode by hydrogenation and electrochemical deposition. *Electrochimica Acta, 166*, pp. 174–182.

Chen, L., Sun, L.J., Luan, F., Liang, Y., Li, Y. and Liu, X.X., 2010. Synthesis and pseudocapacitive studies of composite films of polyaniline and manganese oxide nanoparticles. *Journal of Power Sources, 195*(11), pp. 3742–3747.

Chen, L., Yuan, C., Dou, H., Gao, B., Chen, S. and Zhang, X., 2009. Synthesis and electrochemical capacitance of core—shell poly (3, 4-ethylenedioxythiophene)/poly (sodium 4-styrenesulfonate)-modified multiwalled carbon nanotube nanocomposites. *Electrochimica Acta, 54*(8), pp. 2335–2341.

Chen, Y., Du, L., Yang, P., Sun, P., Yu, X. and Mai, W., 2015. Significantly enhanced robustness and electrochemical performance of flexible carbon nanotube-based supercapacitors by electrodepositing polypyrrole. *Journal of Power Sources, 287*, pp. 68–74.

Chen, Y., Han, M., Tang, Y., Bao, J., Li, S., Lan, Y. and Dai, Z., 2015. Polypyrrole—polyoxometalate/reduced graphene oxide ternary nanohybrids for flexible, all-solid-state supercapacitors. *Chemical Communications, 51*(62), pp. 12377–12380.

Chen, Y., Liu, B., Liu, Q., Wang, J., Liu, J., Zhang, H., Hu, S. and Jing, X., 2015. Flexible all-solid-state asymmetric supercapacitor assembled using coaxial NiMoO4 nanowire arrays with chemically integrated conductive coating. *Electrochimica Acta, 178*, pp. 429–438.

Cheng, X., Zhang, J., Ren, J., Liu, N., Chen, P., Zhang, Y., Deng, J., Wang, Y. and Peng, H., 2016. Design of a hierarchical ternary hybrid for a fiber-shaped asymmetric supercapacitor with high volumetric energy density. *The Journal of Physical Chemistry C, 120*(18), pp. 9685–9691.

Choi, K.S., Liu, F., Choi, J.S. and Seo, T.S., 2010. Fabrication of free-standing multilayered graphene and poly (3, 4-ethylenedioxythiophene) composite films with enhanced conductive and mechanical properties. *Langmuir, 26*(15), pp. 12902–12908.

Choudhary, R.B., Ansari, S. and Purty, B., 2020. Robust electrochemical performance of polypyrrole (PPy) and polyindole (PIn) based hybrid electrode materials for supercapacitor application: A review. *Journal of Energy Storage, 29*, p. 101302.

Chu, C.Y., Tsai, J.T. and Sun, C.L., 2012. Synthesis of PEDOT-modified graphene composite materials as flexible electrodes for energy storage and conversion applications. *International Journal of Hydrogen Energy, 37*(18), pp. 13880–13886.

Cong, H.P., Ren, X.C., Wang, P. and Yu, S.H., 2013. Flexible graphene—polyaniline composite paper for high-performance supercapacitor. *Energy & Environmental Science, 6*(4), pp. 1185–1191.

Das, A.K., Karan, S.K. and Khatua, B.B., 2015. High energy density ternary composite electrode material based on polyaniline (PANI), molybdenum trioxide (MoO3) and graphene nanoplatelets (GNP) prepared by sono-chemical method and their synergistic contributions in superior supercapacitive performance. *Electrochimica Acta, 180*, pp. 1–15.

Deng, X., Li, J., Zhu, S., He, F., He, C., Liu, E., Shi, C., Li, Q. and Zhao, N., 2017. Metal—organic frameworks-derived honeycomb-like Co3O4/three-dimensional graphene networks/Ni foam hybrid as a binder-free electrode for supercapacitors. *Journal of Alloys and Compounds, 693*, pp. 16–24.

Deshmukh, P.R., Patil, S.V., Bulakhe, R.N., Sartale, S.D. and Lokhande, C.D., 2014. Inexpensive synthesis route of porous polyaniline—ruthenium oxide composite for supercapacitor application. *Chemical Engineering Journal*, 257, pp. 82–89.

Dhibar, S. and Das, C.K., 2015. Electrochemical performances of silver nanoparticles decorated polyaniline/graphene nanocomposite in different electrolytes. *Journal of Alloys and Compounds*, 653, pp. 486–497.

Dong, Z.H., Wei, Y.L., Shi, W. and Zhang, G.A., 2011. Characterisation of doped polypyrrole/manganese oxide nanocomposite for supercapacitor electrodes. *Materials Chemistry and Physics*, 131(1–2), pp. 529–534.

Ensafi, A.A., Ahmadi, N. and Rezaei, B., 2015. Electrochemical preparation and characterization of a polypyrrole/nickel-cobalt hexacyanoferrate nanocomposite for supercapacitor applications. *RSC Advances*, 5(111), pp. 91448–91456.

Fan, L.Q., Liu, G.J., Wu, J.H., Liu, L., Lin, J.M. and Wei, Y.L., 2014. Asymmetric supercapacitor based on graphene oxide/polypyrrole composite and activated carbon electrodes. *Electrochimica Acta*, 137, pp. 26–33.

Forouzandeh, P., Kumaravel, V. and Pillai, S.C., 2020. Electrode materials for supercapacitors: A review of recent advances. *Catalysts*, 10(9), p. 969.Fu, H., Du, Z.J., Zou, W., Li, H.Q. and Zhang, C., 2013. Carbon nanotube reinforced polypyrrole nanowire network as a high-performance supercapacitor electrode. *Journal of Materials Chemistry A*, 1(47), pp. 14943–14950.

Gan, J.K., Lim, Y.S., Pandikumar, A., Huang, N.M. and Lim, H.N., 2015. Graphene/polypyrrole-coated carbon nanofiber core—shell architecture electrode for electrochemical capacitors. *RSC Advances*, 5(17), pp. 12692–12699.

Gao, B., He, D., Yan, B., Suo, H. and Zhao, C., 2015. Flexible carbon cloth based polypyrrole for an electrochemical supercapacitor. *Journal of Materials Science: Materials in Electronics*, 26(9), pp. 6373–6379.

Garcia-Torres, J. and Crean, C., 2018. Ternary composite solid-state flexible supercapacitor based on nanocarbons/manganese dioxide/PEDOT:PSS fibres. *Materials & Design*, 155, pp. 194–202.

Gómez, H., Ram, M.K., Alvi, F., Villalba, P., Stefanakos, E.L. and Kumar, A., 2011. Graphene-conducting polymer nanocomposite as novel electrode for supercapacitors. *Journal of Power Sources*, 196(8), pp. 4102–4108.

Gottam, R. and Srinivasan, P., 2015. One-step oxidation of aniline by peroxotitanium acid to polyaniline—titanium dioxide: A highly stable electrode for a supercapacitor. *Journal of Applied Polymer Science*, 132(13).

Groenendaal, L., Jonas, F., Freitag, D., Pielartzik, H. and Reynolds, J.R., 2000. Poly (3,4-ethylenedioxythiophene) and its derivatives: Past, present, and future. *Advanced Materials*, 12(7), 481–494.

Gupta, V. and Miura, N., 2006. Polyaniline/single-wall carbon nanotube (PANI/SWCNT) composites for high performance supercapacitors. *Electrochimica Acta*, 52(4), pp. 1721–1726.

Han, J., Li, L., Fang, P. and Guo, R., 2012. Ultrathin MnO_2 nanorods on conducting polymer nanofibers as a new class of hierarchical nanostructures for high-performance supercapacitors. *The Journal of Physical Chemistry C*, 116(30), pp. 15900–15907.

Han, Y., Shen, M., Wu, Y., Zhu, J., Ding, B., Tong, H. and Zhang, X., 2013. Preparation and electrochemical performances of PEDOT/sulfonic acid-functionalized graphene composite hydrogel. *Synthetic Metals*, 172, pp. 21–27.

Hareesh, K., Shateesh, B., Joshi, R.P., Dahiwale, S.S., Bhoraskar, V.N., Haram, S.K. and Dhole, S.D., 2016. PEDOT: PSS wrapped NiFe2O4/rGO tertiary nanocomposite for the super-capacitor applications. *Electrochimica Acta*, 201, pp. 106–116.

Hareesh, K., Shateesh, B., Joshi, R.P., Williams, J.F., Phase, D.M., Haram, S.K. and Dhole, S.D., 2017. Ultra high stable supercapacitance performance of conducting polymer coated MnO 2 nanorods/rGO nanocomposites. *RSC Advances*, 7(32), pp. 20027–20036.

Hatchett, D.W., Josowicz, M. and Janata, J., 1999. Comparison of chemically and electrochemically synthesized polyaniline films. *Journal of the Electrochemical Society*, 146(12), p. 4535.

He, X., Liu, G., Yan, B., Suo, H. and Zhao, C., 2016. Significant enhancement of electrochemical behaviour by incorporation of carboxyl group functionalized carbon nanotubes into polyaniline based supercapacitor. *European Polymer Journal*, 83, pp. 53–59.

Heeger, A.J., 2010. Semiconducting polymers: The third generation. *Chemical Society Reviews*, 39(7), pp. 2354–2371.

Hong, J.I., Yeo, I.H. and Paik, W.K., 2001. Conducting polymer with metal oxide for electrochemical capacitor: Poly (3, 4-ethylenedioxythiophene) RuO x electrode. *Journal of The Electrochemical Society*, 148(2), p. A156.

Hou, Y., Cheng, Y., Hobson, T. and Liu, J., 2010. Design and synthesis of hierarchical MnO_2 nanospheres/carbon nanotubes/conducting polymer ternary composite for high performance electrochemical electrodes. *Nano Letters*, 10(7), pp. 2727–2733.

Hu, C.C. and Wang, C.C., 2003. Nanostructures and capacitive characteristics of hydrous manganese oxide prepared by electrochemical deposition. *Journal of the Electrochemical Society*, 150(8), p. A1079.

Hu, Z.A., Xie, Y.L., Wang, Y.X., Mo, L.P., Yang, Y.Y. and Zhang, Z.Y., 2009. Polyaniline/SnO2 nanocomposite for supercapacitor applications. *Materials Chemistry and Physics*, *114*(2–3), pp. 990–995.

Huang, Y., Li, H., Wang, Z., Zhu, M., Pei, Z., Xue, Q., Huang, Y. and Zhi, C., 2016. Nanostructured polypyrrole as a flexible electrode material of supercapacitor. *Nano Energy*, *22*, pp. 422–438.

Imani, A. and Farzi, G., 2015. Facile route for multi-walled carbon nanotube coating with polyaniline: Tubular morphology nanocomposites for supercapacitor applications. *Journal of Materials Science: Materials in Electronics*, *26*(10), pp. 7438–7444.

Ji, J., Zhang, X., Liu, J., Peng, L., Chen, C., Huang, Z., Li, L., Yu, X. and Shang, S., 2015. Assembly of polypyrrole nanotube@ MnO$_2$ composites with an improved electrochemical capacitance. *Materials Science and Engineering: B*, *198*, pp. 51–56.

Ji, W., Ji, J., Cui, X., Chen, J., Liu, D., Deng, H. and Fu, Q., 2015. Polypyrrole encapsulation on flower-like porous NiO for advanced high-performance supercapacitors. *Chemical Communications*, *51*(36), pp. 7669–7672.

Jiang, F., Yao, Z., Yue, R., Du, Y., Xu, J., Yang, P. and Wang, C., 2012. Electrochemical fabrication of long-term stable Pt-loaded PEDOT/graphene composites for ethanol electrooxidation. *International Journal of Hydrogen Energy*, *37*(19), pp. 14085–14093.

Jiang, H., Ma, J. and Li, C., 2012. Polyaniline—MnO 2 coaxial nanofiber with hierarchical structure for high-performance supercapacitors. *Journal of Materials Chemistry*, *22*(33), pp. 16939–16942.

Jiang, J. and Kucernak, A., 2002. Electrochemical supercapacitor material based on manganese oxide: Preparation and characterization. *Electrochimica Acta*, *47*(15), pp. 2381–2386.

Jiang, L.L., Lu, X., Xie, C.M., Wan, G.J., Zhang, H.P. and Youhong, T., 2015. Flexible, free-standing TiO2—graphene—polypyrrole composite films as electrodes for supercapacitors. *The Journal of Physical Chemistry C*, *119*(8), pp. 3903–3910.

Jin, Y. and Jia, M., 2015. Design and synthesis of nanostructured graphene-SnO2-polyaniline ternary composite and their excellent supercapacitor performance. *Colloids and Surfaces A: Physicochemical and Engineering Aspects*, *464*, pp. 17–25.

K Sharma, A., Chaudhary, G., Bhardwaj, P., Kaushal, I. and Duhan, S., 2017. Studies on metal doped polyaniline-carbon nanotubes composites for high performance supercapacitor. *Current Analytical Chemistry*, *13*(4), pp. 277–284.

Kalambate, P.K., Dar, R.A., Karna, S.P. and Srivastava, A.K., 2015. High performance supercapacitor based on graphene-silver nanoparticles-polypyrrole nanocomposite coated on glassy carbon electrode. *Journal of Power Sources*, *276*, pp. 262–270.

Keskinen, J., Tuurala, S., Sjödin, M., Kiri, K., Nyholm, L., Flyktman, T., Strømme, M. and Smolander, M., 2015. Asymmetric and symmetric supercapacitors based on polypyrrole and activated carbon electrodes. *Synthetic Metals*, *203*, pp. 192–199.

Khosrozadeh, A., Xing, M. and Wang, Q., 2015. A high-capacitance solid-state supercapacitor based on free-standing film of polyaniline and carbon particles. *Applied Energy*, *153*, pp. 87–93.

Kulandaivalu, S. and Sulaiman, Y., 2019. Recent advances in layer-by-layer assembled conducting polymer based composites for supercapacitors. *Energies*, *12*(11), p. 2107.

Kumar, D., Banerjee, A., Patil, S. and Shukla, A.K., 2015. A 1 V supercapacitor device with nanostructured graphene oxide/polyaniline composite materials. *Bulletin of Materials Science*, *38*(6), pp. 1507–1517.

Li, D., Huang, J. and Kaner, R.B., 2009. Polyaniline nanofibers: A unique polymer nanostructure for versatile applications. *Accounts of Chemical Research*, *42*(1), pp. 135–145.

Li, H., He, Y., Pavlinek, V., Cheng, Q., Saha, P. and Li, C., 2015. MnO 2 nanoflake/polyaniline nanorod hybrid nanostructures on graphene paper for high-performance flexible supercapacitor electrodes. *Journal of Materials Chemistry A*, *3*(33), pp. 17165–17171.

Li, H., Wang, J., Chu, Q., Wang, Z., Zhang, F. and Wang, S., 2009. Theoretical and experimental specific capacitance of polyaniline in sulfuric acid. *Journal of Power Sources*, *190*(2), pp. 578–586.

Li, M. and Yang, L., 2015. Intrinsic flexible polypyrrole film with excellent electrochemical performance. *Journal of Materials Science: Materials in Electronics*, *26*(7), pp. 4875–4879.

Li, X., Chai, Y., Zhang, H., Wang, G. and Feng, X., 2012. Synthesis of polyaniline/tin oxide hybrid and its improved electrochemical capacitance performance. *Electrochimica acta*, *85*, pp. 9–15.

Li, Y., Wang, B., Chen, H. and Feng, W., 2010. Improvement of the electrochemical properties via poly (3, 4-ethylenedioxythiophene) oriented micro/nanorods. *Journal of Power Sources*, *195*(9), pp. 3025–3030.

Li, Y., Zhu, C., Lu, T., Guo, Z., Zhang, D., Ma, J. and Zhu, S., 2013. Simple fabrication of a Fe2O3/carbon composite for use in a high-performance lithium ion battery. *Carbon*, *52*, pp. 565–573.

Lian, Y.M., Utetiwabo, W., Zhou, Y., Huang, Z.H., Zhou, L., Faheem, M., Chen, R.J. and Yang, W., 2019. From upcycled waste polyethylene plastic to graphene/mesoporous carbon for high-voltage supercapacitors. *Journal of Colloid and Interface Science*, *557*, pp. 55–64.

Liang, M., Liu, X., Li, W. and Wang, Q., 2016. A tough nanocomposite aerogel of manganese oxide and polyaniline as an electrode for a supercapacitor. *ChemPlusChem*, *81*(1), p. 40.

Liu, J., Sun, J. and Gao, L., 2010. A promising way to enhance the electrochemical behavior of flexible single-walled carbon nanotube/polyaniline composite films. *The Journal of Physical Chemistry C*, *114*(46), pp. 19614–19620.

Liu, K., Hu, Z., Xue, R., Zhang, J. and Zhu, J., 2008. Electropolymerization of high stable poly (3, 4-ethylenedioxythiophene) in ionic liquids and its potential applications in electrochemical capacitor. *Journal of Power Sources*, *179*(2), 858–862.

Liu, R., Duay, J. and Lee, S.B., 2010. Redox exchange induced MnO_2 nanoparticle enrichment in poly (3, 4-ethylenedioxythiophene) nanowires for electrochemical energy storage. *Acs Nano*, *4*(7), pp. 4299–4307.

Liu, Y., Wang, H., Zhou, J., Bian, L., Zhu, E., Hai, J., Tang, J. and Tang, W., 2013. Graphene/polypyrrole intercalating nanocomposites as supercapacitors electrode. *Electrochimica Acta*, *112*, pp. 44–52.

Lota, K., Khomenko, V. and Frackowiak, E., 2004. Capacitance properties of poly (3, 4-ethylenedioxythiophene)/carbon nanotubes composites. *Journal of Physics and Chemistry of Solids*, *65*(2–3), pp. 295–301.

Low, W.H., Khiew, P.S., Lim, S.S., Siong, C.W. and Ezeigwe, E.R., 2019. Recent development of mixed transition metal oxide and graphene/mixed transition metal oxide based hybrid nanostructures for advanced supercapacitors. *Journal of Alloys and Compounds*, *775*, pp. 1324–1356.

Lu, X., Zhang, W., Wang, C., Wen, T.C. and Wei, Y., 2011. One-dimensional conducting polymer nanocomposites: Synthesis, properties and applications. *Progress in Polymer Science*, *36*(5), pp. 671–712.

Lu, X., Zhang, F., Dou, H., Yuan, C., Yang, S., Hao, L., Shen, L., Zhang, L. and Zhang, X., 2012. Preparation and electrochemical capacitance of hierarchical graphene/polypyrrole/carbon nanotube ternary composites. *Electrochimica Acta*, *69*, pp. 160–166.

Lu, Y., Huang, Y., Zhang, M. and Chen, Y., 2014. Nitrogen-doped graphene materials for supercapacitor applications. *Journal of Nanoscience and Nanotechnology*, *14*(2), pp. 1134–1144.

Ma, G., Peng, H., Mu, J., Huang, H., Zhou, X. and Lei, Z., 2013. In situ intercalative polymerization of pyrrole in graphene analogue of MoS2 as advanced electrode material in supercapacitor. *Journal of Power Sources*, *229*, pp. 72–78.

Masikhwa, T.M., Madito, M.J., Bello, A., Dangbegnon, J.K. and Manyala, N., 2017. High performance asymmetric supercapacitor based on molybdenum disulphide/graphene foam and activated carbon from expanded graphite. *Journal of Colloid and Interface Science*, *488*, pp. 155–165.

Meng, C., Liu, C., Chen, L., Hu, C. and Fan, S., 2010. Highly flexible and all-solid-state paperlike polymer supercapacitors. *Nano Letters*, *10*(10), pp. 4025–4031.

Meng, Q., Cai, K., Chen, Y. and Chen, L., 2017. Research progress on conducting polymer based supercapacitor electrode materials. *Nano Energy*, *36*, pp. 268–285.

Mi, H., Zhang, X., An, S., Ye, X. and Yang, S., 2007. Microwave-assisted synthesis and electrochemical capacitance of polyaniline/multi-wall carbon nanotubes composite. *Electrochemistry Communications*, *9*(12), pp. 2859–2862.

Murugan, A.V., Viswanath, A.K., Gopinath, C.S. and Vijayamohanan, K., 2006. Highly efficient organic-inorganic poly (3, 4-ethylenedioxythiophene)-molybdenum trioxide nanocomposite electrodes for electrochemical supercapacitor. *Journal of Applied Physics*, *100*(7), p. 074319.

Muthulakshmi, B., Kalpana, D., Pitchumani, S. and Renganathan, N.G., 2006. Electrochemical deposition of polypyrrole for symmetric supercapacitors. *Journal of Power Sources*, *158*(2), 1533–1537.

Naskar, P., Maiti, A., Chakraborty, P., Kundu, D., Biswas, B. and Banerjee, A., 2021. Chemical supercapacitors: A review focusing on metallic compounds and conducting polymers. *Journal of Materials Chemistry A*, *9*(4), pp. 1970–2017.

Ng, C.H., Lim, H.N., Lim, Y.S., Chee, W.K. and Huang, N.M., 2015. Fabrication of flexible polypyrrole/graphene oxide/manganese oxide supercapacitor. *International Journal of Energy Research*, *39*(3), pp. 344–355.

Niu, Z., Luan, P., Shao, Q., Dong, H., Li, J., Chen, J., Zhao, D., Cai, L., Zhou, W., Chen, X. and Xie, S., 2012. A "skeleton/skin" strategy for preparing ultrathin free-standing single-walled carbon nanotube/polyaniline films for high performance supercapacitor electrodes. *Energy & Environmental Science*, *5*(9), pp. 8726–8733.

Pandiselvi, K. and Thambidurai, S., 2014. Chitosan-ZnO/polyaniline ternary nanocomposite for high-performance supercapacitor. *Ionics*, *20*(4), pp. 551–561.

Peng, C., Snook, G.A., Fray, D.J., Shaffer, M.S. and Chen, G.Z., 2006. Carbon nanotube stabilised emulsions for electrochemical synthesis of porous nanocomposite coatings of poly [3, 4-ethylene-dioxythiophene]. *Chemical Communications* (44), pp. 4629–4631.

Peng, H., Ma, G., Mu, J., Sun, K. and Lei, Z., 2014. Low-cost and high energy density asymmetric supercapacitors based on polyaniline nanotubes and MoO3 nanobelts. *Journal of Materials Chemistry A*, *2*(27), pp. 10384–10388.

Peng, H., Ma, G., Sun, K., Mu, J., Wang, H. and Lei, Z., 2014. High-performance supercapacitor based on multi-structural CuS@ polypyrrole composites prepared by in situ oxidative polymerization. *Journal of Materials Chemistry A*, *2*(10), pp. 3303–3307.

Pettersson, L.A., Carlsson, F., Inganäs, O. and Arwin, H., 1998. Spectroscopic ellipsometry studies of the optical properties of doped poly (3, 4-ethylenedioxythiophene): An anisotropic metal. *Thin Solid Films*, *313*, pp. 356–361.

Prasad, K.R. and Miura, N., 2004. Polyaniline-MnO₂ composite electrode for high energy density electrochemical capacitor. *Electrochemical and Solid State Letters*, *7*(11), p.A425.

Qian, T., Zhou, J., Xu, N., Yang, T., Shen, X., Liu, X., Wu, S. and Yan, C., 2015. On-chip supercapacitors with ultrahigh volumetric performance based on electrochemically co-deposited CuO/polypyrrole nanosheet arrays. *Nanotechnology*, *26*(42), p. 425402.

Qu, Y., Deng, Y., Li, Q., Zhang, Z., Zeng, F., Yang, Y. and Xu, K., 2017. Core—shell-structured hollow carbon nanofiber@ nitrogen-doped porous carbon composite materials as anodes for advanced sodium-ion batteries. *Journal of Materials Science*, *52*(4), pp. 2356–2365.

Radhakrishnan, S., Rao, C.R. and Vijayan, M., 2011. Performance of conducting polyaniline-DBSA and polyaniline-DBSA/Fe3O4 composites as electrode materials for aqueous redox supercapacitors. *Journal of Applied Polymer Science*, *122*(3), pp. 1510–1518.

Raj, C.J., Kim, B.C., Cho, W.J., Lee, W.G., Jung, S.D., Kim, Y.H., Park, S.Y. and Yu, K.H., 2015. Highly flexible and planar supercapacitors using graphite flakes/polypyrrole in polymer lapping film. *ACS Applied Materials & Interfaces*, *7*(24), pp. 13405–13414.

Rajesh, M., Raj, C.J., Kim, B.C., Cho, B.B., Ko, J.M. and Yu, K.H., 2016. Supercapacitive studies on electropolymerized natural organic phosphate doped polypyrrole thin films. *Electrochimica Acta*, *220*, pp. 373–383.

Rajesh, M., Raj, C.J., Manikandan, R., Kim, B.C., Park, S.Y. and Yu, K.H., 2017. A high performance PEDOT/ PEDOT symmetric supercapacitor by facile in situ hydrothermal polymerization of PEDOT nanostructures on flexible carbon fibre cloth electrodes. *Materials Today Energy*, *6*, pp. 96–104.

Ranjusha, R., Sajesh, K.M., Roshny, S., Lakshmi, V., Anjali, P., Sonia, T.S., Nair, A.S., Subramanian, K.R.V., Nair, S.V., Chennazhi, K.P. and Balakrishnan, A., 2014. Supercapacitors based on freeze dried MnO₂ embedded PEDOT:PSS hybrid sponges. *Microporous and Mesoporous Materials*, *186*, pp. 30–36.

Renault, S., Gottis, S., Barrès, A.L., Courty, M., Chauvet, O., Dolhem, F. and Poizot, P., 2013. A green Li—organic battery working as a fuel cell in case of emergency. *Energy & Environmental Science*, *6*(7), pp. 2124–2133.

Ryu, K.S., Lee, Y.G., Hong, Y.S., Park, Y.J., Wu, X., Kim, K.M., Kang, M.G., Park, N.G. and Chang, S.H., 2004. Poly (ethylenedioxythiophene)(PEDOT) as polymer electrode in redox supercapacitor. *Electrochimica Acta*, *50*(2–3), pp. 843–847.

Sen, P. and De, A., 2010. Electrochemical performances of poly (3, 4-ethylenedioxythiophene)—NiFe2O4 nanocomposite as electrode for supercapacitor. *Electrochimica Acta*, *55*(16), pp. 4677–4684.

Sen, P., De, A., Chowdhury, A.D., Bandyopadhyay, S.K., Agnihotri, N. and Mukherjee, M., 2013. Conducting polymer based manganese dioxide nanocomposite as supercapacitor. *Electrochimica Acta*, *108*, pp. 265–273.

Shao, L., Jeon, J.W. and Lutkenhaus, J.L., 2012. Polyaniline/vanadium pentoxide layer-by-layer electrodes for energy storage. *Chemistry of Materials*, *24*(1), pp. 181–189.

Shayeh, J.S., Ehsani, A., Ganjali, M.R., Norouzi, P. and Jaleh, B.J.A.S.S., 2015. Conductive polymer/reduced graphene oxide/Au nano particles as efficient composite materials in electrochemical supercapacitors. *Applied Surface Science*, *353*, pp. 594–599.

Shen, F., Pankratov, D. and Chi, Q., 2017. Graphene-conducting polymer nanocomposites for enhancing electrochemical capacitive energy storage. *Current Opinion in Electrochemistry*, *4*(1), pp. 133–144.

Shen, K., Ran, F., Zhang, X., Liu, C., Wang, N., Niu, X., Liu, Y., Zhang, D., Kong, L., Kang, L. and Chen, S., 2015. Supercapacitor electrodes based on nano-polyaniline deposited on hollow carbon spheres derived from cross-linked co-polymers. *Synthetic Metals*, *209*, pp. 369–376.

Shi, X., Zheng, S., Wu, Z.S. and Bao, X., 2018. Recent advances of graphene-based materials for high-performance and new-concept supercapacitors. *Journal of Energy Chemistry*, *27*(1), pp. 25–42.

Sivakkumar, S.R., Kim, W.J., Choi, J.A., MacFarlane, D.R., Forsyth, M. and Kim, D.W., 2007. Electrochemical performance of polyaniline nanofibres and polyaniline/multi-walled carbon nanotube composite as an electrode material for aqueous redox supercapacitors. *Journal of Power Sources*, *171*(2), pp. 1062–1068.

Sk, M.M., Yue, C.Y. and Jena, R.K., 2015. Non-covalent interactions and supercapacitance of pseudo-capacitive composite electrode materials (MWCNTCOOH/MnO₂/PANI). *Synthetic Metals*, *208*, pp. 2–12.

Song, H., Cai, K., Wang, J. and Shen, S., 2016. Influence of polymerization method on the thermoelectric properties of multi-walled carbon nanotubes/polypyrrole composites. *Synthetic Metals*, *211*, pp. 58–65.

Su, H., Wang, T., Zhang, S., Song, J., Mao, C., Niu, H., Jin, B., Wu, J. and Tian, Y., 2012. Facile synthesis of polyaniline/TiO2/graphene oxide composite for high performance supercapacitors. *Solid State Sciences*, *14*(6), pp. 677–681.

Su, Z., Yang, C., Xu, C., Wu, H., Zhang, Z., Liu, T., Zhang, C., Yang, Q., Li, B. and Kang, F., 2013. Co-electro-deposition of the MnO 2—PEDOT:PSS nanostructured composite for high areal mass, flexible asymmetric supercapacitor devices. *Journal of Materials Chemistry A*, *1*(40), pp. 12432–12440.

Sulaiman, Y., Azmi, M.K.S., Abdah, M.A.A.M. and Azman, N.H.N., 2017. One step electrodeposition of poly-(3, 4-ethylenedioxythiophene)/graphene oxide/cobalt oxide ternary nanocomposite for high performance supercapacitor. *Electrochimica Acta*, *253*, pp. 581–588.

Sun, D., Jin, L., Chen, Y., Zhang, J.R. and Zhu, J.J., 2013. Microwave-assisted in situ synthesis of graphene/PEDOT hybrid and its application in supercapacitors. *ChemPlusChem*, *78*(3), p. 227.

Sun, D., Wang, Z., Huang, K., Wang, X., Wang, H., Qing, C., Wang, B. and Tang, Y., 2015. A sandwich-structured porous MnO2/polyaniline/MnO2 thin film for supercapacitor applications. *Chemical Physics Letters*, *638*, pp. 38–42.

Sun, H., She, P., Xu, K., Shang, Y., Yin, S. and Liu, Z., 2015. A self-standing nanocomposite foam of polyaniline@ reduced graphene oxide for flexible super-capacitors. *Synthetic Metals*, *209*, pp. 68–73.

Sun, K., Peng, H., Mu, J., Ma, G., Zhao, G. and Lei, Z., 2015. High energy density asymmetric supercapacitors based on polyaniline nanotubes and tungsten trioxide rods. *Ionics*, *21*(8), pp. 2309–2317.

Sun, W., Chen, L., Wang, Y., Zhou, Y., Meng, S., Li, H. and Luo, Y., 2016. Synthesis of highly conductive PPy/graphene/MnO2 composite using ultrasonic irradiation. *Synthesis and Reactivity in Inorganic, Metal-Organic, and Nano-Metal Chemistry*, *46*(3), pp. 437–444.

Sun, W. and Chen, X., 2009. Preparation and characterization of polypyrrole films for three-dimensional micro supercapacitor. *Journal of Power Sources*, *193*(2), pp. 924–929.

Sun, X., Gan, M., Ma, L., Wang, H., Zhou, T., Wang, S., Dai, W. and Wang, H., 2015. Fabrication of PANI-coated honeycomb-like MnO2 nanospheres with enhanced electrochemical performance for energy storage. *Electrochimica Acta*, *180*, pp. 977–982.

Sun, X., Li, Q. and Mao, Y., 2015. Understanding the influence of polypyrrole coating over V2O5 nanofibers on electrochemical properties. *Electrochimica Acta*, *174*, pp. 563–573.

Syed Zainol Abidin, S.N.J., Mamat, S., Abdul Rasyid, S., Zainal, Z. and Sulaiman, Y., 2018. Fabrication of poly (vinyl alcohol)-graphene quantum dots coated with poly (3, 4-ethylenedioxythiophene) for supercapacitor. *Journal of Polymer Science Part A: Polymer Chemistry*, *56*(1), pp. 50–58.

Tang, H., Wang, J., Yin, H., Zhao, H., Wang, D. and Tang, Z., 2015. Growth of polypyrrole ultrathin films on MoS2 monolayers as high-performance supercapacitor electrodes. *Advanced Materials*, *27*(6), pp. 1117–1123.

Tang, P., Han, L. and Zhang, L., 2014. Facile synthesis of graphite/PEDOT/MnO2 composites on commercial supercapacitor separator membranes as flexible and high-performance supercapacitor electrodes. *ACS Applied Materials & Interfaces*, *6*(13), pp. 10506–10515.

Tang, P., Zhao, Y. and Xu, C., 2013. Step-by-step assembled poly (3, 4-ethylenedioxythiophene)/manganese dioxide composite electrodes: Tuning the structure for high electrochemical performance. *Electrochimica Acta*, *89*, pp. 300–309.

Tang, Q., Chen, M., Yang, C., Wang, W., Bao, H. and Wang, G., 2015. Enhancing the energy density of asymmetric stretchable supercapacitor based on wrinkled CNT@ MnO2 cathode and CNT@ polypyrrole anode. *ACS Applied Materials & Interfaces*, *7*(28), pp. 15303–15313.

Tao, J., Ma, W., Liu, N., Ren, X., Shi, Y., Su, J. and Gao, Y., 2015. High-performance solid-state supercapacitors fabricated by pencil drawing and polypyrrole depositing on paper substrate. *Nano-Micro Letters*, *7*(3), pp. 276–281.

Tong, L., Skorenko, K.H., Faucett, A.C., Boyer, S.M., Liu, J., Mativetsky, J.M., Bernier, W.E. and Jones Jr, W.E., 2015. Vapor-phase polymerization of poly (3, 4-ethylenedioxythiophene)(PEDOT) on commercial carbon coated aluminum foil as enhanced electrodes for supercapacitors. *Journal of Power Sources*, *297*, pp. 195–201.

Tran, C., Singhal, R., Lawrence, D. and Kalra, V., 2015. Polyaniline-coated freestanding porous carbon nanofibers as efficient hybrid electrodes for supercapacitors. *Journal of Power Sources*, *293*, pp. 373–379.

Tran, H.D., D'Arcy, J.M., Wang, Y., Beltramo, P.J., Strong, V.A. and Kaner, R.B., 2011. The oxidation of aniline to produce "polyaniline": A process yielding many different nanoscale structures. *Journal of Materials Chemistry*, *21*(11), pp. 3534–3550.

Vellakkat, M. and Hundekal, D., 2017. Electrical conductivity and supercapacitor properties of polyaniline/chitosan/nickel oxide honeycomb nanocomposite. *Journal of Applied Polymer Science*, *134*(9).

Villers, D., Jobin, D., Soucy, C., Cossement, D., Chahine, R., Breau, L. and Bélanger, D., 2003. The influence of the range of electroactivity and capacitance of conducting polymers on the performance of carbon conducting polymer hybrid supercapacitor. *Journal of the Electrochemical Society*, *150*(6), p.A747.

Wang, F., Zhan, X., Cheng, Z., Wang, Z., Wang, Q., Xu, K., Safdar, M. and He, J., 2015. Tungsten oxide@ polypyrrole core—shell nanowire arrays as novel negative electrodes for asymmetric supercapacitors. *Small*, *11*(6), pp. 749–755.

Wang, G., Zhang, L. and Zhang, J., 2012. A review of electrode materials for electrochemical supercapacitors. *Chemical Society Reviews*, *41*(2), pp. 797–828.

Wang, H., Hao, Q., Yang, X., Lu, L. and Wang, X., 2009. Graphene oxide doped polyaniline for supercapacitors. *Electrochemistry Communications*, *11*(6), pp. 1158–1161.

Wang, J., Cai, K., Shen, S. and Yin, J., 2014. Preparation and thermoelectric properties of multi-walled carbon nanotubes/polypyrrole composites. *Synthetic Metals*, *195*, pp. 132–136.

Wang, J., Li, X., Du, X., Wang, J., Ma, H. and Jing, X., 2017. Polypyrrole composites with carbon materials for supercapacitors. *Chemical Papers*, *71*(2), pp. 293–316

Wang, S., Ma, L., Gan, M., Fu, S., Dai, W., Zhou, T., Sun, X., Wang, H. and Wang, H., 2015. Free-standing 3D graphene/polyaniline composite film electrodes for high-performance supercapacitors. *Journal of Power Sources*, *299*, pp. 347–355.

Wang, Z., Carlsson, D.O., Tammela, P., Hua, K., Zhang, P., Nyholm, L. and Strømme, M., 2015. Surface modified nanocellulose fibers yield conducting polymer-based flexible supercapacitors with enhanced capacitances. *ACS Nano*, *9*(7), pp. 7563–7571.

Warren, R., Sammoura, F., Teh, K.S., Kozinda, A., Zang, X. and Lin, L., 2015. Electrochemically synthesized and vertically aligned carbon nanotube—polypyrrole nanolayers for high energy storage devices. *Sensors and Actuators A: Physical*, *231*, pp. 65–73.

Wen, J., Jiang, Y., Yang, Y. and Li, S., 2014. Conducting polymer and reduced graphene oxide Langmuir—Blodgett films: A hybrid nanostructure for high performance electrode applications. *Journal of Materials Science: Materials in Electronics*, *25*(2), pp. 1063–1071.

Weng, Y.T., Pan, H.A., Wu, N.L. and Chen, G.Z., 2015. Titanium carbide nanocube core induced interfacial growth of crystalline polypyrrole/polyvinyl alcohol lamellar shell for wide-temperature range supercapacitors. *Journal of Power Sources*, *274*, pp. 1118–1125.

White, A.M. and Slade, R.C., 2004. Electrochemically and vapour grown electrode coatings of poly (3, 4-ethylenedioxythiophene) doped with heteropolyacids. *Electrochimica Acta*, *49*(6), pp. 861–865.

Wong, S.I., Sunarso, J., Wong, B.T., Lin, H., Yu, A. and Jia, B., 2018. Towards enhanced energy density of graphene-based supercapacitors: Current status, approaches, and future directions. *Journal of Power Sources*, *396*, pp. 182–206.

Wu, K., Xu, S.Z., Zhou, X.J. and Wu, H.X., 2013. Graphene quantum dots enhanced electrochemical performance of polypyrrole as supercapacitor electrode. *J. Electrochem*, *19*(4), pp. 361–370.

Wu, K., Zhao, J., Zhang, X., Zhou, H. and Wu, M., 2019. Hierarchical mesoporous MoO2 sphere as highly effective supercapacitor electrode. *Journal of the Taiwan Institute of Chemical Engineers*, *102*, pp. 212–217.

Wu, W., Yang, L., Chen, S., Shao, Y., Jing, L., Zhao, G. and Wei, H., 2015. Core—shell nanospherical polypyrrole/graphene oxide composites for high performance supercapacitors. *RSC Advances*, *5*(111), pp. 91645–91653.

Xia, X., Chao, D., Fan, Z., Guan, C., Cao, X., Zhang, H. and Fan, H.J., 2014. A new type of porous graphite foams and their integrated composites with oxide/polymer core/shell nanowires for supercapacitors: Structural design, fabrication, and full supercapacitor demonstrations. *Nano letters*, *14*(3), pp. 1651–1658.

Xie, Y. and Wang, D., 2016. Supercapacitance performance of polypyrrole/titanium nitride/polyaniline coaxial nanotube hybrid. *Journal of Alloys and Compounds*, *665*, pp. 323–332.

Xu, C., Sun, J. and Gao, L., 2011. Synthesis of novel hierarchical graphene/polypyrrole nanosheet composites and their superior electrochemical performance. *Journal of Materials Chemistry*, *21*(30), pp. 11253–11258.

Xu, H., Wu, J.X., Chen, Y., Zhang, J.L. and Zhang, B.Q., 2015. Facile synthesis of polyaniline/NiCo 2 O 4 nanocomposites with enhanced electrochemical properties for supercapacitors. *Ionics*, *21*(9), pp. 2615–2622.

Xu, J., Wang, D., Fan, L., Yuan, Y., Wei, W., Liu, R., Gu, S. and Xu, W., 2015. Fabric electrodes coated with polypyrrole nanorods for flexible supercapacitor application prepared via a reactive self-degraded template. *Organic Electronics*, *26*, pp. 292–299.

Xu, J., Wang, K., Zu, S.Z., Han, B.H. and Wei, Z., 2010. Hierarchical nanocomposites of polyaniline nanowire arrays on graphene oxide sheets with synergistic effect for energy storage. *ACS Nano*, *4*(9), pp. 5019–5026.

Yan, Y., Cheng, Q., Pavlinek, V., Saha, P. and Li, C., 2012. Fabrication of polyaniline/mesoporous carbon/MnO2 ternary nanocomposites and their enhanced electrochemical performance for supercapacitors. *Electrochimica acta*, *71*, pp. 27–32.

Yan, D., Liu, Y., Li, Y., Zhuo, R., Wu, Z., Ren, P., Li, S., Wang, J., Yan, P. and Geng, Z., 2014. Synthesis and electrochemical properties of MnO2/rGO/PEDOT:PSS ternary composite electrode material for supercapacitors. *Materials Letters*, *127*, pp. 53–55.

Yang, L., Shi, Z. and Yang, W., 2015. Polypyrrole directly bonded to air-plasma activated carbon nanotube as electrode materials for high-performance supercapacitor. *Electrochimica Acta*, *153*, pp. 76–82.

Yang, Q., Hou, Z. and Huang, T., 2015. Self-assembled polypyrrole film by interfacial polymerization for supercapacitor applications. *Journal of Applied Polymer Science*, *132*(11).

Yang, X., Shi, K., Zhitomirsky, I. and Cranston, E.D., 2015. Cellulose nanocrystal aerogels as universal 3D lightweight substrates for supercapacitor materials. *Advanced Materials*, *27*(40), pp. 6104–6109.

Yoon, S.B. and Kim, K.B., 2013. Effect of poly (3, 4-ethylenedioxythiophene)(PEDOT) on the pseudocapacitive properties of manganese oxide (MnO_2) in the PEDOT/MnO_2/multiwall carbon nanotube (MWNT) composite. *Electrochimica Acta*, *106*, pp. 135–142.

Yu, L., Gan, M., Ma, L., Huang, H., Hu, H., Li, Y., Tu, Y., Ge, C., Yang, F. and Yan, J., 2014. Facile synthesis of MnO_2/polyaniline nanorod arrays based on graphene and its electrochemical performance. *Synthetic Metals*, *198*, pp. 167–174.

Yu, M., Zeng, Y., Han, Y., Cheng, X., Zhao, W., Liang, C., Tong, Y., Tang, H. and Lu, X., 2015. Valence-optimized vanadium oxide supercapacitor electrodes exhibit ultrahigh capacitance and super-long cyclic durability of 100 000 cycles. *Advanced Functional Materials*, *25*(23), pp. 3534–3540.

Yu, Z., Tetard, L., Zhai, L. and Thomas, J., 2015. Supercapacitor electrode materials: Nanostructures from 0 to 3 dimensions. *Energy & Environmental Science*, *8*(3), pp. 702–730.

Yun, T.G., Hwang, B.I., Kim, D., Hyun, S. and Han, S.M., 2015. Polypyrrole—MnO_2-coated textile-based flexible-stretchable supercapacitor with high electrochemical and mechanical reliability. *ACS Applied Materials & Interfaces*, *7*(17), pp. 9228–9234.

Zhang, J., Shi, L., Liu, H., Deng, Z., Huang, L., Mai, W., Tan, S. and Cai, X., 2015. Utilizing polyaniline to dominate the crystal phase of Ni (OH) 2 and its effect on the electrochemical property of polyaniline/Ni (OH) 2 composite. *Journal of Alloys and Compounds*, *651*, pp. 126–134.

Zhang, J., Shu, D., Zhang, T., Chen, H., Zhao, H., Wang, Y., Sun, Z., Tang, S., Fang, X. and Cao, X., 2012. Capacitive properties of PANI/MnO_2 synthesized via simultaneous-oxidation route. *Journal of Alloys and Compounds*, *532*, pp. 1–9.

Zhang, J. and Zhao, X.S., 2012. Conducting polymers directly coated on reduced graphene oxide sheets as high-performance supercapacitor electrodes. *The Journal of Physical Chemistry C*, *116*(9), pp. 5420–5426.

Zhang, K., Zhang, L.L., Zhao, X.S. and Wu, J., 2010. Graphene/polyaniline nanofiber composites as supercapacitor electrodes. *Chemistry of Materials*, *22*(4), pp. 1392–1401.

Zhang, X., Ji, L., Zhang, S. and Yang, W., 2007. Synthesis of a novel polyaniline-intercalated layered manganese oxide nanocomposite as electrode material for electrochemical capacitor. *Journal of Power Sources*, *173*(2), pp. 1017–1023.

Zhao, Z., Richardson, G.F., Meng, Q., Zhu, S., Kuan, H.C. and Ma, J., 2015. PEDOT-based composites as electrode materials for supercapacitors. *Nanotechnology*, *27*(4), p. 042001.

Zhou, C., Zhang, Y., Li, Y. and Liu, J., 2013. Construction of high-capacitance 3D CoO@ polypyrrole nanowire array electrode for aqueous asymmetric supercapacitor. *Nano Letters*, *13*(5), pp. 2078–2085.

Zhou, H., Han, G., Xiao, Y., Chang, Y. and Zhai, H.J., 2015a. A comparative study on long and short carbon nanotubes-incorporated polypyrrole/poly (sodium 4-styrenesulfonate) nanocomposites as high-performance supercapacitor electrodes. *Synthetic Metals*, *209*, pp. 405–411.

Zhou, J., Zhao, H., Mu, X., Chen, J., Zhang, P., Wang, Y., He, Y., Zhang, Z., Pan, X. and Xie, E., 2015b. Importance of polypyrrole in constructing 3D hierarchical carbon nanotube@ MnO_2 perfect core—shell nanostructures for high-performance flexible supercapacitors. *Nanoscale*, *7*(35), pp. 14697–14706.

Zhou, Y., Qin, Z.Y., Li, L., Zhang, Y., Wei, Y.L., Wang, L.F. and Zhu, M.F., 2010. Polyaniline/multi-walled carbon nanotube composites with core—shell structures as supercapacitor electrode materials. *Electrochimica Acta*, *55*(12), pp. 3904–3908.

Zhu, J., Sun, W., Yang, D., Zhang, Y., Hoon, H.H., Zhang, H. and Yan, Q., 2015. Multifunctional architectures constructing of PANI nanoneedle arrays on MoS2 thin nanosheets for high-energy supercapacitors. *Small*, *11*(33), pp. 4123–4129.

Zhu, J., Xu, Y., Wang, J., Wang, J., Bai, Y. and Du, X., 2015. Morphology controllable nano-sheet polypyrrole—graphene composites for high-rate supercapacitor. *Physical Chemistry Chemical Physics*, *17*(30), pp. 19885–19894.

Zhu, M., Huang, Y., Deng, Q., Zhou, J., Pei, Z., Xue, Q., Huang, Y., Wang, Z., Li, H., Huang, Q. and Zhi, C., 2016. Highly flexible, freestanding supercapacitor electrode with enhanced performance obtained by hybridizing polypyrrole chains with MXene. *Advanced Energy Materials*, *6*(21), p. 1600969.

Zhu, S., Wu, M., Ge, M.H., Zhang, H., Li, S.K. and Li, C.H., 2016. Design and construction of three-dimensional CuO/polyaniline/rGO ternary hierarchical architectures for high performance supercapacitors. *Journal of Power Sources*, *306*, pp. 593–601.

12 Theory, Modelling, and Simulation in Supercapacitors

Kottoly Raveendran Raghi, Daisy Rajaian Sherin, and Thanathu K. Manojkumar

CONTENTS

12.1 INTRODUCTION

Supercapacitors or electrochemical capacitors are electrostatic devices that have attracted great attention due to their fast dynamics of charge propagation, high efficiency, low maintenance, long lifecycle performance (> 10,000 cycles), high power density, safe and reliable advancement in energy or charge storage and conversion, in electric double layers (EDL) at electrode/electrolyte interfaces [1–3]. Since they behave similarly to any other capacitors, they can be fully charged/discharged and recharged very fast with a long life cycle and are widely used in many power-management applications. During the last few decades, supercapacitor-related research has tremendously increased in response to the demand for applications requiring properties like reliability, better life cycle, and higher specific energy [4].

Most theoretical studies on the energy storage mechanism in EDLCs mainly focused on examining the EDL behavior in different geometries of the electrolyte/electrode components in the particular system, with the help of various simulation techniques such as molecular simulations (molecular dynamics (MD) and Monte Carlo (MC) simulations) and classical density functional theory (DFT) [5]. In comparison with the in situ experimental techniques, these simulations are very useful for a better understanding into the charge storage mechanisms in the system. Also, these techniques play an effective and vital role in the design and estimation of its performances. As an

DOI: 10.1201/9781003174646-12

example, the rational optimization of the electrostatic interaction between the electrode and electrolyte in supercapacitors have fundamental importance to enhance its performance for practical applications [5–6]. Besides, in order to understand the mechanisms of the functioning of supercapacitors, it's essential to model them at the molecular scale.

Theory, modelling, and simulation of the supercapacitors can complement the experimental results in this research area and can give insights into the energy storage mechanism, predict outcome of new innovative materials, novel electrolytes, and electrodes. Thus, in the future we can expect the multiscale modeling and simulation methods to become well-established ways to get a deep knowledge of ion transport in electrolytes, phase transitions in electrode materials, and charge transfer processes. The most effective goal for theory is the development of novel materials (electrode/electrolyte) with better performance for supercapacitors [7].

In this chapter an overview of the modeling techniques like the electrochemical model, equivalent circuit model, intelligent model, transmission line and fractional-order model, self-discharge, simplified analytical model and thermal model, and simulation techniques of supercapacitors are presented.

12.2 THEORIES OF SUPERCAPACITORS

The understanding of electric double layer capacitance (EDLC) and pseudocapacitance is significant to realize the origination of electrochemical capacitance in supercapacitors, which in turn helps to comprehend hybrid or asymmetric capacitors as they are formed by combining EDL and pseudocapacitive materials. Several theories have been suggested to realize the exact methodology involved in the origination of EDLC and these theories are revised occasionally to integrate the pseudocapacitive materials. As these are only elementary concepts, the properties of electrode and electrolyte should be considered to envisage the real performance of a supercapacitor [8].

12.2.1 HELMHOLTZ THEORY

Hermann von Helmholtz first theorized the concept of the supercapacitor and provided a rough idea of arrangements of electrodes and ionic layers in a supercapacitor [9]. He noted the formation of the electrical double-layer (EDL) of polarized ions at the interface of electrode and electrolyte (Figure 12.1). This phenomenon is modeled as a conventional capacitor, in which the radius of solvated ions is taken as charge separation H, as shown in Figure 12.2a [9].

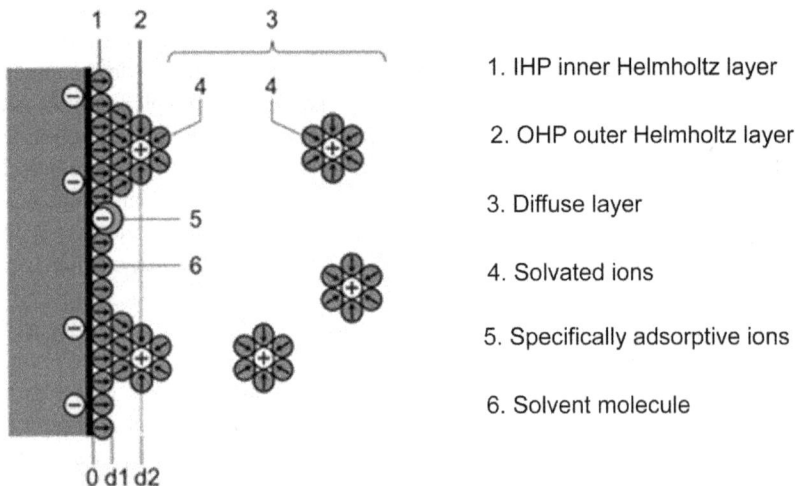

1. IHP inner Helmholtz layer

2. OHP outer Helmholtz layer

3. Diffuse layer

4. Solvated ions

5. Specifically adsorptive ions

6. Solvent molecule

FIGURE 12.1   Helmholtz electrical double layer [10].

One of the layers is attached to the electrode surface and the other one is a solvated, made up of dissolved electrolytic ions. The inner Helmholtz plane (IHP) – a monolayer of solvents, which separates these two layers. This single layer act as a molecular dielectric between two oppositely charged polarized entities. The polarized ions of the electrolyte are received at outer Helmholtz plane (OHP) and the net charge on the electrode is neutralized by the oppositely charged ions present in this plane. The surface area of electrodes and the number of the ions adsorbed determine the electric charge gathered in these layers. The EDLC (C_d) varies with distance between the layers δ and dielectric constant ε, and can be evaluated by the following formula.

$$C_d = \frac{\in}{4\pi\delta}$$

But this theory fails to clarify the interactions that arise farther from the electrode, which is responsible for pseudocapacitance.

12.2.2 GOUY-CHAPMAN THEORY

To overcome the limitations of Helmholtz theory, L. G. Guoy [11] and D. L. Chapman [12] considered the ion mobility as one of the important factor and they independently generated a model for the double layer. In this diffuse model Figure 12.2b, ions in electrolytic region were treated as point charges and a diffuse layer. The ion mobility can be explained by the combined activity of diffusion and electrostatic forces [13]. The capacitance varies with the ionic concentration of the electrolyte and the potential applied. This model assumes the dispersal of ionic charge depends on their distance from the electrode surface by applying Maxwell Boltzmann statistics. As the distance from the surface of the bulk increases, electric potential decreases exponentially. However, the capacitance values measured by this model are higher than actual quantities, as the ions are treated as point charges [14]. At equilibrium, the concentration is calculated using Boltzmann distribution as,

$$c_i = c_{i\infty} exp\left(\frac{-z_i e\varnothing}{k_b T}\right)$$

Here, z_i is the valency, $c_{i\infty}$ is the bulk concentration of ions i, k_b is Boltzmann constant, T is the absolute temperature and e is the electron charge. In this model, electric potential is calculated using Poisson-Boltzmann Equation, which is expressed as,

$$\nabla.\left(\in_0\in_0 \nabla\varnothing\right) = 2zeN_A c_\infty sinh\left(\frac{ze\varnothing}{k_b T}\right)$$

Here, N_A is the Avogadro's number. This equation gives an exact solution when electrodes are planar and electrolyte properties are constant. The boundary conditions defined should be potential, $\varnothing(0) = \varnothing(D)$ and $\varnothing(\infty) = 0$. Then the specific capacitance is calculated for the diffuse layer as,

$$C_S^D = \frac{q_s}{\varnothing_D} = \frac{4zeN_A c_\infty \lambda_d}{\varnothing_D} sinh\left(\frac{ze\varnothing_D}{2k_b T}\right)$$

Here, q_s is the surface charge density and λ_d is the Debye length [15–18]

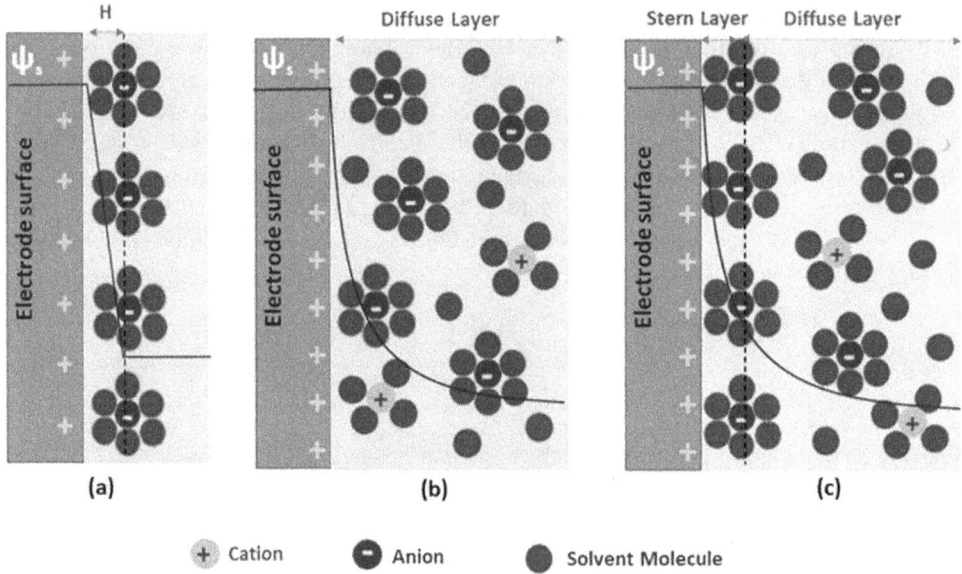

FIGURE 12.2 Schematic representation of (a) Helmholtz (b) Gouy-Chapman and (c)Gouy-Chapman-Stern models [9].

12.2.3 STERN THEORY

Otto Stern defined the double layer as a blend of Helmholtz and Gouy-Chapman models Figure 12.2c [19]. He proposed an internal stern layer, similar to the Helmholtz layer, by considering the effect of the fixed size of the ions and a second diffuse layer as in Gouy—Chapman model [13, 19]. This model is based on the assumption that most of the activities inside the second layer are coulombic in nature. In this model ions are treated as point charges. Stern assume the viscosity of the fluid lies in a constant plane and the dielectric permittivity across the EDL persist as same.

12.2.4 GRAHAME THEORY

By modifying the Stern model [20] D C Grahame proposed that if ions lose their solvation shell while approaching the electrode, most of the solvated molecules persist closer to the electrode. This model permits the movement of some of the charged/neutral particles through the Stern layer. The ions responsible for pseudocapacitance are named as specifically adsorbed ions, are very close to the electrode. Grahame further divided the Helmholtz region in two planes—the inner Helmholtz plane (IHP) and outer Helmholtz plane (OHP). The IHP crosses through the centres of specifically absorbed ions and OHP lies closest to solvated ions. The diffuse layer comprises the region outside the OHP, involves the non-specifically adsorbed solvated ions. On heating the electrolyte non-specifically adsorbed ions of OHP circulate to the bulk liquid, results in the formation of a 3D diffuse layer. Width of diffuse layer may calculated using the total ionic concentration of the solution. The capacitance of this diffuse layer, C_{diff} and Helmholtz type capacitance, C_H are used for calculation of total capacitance.

$$\frac{1}{C_{dl}} = \frac{1}{C_{diff}} + \frac{1}{C_H}$$

Contradictory to EDLC, in pseudocapacitors the charge transferred becomes voltage dependent (dQ/dV) as a result of faradaic interactions between the electrolyte and solid material. Mainly three types of electrochemical processes are involved in pseudocapacitors-redox reaction of ions, adsorption of these ions present and electrode doping. The average capacitance is estimated by the following equation.

$$C_{av} = \frac{Q_{tot}}{V_{tot}}$$

This may further used to compute the specific capacitance ($C_{avg}-1$). The total charge of the system is given by Q_{tot}, where as V_{tot} represents the charge in voltage either electrode charging or discharging.

12.3   MODELING OF SUPERCAPACITORS

The mathematical modeling of supercapacitor is a convenient method for the design, condition monitoring, analysis of material properties and control synthesis. It is possible to predict the dispersal and alignment of electrolyte ions, the alteration in the electrode morphology [21], impact of the ionic resistance of the separator and the ionic and electronic resistance of the porous electrode [22]. The literature review shows numerous SC models suggested for various purposes, includes electrical and thermal properties, self-discharge, aging etc. Models such as electrochemical or equivalent circuit models or fractional-order models are frequently applied in electrical behavior modeling. Even though electrochemical models have high accuracy, their low calculation efficiency makes their use limited in systems in real-time energy management as compared to the equivalent circuit models.

The molecular models of SCs were originated by considering the equilibrium and dynamic states, applying the physicochemical rules, sophisticated simulation techniques like spectral element methods [23] or Monte Carlo methods [24–25]. The reliability and computational cost of these models relied on the electrode/electrolyte models. The computational cost may be reduced by accepting primitive models, where ions are considered as hard spheres and electrodes are represented as simple walls. But this unrealistic modeling reduces the accuracy as it was unable to compute electrostatic parameters. Another option for increasing the accuracy and reducing computational cost is by modeling solvent molecules as hard spheres with no charge on it. But this model has limited application, mainly in simulations of large systems as the computational cost is very high as it considers all atom electrolytes [26]. To overcome these limitations, reduced-order [27] and coarse-grained models [21]are proposed. Also, for the electrode models, two main strategies are considered. One stage is where the voltage on each electrode atom at each molecular dynamic step can be supposed to be equivalent to a specified value [28–29], while in the second approach, fixed partial charges are assigned to each atom [30–31]. The use of persistent charge simulations changes both the structure of the adsorbed fluid at the interface and the time scales over which relaxation phenomena occur [32].

Some of the most realistic modeling attempts are described here.

12.3.1   ELECTROCHEMICAL MODEL

Helmholtz [33] described a model which is similar to the classical structures with dielectric capacitors [13]. Later Gouy [34] and Chapman [12] redefined this by adopting the Boltzmann distribution equation to represent how the concentration of ions depend on the diffuse layers of electrochemical potential. They interpret the ion mobility in the electrolyte solutions as a result of diffusion and electrostatic forces. Further Stern [35] combined these two models and he considered the EDL is having two separate layers, the Stern layer (also known as Helmholtz layer) and the Gouy-Chapman diffuse layer. The Stern layer controls the total capacitance of EDL by selective absorption on the electrode surface. But this model considers the ions as point charges and this is good only for low concentration of ions as well as low potential. This leads to an unrealistic condition of ion concentration by applying PB equation [14, 36]. Bikerman [37] put forward the idea of the PB model by considering fixed ion size under equilibrium conditions, in which both the positively and negatively charged ions of the electrolyte may have different size even if they are in same valence state. Verbrugge and coworkers [38] came forward with idea of a 1D one-domain model, where they considered dilute solution theory and porous electrode. The SC was modelled as a continuum entity having physical properties which are highly homogeneous and isotropic. Allu et al. [39] modified this to a 3D model where they considered the electrode-electrolyte system got uniform formulation. This model captures uneven conformation, charge transport, associated performance in 3D and spatio-temporal discrepancies, important physical properties, etc., into simulations. Later Wang and coworkers [40] established a 3D model for SCs which reflects 3D electrode, ion size and dialectic permittivity.

12.3.2 EQUIVALENT CIRCUIT MODEL

This model use capacitor-resistor (RC) grids to imitate the electrical performance of SCs. Ordinary differential equations are used in this modeling and hence they are easy to implement [41]. The accuracy of the model varies with configuration of the circuit and number of elements, and it can be increased by increasing circuit sophistication. The overall resistance is represented by the series resistor and the capacitor represents normal capacitance of SCs. Spyker and Nelms [42] interpret the classical equivalent circuit model by adding another parallel resistor to aid the self-discharge phenomenon as shown in Figure 12.3a. Zubieta and Boner [43] established a model, comprised of three capacitor-resistor branches- immediate branch which incorporates a nonlinear capacitance, delayed branch, and long-term branch. Liu et al. [44] developed a synthetic route for a three-branch model, in which model parameters are relied on temperature. In order to characterize the self-discharge process, Zhang and coworkers [45] exploited a variable resistor in three-branch model as shown in Figure 12.3b. Later Buller et al. [46] anticipated a dynamic model, which comprises of resistor in series, a bulk capacitor and two RC networks (parallel), Figure 12.3c. Targeting at unfolding the full frequency-range performance, Musolino and coworkers [47] proposed a dynamic model by substituting the immediate branch with a parallel leakage resistor. Gualous et al. [48] came forward with a temperature-dependent model based on these laboratory studies on SC capacitance variation with respect to serial resistance. Rafika et al. [49] proposed a model with 14 elements varies with voltage and/or temperature. To mimic dispersed capacitance and electrolyte resistance by the porous electrodes, transmission line models were introduced, taking transient and long-term behavior into consideration, as shown in Figure 12.3d. The complexity of the model depends on the number of the active RC networks [50–51]. Usually, the reliability of the model increases by increasing RC networks at the expense of computational efficacy.

12.3.3 INTELLIGENT MODEL

Artificial neural networks (ANN) found to be highly effective in investigating the properties of energy storage systems [52–53]. Generally, these models have the ability to define the complex nonlinear relationship amongst the performance and the features affecting them, without a deep knowledge of underlying mechanisms [54]. The model accuracy depends on the quantity and quality of training data. These type of intelligent methods found to be useful in designing of efficient energy storage system apart from the prediction of performance of SC. Farsi and Gobal [55] made an ANN based model for investigating the influences of several features on the performance of SC performance, especially factors affecting the utilization as well as energy and power density. In the model they employed the most important factors which may influence the design of a good SC such as size of the crystal, dimensions of the surface lattice especially the length, effective cell current

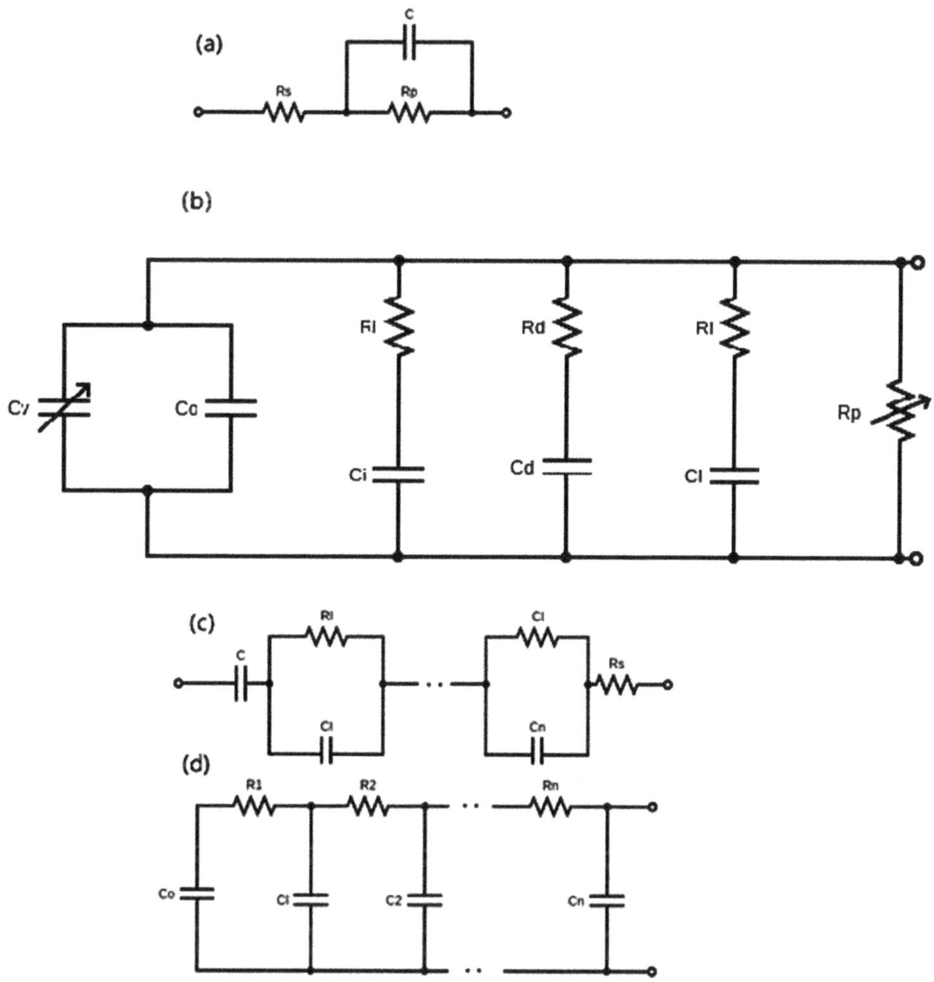

FIGURE 12.3 Equivalent circuit models [8].

and active materials' exchange current density. A simulation model of SC behavior was developed by Wu et al. [56], in which a well-known ANN model was used to predict its parameters by applying terminal voltage and temperature as inputs. Later scientists came forward with a feed-forward ANN to represent properties of SC in which they employed underlying chemistry, temperature of the system and rate of the current along with the historical data [57]. The model performance was tested using power cycling and the model was found to be useful in controlling voltage in SC. Weigert et al. [58] came forward with a ANN network model for estimating the state-of-charge (SOC) in hybrid energy storage battery-ultracapacitor device. An ANN model for estimating the output voltage as a function of voltage variations, temperature of the system and current was developed by Francoise et al. [59].

12.3.4 Transmission Line and Fractional-Order Model

De Levie [60–61] introduced a transmission line model for the electrical impedance of an SC by applying macroscopic scale of an electrode. The overall performance of an electrode represented by a transmission line instead of using all routes that each of the adsorbed species follows. Two electrodes are using transmission lines as represented in Figure 12.4 and the bulk resistance of

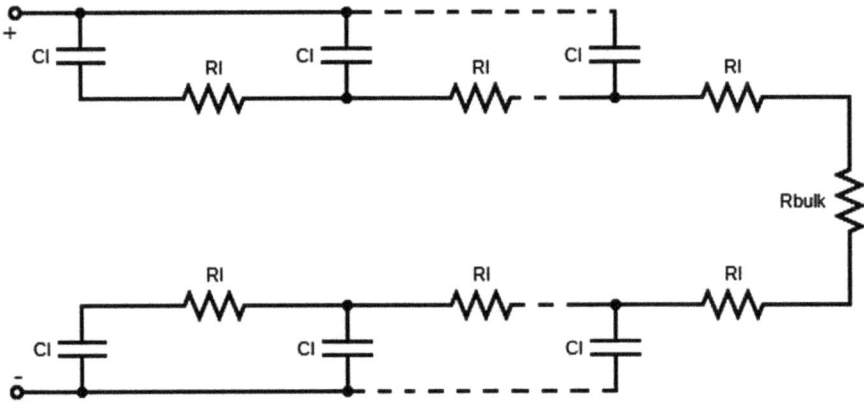

FIGURE 12.4 Equivalent circuit model in the transmission line model. R_{bulk} is the resistance of the electrolyte in the bulk region, while Rl and Cl are the resistance and the capacitance inside the electrodes respectively [9].

the electrolyte (Rbulk) are combined to denote the performance of a SC [62]. The computational time increases with the increase in the branches in a transmission line. There are several models suggested by different research groups having transmission lines with 5 to 15 branches, depending on the application [63]. The computational simulation cost can be reduced without affecting accuracy, by waveform relaxation strategy in a transmission line model [64]. A parameter fixation method for a transmission line model was proposed, merely by comparing molecular simulations and electrochemical impedance experiments [62]. The designing of best mesoporous material for SCs having energy and power density can be done by applying a transmission line model along with the details of distribution of pore size of the material [65]. Novel methods to improve the durability and electrochemical properties of SCs was put forward by combining a transmission line model with spectroscopic techniques such as impedance and density functional computational methods. One of the main drawbacks of transmission line model is the difficulty when many cells are linked either in parallel or in series [66]. Fractional models are extension of transmission-line models, in which the dispersal of relaxation time constants are denoted as elements such as "Constant Phase Elements" (CPE) or "Warburg Impedances". Actually, a RC tree with infinite branches is comparable to a one CPE and the components used in time domain is represented by a differential equation of fractional-order.

12.3.5 SIMPLIFIED ANALYTICAL MODEL

Simplified analytical model is a good method in evaluating the electrical performance of SCs, which considers the "coulombic efficiency", "self-discharge", and "parasitic inductances" represented as equivalent electric circuits [67–68]. The simplification of the model lowers its mathematical complexity and accuracy. Most of the simplified analytical models consider the voltage dependence of capacitance of SC. Many scientists proposed tangential expression [66] or linear formulation to predict the voltage as a function of capacitance [43, 69–70]:

$$C = C_a + C_b tanh\left(\frac{v}{U_x} - U_x\right) \text{ and } C = C_0 + kv$$

where U_x represents the voltage at the inflexion point of the hyperbolic tangent term, v is the "supercapacitor voltage" and C_a and C_b coefficient of fitting.

In addition to this variable capacitance, simplified analytical models can account for other observables which are having effect on the performance of SC. The widely employed electric circuits for this purpose are RC circuit [42, 71–72], multi-branch models [73–75] and dynamic models [76]. Multipurpose models are also proposed by combining the features of these three types [51, 69, 77]

12.3.6 Thermal Model

SCs are strongly sensitive to temperature mainly because it is operated in high-rate cycling [78–79]. This necessitates an accurate prediction of SC thermal behavior for scheming a effective cooling management, finding best temperature dependent features of electrical circuit models, and also for estimating the aging factors [80]. Schiffer et al. [81] measured the thermal properties of a SC. They observed that the heat generation is stimulated by the entropy change of ion movement between different charged state of the system. Dandeville et al. [82] verified the Schiffer's conclusion by acquiring heat profiles of a SC which depend on time using calorimetric methods. Various models have been suggested to forecast temperature performance, which can be commonly grouped into two classes, i.e., first principle models and comprehensive models. The first principle models employ partial differential equations to represent the thermal dynamics of SCs, and numerical discretization methods are good for solving these equations [83–87]. The model is good for evaluating heat generation as a function of ion diffusion, steric effect, and changes in entropy. There exist several comprehensive models to evaluate the SC thermal dynamics [69].

12.4 SIMULATION TECHNIQUES OF SUPERCAPACITORS

Over the past decades, the development of advanced electrode active materials through simulation has highly enhanced the performance of electrochemical capacitors or supercapacitors [88]. The most commonly used theoretical approaches to simulate supercapacitors are "Molecular dynamics" (MD), and "Monte Carlo (MC) simulations", and classical DFT methods [6]. Of these, in CDFT and MC generally coarse-grained models are employed for the electrolyte, whereas in MD usually all-atom models are used [5].

12.4.1 Molecular Simulations

The most suitable molecular simulation techniques employed for the modelling of supercapacitors are molecular dynamics (MD) and Monte Carlo (MC) simulations [89]. The computational resources and correctness of these methods mainly depend on the modeling of its electrode and electrolyte [9]. Compared to other simulation techniques, mainly these methods have two unique advantages one is they can provide direct insights on the microstructure (very difficult to determine with any experimental technique) and effective macroscopic properties (like capacitance) of the EDLs. This paves the way for the researchers to determine the microscopic origins of the capacitance of SC and thus gives the guidance of the selection and design of electrode and electrolyte materials for SC [6].

Among the computational simulation techniques, MD methods have the unique nature of allowing to apply controllable external potential conditions in the system, thus it provides real-time simulations and the simulation results are directly compared with the extensive in situ experimental analysis [89]. MD simulations have been broadly and effectively used to describe the electrostatic and van der walls interactions between the atoms or molecules, and it reproduce the dynamical behaviors of the system averaged within a short simulation period, which are characterized by solving the equation of motion defined in classical (i.e., Newtonian) mechanics [21, 90–91].

With the help of specified force fields (FFs) or interatomic potentials, the force imposed on the atoms or molecules are measured in the MD system during a period of time. In MD simulations the iterations on the calculating the instantaneous forces and the consequential movement of particles in the system reveals the precise information on the position, and velocity of atoms/molecules and also the trajectories of the entire process. By monitoring these trajectories, MD can provide information on thermodynamic and dynamic properties of the molecules [92]. In the case of EDLCs, based on the specified interaction between the ions in the system and ion pathways obtained from the MD simulation results, the distribution and motions of the electrolyte ions can be obtained, which respectively reflect their energy storage and dynamics behavior [6].

Monte Carlo (MC) simulation technique is based on statistical mechanics, widely used approach for analyzing the molecular behaviors and have been greatly employed in electro-osmotic flow, as well as handling higher charge densities, which are of great importance in EDLCs [93]. MC methods are very useful for the EDLCs design with significant uncertainty in inputs and/or systems containing a large degrees of freedom, which are often the only practical way to sample governed by the density of a continuous random variable.

The accuracy of molecular simulations of EDLs using MD methods is not much different from other atomistic simulations. Generally, the molecular simulations of electronically polarizable objects are challenging, therefore most of the EDL simulations adopt an approximation technique, i.e., the partial charges in the system are distributed uniformly among the surface atoms of electrodes. Idealized electrodes such as spherical, planar, or cylindrical surface electrodes are effectively used this method. But in the case of complex shapes electrodes this technique is very difficult because the accumulation of charge on different electrodes are different, even in the time-averaged sense. If so a constant electric potential will impose on the surface of the electrode. Two methods have been used for this, (i) by modifying the electrical charges on the surface atoms of each electrodes, the electrical potential is kept as a constant, and (ii) by solving an auxiliary Laplace equation, the electrical potential is implemented on numerical grid points lying over the electrode surface. Both these methods are effectively used to model nanopore electrodes.

12.4.2 CLASSICAL DENSITY FUNCTIONAL THEORY

Classical density functional theory (CDFT) is a powerful computational quantum mechanical technique that has been widely and successfully used to simulate the equilibrium properties of soft matters and complex liquids. Compared with other simulation techniques such as MC or MD, DFT calculations requires less computational resources, and also it allows one to accurately steer the important parameters such as electrode surface charge density/potential, and ion size in the system. For supercapacitor simulations, the solvent molecules, ionic species, and impurities in the electrolyte solution is specified by using both primitive as well as non-primitive coarse-grained models [5, 8]. The model system consists of charged hard spheres for ionic species (cations and anions), and a hardsphere segment for solvent molecules [94]. For various pore geometries of the supercapacitors, CDFT simulation had been successfully used to examine the EDL structure and capacitance for the electrolyte solution. CDFT predictions can be capture the outcomes from earlier experiments, and simulation studies, and provide detailed insights into the electrochemical properties of ionic liquids (IL) as working electrolytes for supercapacitors [94]. One of the biggest limitations of CDFT is that, it does not handled the issues of the concentration of ions, different size of ions and solvent molecules, and the polarity of solvent, which are very crucial to enhance the properties of EDLCs [95–97].

12.5 SIMULATION SOFTWARE/PROGRAMS

Whenever a supercapacitor model is designed, one has to search a suitable program or software, which can be utilized for the development of that particular model. Complex models resembling the existing supercapacitors have also been developed using this software. Each simulation program or software has a different impact on the development of the designed models and thus have different advantages and disadvantages. The most commonly used software packages for carrying out modeling of supercapacitors are Simulink [98], OrCAD capture, PSCAD, SABER, PLECS, etc. [99].

There are several reports on the use of these software tools in modeling of supercapacitors. The new hybrid system proposed by Sachin and Blaabjerg used MATLAB/Simulink for 6 kW rated power for simulations [100]. Andari et al. used MATLAB/Simulink software to simulate a designed fuel cell hybrid electric vehicle model, which consists of a supercapacitor, proton exchange membrane (PEM) fuel cell, and permanent magnet synchronous (PMS) motor [101]. A new strategy of an energy management between battery and supercapacitors for an urban electric vehicle is proposed by Azizi and Radjeai in 2017. To evaluate the performance of the proposed strategy, a simulation of an urban hybrid electric vehicle is implemented in MATLAB/Simulink [102]. Farcas and his co-workers modeled some supercapacitors and the simulations are made in Simulink 7.5 and Orcad 9.2 to determine their operation are in time and frequency domain [103]. Akash and vijay modeled a hybrid supercapacitor by the combination of battery and supercapacitors through simulations in ORCAD/PSPICE [104]. Vargas et al. in 2019, proposed a flexible extended harmonic domain (FEHD) hybrid model of a stand-alone PV system, that involves a battery-supercapacitor hybrid energy storage system. The proposed model keeps accuracy of harmonics dynamics to the actual components and it also reduced the computational cost compared with the time domain-based model and conventional extended harmonic domain-based models. Moreover, the validity of this hybrid system is verified by simulation based on PSCAD/EMTDC [105]. To describe the electric and thermal behavior of supercapacitor in transportation applications, Goulas and his co-workers proposed an equivalent electric circuit by using Saber and Spice software [106]. Schonberger used another software, PLECS, to model a simplified as well as frequency dependent supercapacitor. He used a lumped parameter circuit to model two supercapacitors [107].

12.6 CONCLUSIONS

This chapter gives a brief review of theoretical advancements in SC modeling and simulations, as well as basic theories involved in the designing of ultracapacitors. The fundamental EDL theory developed by Helmholtz, used in the design of conventional capacitors, were further modified by Gouy-Chapman, Stern and Grahame to incorporate pseudocapacitive materials. Various models such as the electrochemical model, equivalent circuit model, intelligent model, transmission-line model, fractional model, simplified analytical model, and thermal model are briefly discussed to account for the structure of the model, computational complexity, and specificity and accuracy for the simulation of electrical behavior. Then the chapter summarized molecular simulation approaches like molecular dynamics and Monte Carlo techniques, and density functional theory, as well as the software used in these simulations.

As a counterpart to experimental efforts, different models and simulation methods are helpful in predicting the use of novel electrode and electrolyte materials for better performance, efficient design of SCs, and to understand the energy storage mechanism. In the near future, advances in research will generate large multi-scale computing that integrates methods at various times and different length scales, which will be able to provide a fundamental understanding of processes such as phase transitions in electrode materials, ion transport in electrolytes, charge transfer at interfaces, and electronic transports in electrodes.

REFERENCES

1. Kumar A, Ahmed G, Gupta M, Bocchetta P, Adalati R, Chandra R, Kumar Y. Theories and models of supercapacitors with recent advancements: Impact and interpretations. *Nano Express*. 2021 Apr 30;2(2):022004.
2. Lemine AS, Zagho MM, Altahtamouni TM, Bensalah N. Graphene a promising electrode material for supercapacitors—A review. *International Journal of Energy Research*. 2018 Nov;42(14):4284–300.

3. Ban S, Zhang J, Zhang L, Tsay K, Song D, Zou X. Charging and discharging electrochemical superca-pacitors in the presence of both parallel leakage process and electrochemical decomposition of solvent. *Electrochimica Acta.* 2013 Feb 15;90:542–9.

4. Muzaffar A, Ahamed MB, Deshmukh K, Thirumalai J. A review on recent advances in hybrid superca-pacitors: Design, fabrication and applications. *Renewable and Sustainable Energy Reviews.* 2019 Mar 1;101:123–45.

5. Zhan C, Lian C, Zhang Y, Thompson MW, Xie Y, Wu J, Kent PR, Cummings PT, Jiang DE, Wesolowski DJ. Computational insights into materials and interfaces for capacitive energy storage. *Advanced Science.* 2017;4(7):1700059.

6. Bo Z, Li C, Yang H, Ostrikov K, Yan J, Cen K. Design of supercapacitor electrodes using molecular dynamics simulations. *Nano-Micro Letters.* 2018 Apr;10(2):1–23.

7. Huang S, Zhu X, Sarkar S, Zhao Y. Challenges and opportunities for supercapacitors. *APL Materials.* 2019 Oct 1;7(10):100901.

8. Zhang L, Hu X, Wang Z, Sun F, Dorrell DG. A review of supercapacitor modeling, estimation, and applications: A control/management perspective. *Renewable and Sustainable Energy Reviews.* 2018 Jan 1;81:1868–78.

9. Berrueta A, Ursúa A, San Martín I, Eftekhari A, Sanchis P. Supercapacitors: Electrical characteristics, modeling, applications, and future trends. *Ieee Access.* 2019 Apr 22;7:50869–96.

10. Supercapacitors vs. Batteries | Engineering Center (kemet.com)

11. Guoy G. Constitution of the electric charge at the surface of an electrolyte. *Journal de Physique.* 1910;9:457–67.

12. Chapman DL. LI. A contribution to the theory of electrocapillarity. *The London, Edinburgh, and Dublin Philosophical Magazine and Journal of Science.* 1913 Apr 1;25(148):475–81.

13. Wang H, Pilon L. Accurate simulations of electric double layer capacitance of ultramicroelectrodes. *The Journal of Physical Chemistry C.* 2011 Aug 25;115(33):16711–9.

14. Bard AJ, Faulkner LR. Fundamentals and applications. *Electrochemical Methods.* 2001;2(482):580–632.

15. Faranda R, Gallina M, Son DT. A new simplified model of double-layer capacitors. In *2007 International Conference on Clean Electrical Power 2007 May 21* (pp. 706–710). IEEE.

16. Karlsson A. Evaluation of Simulink/SimPowerSystems and other commercial simulation tools for the simulation of machine system transients. Master's Degree Project, Stocholm, Sweden 2005.

17. Belhachemi F, Rael S, Davat B. A physical based model of power electric double-layer supercapacitors. In *Conference Record of the 2000 IEEE Industry Applications Conference. Thirty-Fifth IAS Annual Meeting and World Conference on Industrial Applications of Electrical Energy* (Cat. No. 00CH37129) 2000 Oct 8 (Vol. 5, pp. 3069–3076). IEEE.

18. Van der Sluis L. *Transients in Power Systems.* John Wiley & Sons Ltd, 2001, p. 153.

19. Stern O. The theory of the electrolytic double-layer. *Z. Elektrochem.* 1924;30(508):1014–20.

20. Grahame DC. The electrical double layer and the theory of electrocapillarity. *Chemical Reviews.* 1947 Dec 1;41(3):441–501.

21. Burt R, Birkett G, Zhao XS. A review of molecular modelling of electric double layer capacitors. *Physical Chemistry Chemical Physics.* 2014;16(14):6519–38.

22. Srinivasan V, Weidner JW. Mathematical modeling of electrochemical capacitors. *Journal of the Electrochemical Society.* 1999 May 1;146(5):1650.

23. Drummond R, Howey DA, Duncan SR. Low-order mathematical modelling of electric double layer supercapacitors using spectral methods. *Journal of Power Sources.* 2015 Mar 1;277:317–28.

24. Allen MP, Tildesley DJ. *Computer Simulation of Liquids.* Oxford University Press, 1989.

25. Frenkel D, Smit B. *Understanding Molecular Simulation: From Algorithms to Applications.* Elsevier, 2001 Oct 19.

26. Wang Y, Jiang WE, Yan T, Voth GA. Understanding ionic liquids through atomistic and coarse-grained molecular dynamics simulations. *Accounts of Chemical Research.* 2007 Nov 20;40(11):1193–9.

27. Mundy A, Plett GL. Reduced-order physics-based modeling and experimental parameter identification for non-faradaic electrical double-layer capacitors. *Journal of Energy Storage.* 2016 Aug 1;7:167–80.

28. Vatamanu J, Borodin O, Smith GD. Molecular insights into the potential and temperature dependences of the differential capacitance of a room-temperature ionic liquid at graphite electrodes. *Journal of the American Chemical Society.* 2010 Oct 27;132(42):14825–33.

29. Xing L, Vatamanu J, Borodin O, Bedrov D. On the atomistic nature of capacitance enhancement gen-erated by ionic liquid electrolyte confined in subnanometer pores. *The Journal of Physical Chemistry Letters.* 2013 Jan 3;4(1):132–40.

30. Li S, Feng G, Fulvio PF, Hillesheim PC, Liao C, Dai S, Cummings PT. Molecular dynamics simulation study of the capacitive performance of a binary mixture of ionic liquids near an onion-like carbon electrode. *The Journal of Physical Chemistry Letters*. 2012 Sep 6;3(17):2465–9.

31. Paek E, Pak AJ, Hwang GS. Erratum: A computational study of the interfacial structure and capacitance of graphene in [BMIM][PF6] ionic liquid [j. electrochem. soc., 160, a1 (2013)]. *Journal of The Electrochemical Society*. 2014 Jan;161(10):X15-.

32. Merlet C, Péan C, Rotenberg B, Madden PA, Simon P, Salanne M. Simulating supercapacitors: Can we model electrodes as constant charge surfaces? *The Journal of Physical Chemistry Letters*. 2013 Jan 17;4(2):264–8.

33. Helmholtz HV. Studien über electrische Grenzschichten. *Annalen der Physik*. 1879;243(7):337–82.

34. Guoy G. Constitution of the electric charge at the surface of an electrolyte. *Journal de Physique*. 1910;9:457–67.

35. Stern O. The theory of the electrolytic double-layer. *Z. Elektrochem.* 1924;30(508):1014–20.

36. Bagotsky VS, editor. *Fundamentals of Electrochemistry*. John Wiley & Sons, 2005 Dec 2.

37. Bikerman JJ. XXXIX. Structure and capacity of electrical double layer. *The London, Edinburgh, and Dublin Philosophical Magazine and Journal of Science*. 1942 May 1;33(220):384–97.

38. Verbrugge MW, Liu P. Microstructural analysis and mathematical modeling of electric double-layer supercapacitors. *Journal of the Electrochemical Society*. 2005 Mar 25;152(5):D79.

39. Allu S, Asokan BV, Shelton WA, Philip B, Pannala S. A generalized multi-dimensional mathematical model for charging and discharging processes in a supercapacitor. *Journal of Power Sources*. 2014 Jun 15;256:369–82.

40. Wang H, Pilon L. Mesoscale modeling of electric double layer capacitors with three-dimensional ordered structures. *Journal of Power Sources*. 2013 Jan 1;221:252–60.

41. Hu X, Li S, Peng H. A comparative study of equivalent circuit models for Li-ion batteries. *Journal of Power Sources*. 2012 Jan 15;198:359–67.

42. Spyker RL, Nelms RM. Classical equivalent circuit parameters for a double-layer capacitor. *IEEE Transactions on Aerospace and Electronic Systems*. 2000 Jul;36(3):829–36.

43. Zubieta L, Bonert R. Characterization of double-layer capacitors for power electronics applications. *IEEE Transactions on Industry Applications*. 2000 Jan;36(1):199–205.

44. Liu K, Zhu C, Lu R, Chan CC. Improved study of temperature dependence equivalent circuit model for supercapacitors. *IEEE Transactions on Plasma Science*. 2013 Mar 25;41(5):1267–71.

45. Zhang Y, Yang H. Modeling and characterization of supercapacitors for wireless sensor network applications. *Journal of Power Sources*. 2011 Apr 15;196(8):4128–35.

46. Buller S, Karden E, Kok D, De Doncker RW. Modeling the dynamic behavior of supercapacitors using impedance spectroscopy. *IEEE Transactions on Industry Applications*. 2002 Dec 10;38(6):1622–6.

47. Musolino V, Piegari L, Tironi E. New full-frequency-range supercapacitor model with easy identification procedure. *IEEE Transactions on Industrial Electronics*. 2012 Feb 10;60(1):112–20.

48. Gualous H, Bouquain D, Berthon A, Kauffmann JM. Experimental study of supercapacitor serial resistance and capacitance variations with temperature. *Journal of Power Sources*. 2003 Sep 15;123(1):86–93.

49. Rafik F, Gualous H, Gallay R, Crausaz A, Berthon A. Frequency, thermal and voltage supercapacitor characterization and modeling. *Journal of Power Sources*. 2007 Mar 20;165(2):928–34.

50. Belhachemi F, Rael S, Davat B. A physical based model of power electric double-layer supercapacitors. In *Conference Record of the 2000 IEEE Industry Applications Conference. Thirty-Fifth IAS Annual Meeting and World Conference on Industrial Applications of Electrical Energy* (Cat. No. 00CH37129) 2000 Oct 8 (Vol. 5, pp. 3069–3076). IEEE.

51. Torregrossa D, Bahramipanah M, Namor E, Cherkaoui R, Paolone M. Improvement of dynamic modeling of supercapacitor by residual charge effect estimation. *IEEE Transactions on Industrial Electronics*. 2013 Apr 24;61(3):1345–54.

52. Hu X, Li SE, Yang Y. Advanced machine learning approach for lithium-ion battery state estimation in electric vehicles. *IEEE Transactions on Transportation Electrification*. 2015 Dec 24;2(2):140–9.

53. Soualhi A, Sari A, Razik H, Venet P, Clerc G, German R, Briat O, Vinassa JM. Supercapacitors ageing prediction by neural networks. In *IECON 2013–39th Annual Conference of the IEEE Industrial Electronics Society 2013 Nov 10* (pp. 6812–6818). IEEE.

54. Haykin S. *Neural Networks: A Comprehensive Foundation*, MacMillan College Publishing Co, 1994.

55. Farsi H, Gobal F. Artificial neural network simulator for supercapacitor performance prediction. *Computational Materials Science*. 2007 May 1;39(3):678–83.

56. Wu CH, Hung YH, Hong CW. On-line supercapacitor dynamic models for energy conversion and management. *Energy Conversion and Management*. 2012 Jan 1;53(1):337–45.

57. Eddahech A, Briat O, Ayadi M, Vinassa JM. Modeling and adaptive control for supercapacitor in automotive applications based on artificial neural networks. *Electric Power Systems Research*. 2014 Jan 1;106:134–41.

58. Weigert T, Tian Q, Lian K. State-of-charge prediction of batteries and battery—supercapacitor hybrids using artificial neural networks. *Journal of Power Sources*. 2011 Apr 15;196(8):4061–6.

59. Marie-Francoise JN, Gualous H, Berthon A. Supercapacitor thermal-and electrical-behaviour modelling using ANN. *IEE Proceedings-Electric Power Applications*. 2006 Mar 1;153(2):255–62.

60. De Levie R. On porous electrodes in electrolyte solutions: I. Capacitance effects. *Electrochimica Acta*. 1963 Oct 1;8(10):751–80.

61. De Levie R. On the impedance of electrodes with rough interfaces. *Journal of Electroanalytical Chemistry and Interfacial Electrochemistry*. 1989 Mar 24;261(1):1–9.

62. Péan C, Rotenberg B, Simon P, Salanne M. Multi-scale modelling of supercapacitors: From molecular simulations to a transmission line model. *Journal of Power Sources*. 2016 Sep 15;326:680–5.

63. Logerais PO, Camara MA, Riou O, Djellad A, Omeiri A, Delaleux F, Durastanti JF. Modeling of a supercapacitor with a multibranch circuit. *International Journal of Hydrogen Energy*. 2015 Oct 19;40(39):13725–36.

64. Moayedi S, Cingöz F, Davoudi A. Accelerated simulation of high-fidelity models of supercapacitors using waveform relaxation techniques. *IEEE Transactions on Power Electronics*. 2013 Mar 7;28(11):4903–9.

65. Ghosh A, Le VT, Bae JJ, Lee YH. TLM-PSD model for optimization of energy and power density of vertically aligned carbon nanotube supercapacitor. *Scientific Reports*. 2013 Oct 22;3(1):1–0.

66. Miller J. *Ultracapacitor Applications (Energy Engineering)*. IET, 2011.

67. Meyers JP, Doyle M, Darling RM, Newman J. The impedance response of a porous electrode composed of intercalation particles. *Journal of the Electrochemical Society*. 2000 Aug 1;147(8):2930.

68. Drummond R, Zhao S, Howey DA, Duncan SR. Circuit synthesis of electrochemical supercapacitor models. *Journal of Energy Storage*. 2017 Apr 1;10:48–55.

69. Berrueta A, San Martin I, Hernández A, Ursúa A, Sanchis P. Electro-thermal modelling of a supercapacitor and experimental validation. *Journal of Power Sources*. 2014 Aug 1;259:154–65.

70. Lajnef W, Vinassa JM, Briat O, Azzopardi S, Woirgard E. Characterization methods and modelling of ultracapacitors for use as peak power sources. *Journal of Power Sources*. 2007 Jun 1;168(2):553–60.

71. Goh CT, Cruden A. Bivariate quadratic method in quantifying the differential capacitance and energy capacity of supercapacitors under high current operation. *Journal of Power Sources*. 2014 Nov 1;265:291–8.

72. Eddahech A, Ayadi M, Briat O, Vinassa JM. Online parameter identification for real-time supercapacitor performance estimation in automotive applications. *International Journal of Electrical Power & Energy Systems*. 2013 Oct 1;51:162–7.

73. Fletcher S, Kirkpatrick I, Dring R, Puttock R, Thring R, Howroyd S. The modelling of carbon-based supercapacitors: Distributions of time constants and pascal equivalent circuits. *Journal of Power Sources*. 2017 Mar 31;345:247–53.

74. Sedlakova V, Sikula J, Majzner J, Sedlak P, Kuparowitz T, Buergler B, Vasina P. Supercapacitor equivalent electrical circuit model based on charges redistribution by diffusion. *Journal of Power Sources*. 2015 Jul 15;286:58–65.

75. Nadeau A, Hassanalieragh M, Sharma G, Soyata T. Energy awareness for supercapacitors using Kalman filter state-of-charge tracking. *Journal of Power Sources*. 2015 Nov 20;296:383–91.

76. Parvini Y, Siegel JB, Stefanopoulou AG, Vahidi A. Supercapacitor electrical and thermal modeling, identification, and validation for a wide range of temperature and power applications. *IEEE Transactions on Industrial Electronics*. 2015 Oct 26;63(3):1574–85.

77. Fletcher S, Black VJ, Kirkpatrick I. A universal equivalent circuit for carbon-based supercapacitors. *Journal of Solid State Electrochemistry*. 2014 May 1;18(5):1377–87.

78. Miller JR. Electrochemical capacitor thermal management issues at high-rate cycling. *Electrochimica Acta*. 2006 Dec 1;52(4):1703–8.

79. Liu P, Verbrugge M, Soukiazian S. Influence of temperature and electrolyte on the performance of activated-carbon supercapacitors. *Journal of Power Sources*. 2006 Jun 1;156(2):712–8.

80. Weddell AS, Merrett GV, Kazmierski TJ, Al-Hashimi BM. Accurate supercapacitor modeling for energy harvesting wireless sensor nodes. *IEEE Transactions on Circuits and Systems II: Express Briefs*. 2011 Dec 1;58(12):911–5.

81. Schiffer J, Linzen D, Sauer DU. Heat generation in double layer capacitors. *Journal of Power Sources*. 2006 Sep 29;160(1):765–72.

82. Dandeville Y, Guillemet P, Scudeller Y, Crosnier O, Athouel L, Brousse T. Measuring time-dependent heat profiles of aqueous electrochemical capacitors under cycling. *Thermochimica Acta*. 2011 Nov 10;526(1–2):1–8.

83. Gualous H, Louahlia H, Gallay R. Supercapacitor characterization and thermal modelling with reversible and irreversible heat effect. *IEEE Transactions on Power Electronics*. 2011 Apr 21;26(11):3402–9.
84. Wang K, Zhang L, Ji B, Yuan J. The thermal analysis on the stackable supercapacitor. *Energy*. 2013 Sep 15;59:440–4.
85. d'Entremont A, Pilon L. First-principles thermal modeling of electric double layer capacitors under constant-current cycling. *Journal of Power Sources*. 2014 Jan 15;246:887–98.
86. Al Sakka M, Gualous H, Van Mierlo J, Culcu H. Thermal modeling and heat management of supercapacitor modules for vehicle applications. *Journal of Power Sources*. 2009 Dec 1;194(2):581–7.
87. Sarwar W, Marinescu M, Green N, Taylor N, Offer G. Electrochemical double layer capacitor electrothermal modelling. *Journal of Energy Storage*. 2016 Feb 1;5:10–24.
88. Noori Abolhassan, Maher F. El-Kady, Mohammad S. Rahmanifar, Richard B. Kaner, Mir F. Mousavi. Towards establishing standard performance metrics for batteries, supercapacitors and beyond. *Chemical Society Reviews*. 2019;48(5):1272–341.
89. Xu K, Shao H, Lin Z, Merlet C, Feng G, Zhu J, Simon P. Computational insights into charge storage mechanisms of supercapacitors. *Energy & Environmental Materials*. 2020 Sep;3(3):235–46.
90. Allen MP. Introduction to molecular dynamics simulation. *Computational Soft Matter: From Synthetic Polymers to Proteins*. 2004;23(1):1–28.
91. Hollingsworth SA, Dror RO. Molecular dynamics simulation for all. *Neuron*. 2018 Sep 19;99(6):1129–43.
92. Polanski J. 4.14 *Chemoinformatics*. University of Silesia, 2009.
93. Ike IS, Sigalas I, Iyuke SE. The influences of operating conditions and design configurations on the performance of symmetric electrochemical capacitors. *Physical Chemistry Chemical Physics*. 2016;18(41):28626–47.
94. Lian C, Liu H. Classical density functional theory insights for supercapacitors. *IntechOpen, Rijeka*. 2018 Jun 27, chapter 8, 137–156.
95. Feng Guang, De-en Jiang, Peter T. Cummings. Curvature effect on the capacitance of electric double layers at ionic liquid/onion-like carbon interfaces. *Journal of Chemical Theory and Computation*. 2012;8(3):1058–63.
96. Nguyen Phuong TM, Chunyan Fan DD Do, Nicholson D. On the cavitation-like pore blocking in ink-bottle pore: Evolution of hysteresis loop with neck size. *The Journal of Physical Chemistry C*. 2013;117(10):5475–84.
97. Pizio O, Sokołowski S, Sokołowska Z. Electric double layer capacitance of restricted primitive model for an ionic fluid in slit-like nanopores: A density functional approach. *The Journal of Chemical Physics*. 2012 Dec 21;137(23):234705.
98. Hinov N, Vacheva G, Zlatev Z. Modelling a charging process of a supercapacitor in MATLAB/Simulink for electric vehicles. In *AIP Conference Proceedings 2018 Dec 10* (Vol. 2048, No. 1, p. 060023). AIP Publishing LLC.
99. Johansson P, Andersson B. Comparison of simulation programs for supercapacitor modelling. Master of Science Thesis. Chalmers University of Technology, Sweden. 2008.
100. Şahin ME, Blaabjerg F. A hybrid PV-battery/supercapacitor system and a basic active power control proposal in MATLAB/simulink. *Electronics*. 2020 Jan;9(1):129.
101. Andari W, Ghozzi S, Allagui H, Mami A. Design, modeling and energy management of a PEM fuel cell/supercapacitor hybrid vehicle. *International Journal of Advanced Computer Science and Applications*. 2017 Jan 1;8(1):273–8.
102. Azizi I, Radjeai H. A new strategy for battery and supercapacitor energy management for an urban electric vehicle. *Electrical Engineering*. 2018 Jun;100(2):667–76.
103. Fărcaş C, Petreuş D, Ciocan I, Palaghiţă N. Modeling and simulation of supercapacitors. In *2009 15th International Symposium for Design and Technology of Electronics Packages (SIITME) 2009 Sep 17* (pp. 195–200). IEEE.
104. Chipade AD, Bhagat V. Design and analysis hybrid combination of battery and supercapacitor using ORCAD/PSPICE. *IJRPET*, 2017 Sep;3(9).
105. Vargas U, Lazaroiu GC, Tironi E, Ramirez A. Harmonic modeling and simulation of a stand-alone photovoltaic-battery-supercapacitor hybrid system. *International Journal of Electrical Power & Energy Systems*. 2019 Feb 1;105:70–8.
106. Gualous H, Bouquain D, Berthon A, Kauffmann JM. Experimental study of supercapacitor serial resistance and capacitance variations with temperature. *Journal of Power Sources*. 2003 Sep 15;123(1):86–93.
107. Schönberger J. *Modeling a Supercapacitor using PLECS*. Plexim GmbH, Version, 2010, p. 3

13 Future Perspectives of Polymer Supercapacitors for Advanced Energy Storage Applications

Ajalesh Balachandran Nair, Shasiya Panikkaveettil Shamsudeen, Minu Joys, and Neethumol Varghese

CONTENTS

13.1 INTRODUCTION

The supercapacitor is a novel type of device to store energy and it is an advanced form of conventional capacitor, which contains two electrode materials. They are progressively used for energy transformation and in storing energy. Interest in the field of supercapacitors is mainly owing to their excellent energy capacity, excellent power density, outstanding storage capacity, faster charging and discharge rates, and longer shelf life. These important characteristics of supercapacitors bridge the performance gap between classical capacitors and novel secondary cells/rechargeable batteries, and they have tremendous applications such as in electronic communication, transportation, aerospace, and energy storage fields [1–3].

Conducting polymers (CPs) are considered important pseudocapacitive redox active materials due to their attributes. Some of the prominent and well-known CPs in the field of supercapacitors are polyaniline (PANi), polypyrrole (PPy), and polythiophene (PTh). The solid electrodes fabricated with CPs demonstrate various benefits like high conductivity, excellent flexibility, and ease of preparation methods [4]. The electrochemical capability of these electrodes is undesirable and various techniques were tried to enhance their performance. The polymerization of aniline monomer through different techniques such as chemical or electrochemical exhibited several added attractions such as facile synthesis, basic doping/de-doping systems, and ecological stability [5]. It is among the most prominent materials appropriate for pseudocapacitors. PPy also has many benefits including ease of preparation, excellent capacitivity, and improved cycle stability. The supercapacitor system based on PPy shows superior electrochemical properties in high performance applications [6]. PTh and functionalized PTh finds applications in the field of supercapacitors due to their improved conductivity, ecological stability, and higher absorption [7–8]. The electrochemical performance of supercapacitor electrode developed from pristine PTh was examined by several scientists [9–10] and have attempted various fabricated methods to improve their properties.

DOI: 10.1201/9781003174646-13

Recently, researchers have focused on fabricating nanomaterials to enhance their capacitive performance of supercapacitors. Carbon materials are important class, have high surface area and can be used as electrical double-layer capacitors (EDLCs). Several kinds of carbon materials are available, ID and 2D materials are most important in this category. The physical and chemical properties of outstanding nano-sized carbon-based materials viz. single-walled and multiwalled carbon nanotubes, Graphene (G), reduced graphene oxide (rGO) etc. are usually based on the carbon pore size.

The high electrical conductivity and outstanding mechanical strength exhibited by CNTs also paid consideration as supercapacitor electrodes. A high specific capacitance by nitric acid treated CNTs in H_2SO_4 exhibited was first reported by Niu et al. [11]. Various researches on CNT based supercapacitor electrode have been reported in light of this report. Hata et al. fabricated compactly aligned bulk form of SWCNTs as the supercapacitor electrode applications [12]. Later on, extensive research based on this established that the heteroatoms like nitrogen, oxygen, sulfur etc. make a widespread influence on the electrical capacitively of CNTs [13–14]. One dimensional CNTs with its exceptional performances, excellent conductance measurement, mechanical properties etc. show a significant role in fiber based wearable electrical devices.

Graphene is a well-known two-dimensional monolayer carbon-based compound comprised of sp^2 hybridized carbon atoms. This exhibits a high surface area in addition to high electrical conductivity which makes them suitable as electrode type material for supercapacitors applications [15]. Graphene displays an excessive theoretic capacitance of 550 F/g [16–17]. But the experimental capacitance of graphene—based supercapacitor is poorer. Due to the lower surface area of reduced graphene oxide (rGO) than the theoretical one by sheet restacking, the specific capacitance is reported as 191.0 F/g in Potassium hydroxide [18]. Enhancement in specific capacitance can be accomplished by proper physical outline of the electrode material using different approaches. In recent years, research is mainly focused on functionalization and hybridization of materials for increasing the surface area, packing density, conductivity, reducing restacking [19], and achieving defect control [20].

Solid state supercapacitors have been regarded as promising candidates aimed at an ample selection of applications in compact and elastic electrical and electronic devices. Fiber- shaped supercapacitors displayed high applications in portable and wearable electronics, in comparison with the two-dimensional planar structured supercapacitors, due to their low volume, extreme elasticity, and exceptional deformability [21–22]. Recently, fiber-based supercapacitors have rendered significant advancement in this field [23–25].

This chapter aims to give a comprehensive picture of the CPs based supercapacitors; nanomaterial-based supercapacitors, their binary/ternary systems, fiber-based supercapacitors, as well as a summary of certain selected supercapacitors which we take into account for future developments. In the last section of this chapter, a brief summary of the forthcoming perceptions in fast emerging area of supercapacitors is given. Further developments and challenges in the field of energy and power area are yet required. The appropriate designing of electrode materials, highly conducting electrolyte, high voltage display and security measures are essential. Furthermore, it is extremely necessary to remodel superconductors with numerous functions having good energy density, which make them fascinating for commercialization.

13.2 CONDUCTING POLYMER-BASED SUPERCAPACITORS

Conducting polymers (CPs) are materials with pseudocapacitivity, where the bulk material endures a quick faradaic redox response to furnish the capacitivity and exhibit greater specific energies to the carbon-based supercapacitor. Supercapacitors encompasses electrodes, electrolytes, and a divider. The supercapacitor is classified into two major classes centered on the dissimilarity in the method of storing. They are EDLCs and pseudocapacitors. Energy is stored in electrochemical double-layer capacitors based on the electrostatic interaction between ions of electrode materials and electrolytes on a significant area. The pseudocapacitors store energy based on the transfer of electron charge faradaically between electrode and electrolyte. Electro sorption, redox reactions, and intercalation processes are some of the methods for attaining this.

Conducting Polymers were counted as favorable resources of pseudocapacitivity due to their excellent features. PPy, PANi, and PTh are noteworthy in the field of conducting polymers. Electrodes based on these resources display a wide variety of features, such as good conducting power, excellent elasticity, comparatively economical, and the ease of fabrication [26]. Researchers have reviewed and have attempted a lot of approaches to enrich the electrochemical performance of the CPs based electrodes. The schematic representation and design of CP-based supercapacitors are shown in Figure 13.1. In this section, evaluation of the research development of supercapacitor electrodes based on PPy, PANi, or PTh is incorporated.

PANi is one of the mainly auspicious dynamic materials appropriate for pseudocapacitor electrode applications. PANi, which can be synthesized by polymerization of aniline monomer and exhibit a wide variety of benefits such as ease of processing and preparation, ingenuous chemistry of doping and de-doping of acid-base, and ecological firmness [27]. The surface morphology of PANi exhibit an immense effect on their electrical and chemical properties. So, the adoption of a suitable and high-productivity synthesis method is vital for synthesizing PANi using an appropriate nanostructure. PANi developed as nanofibers in an aqueous medium in chemical oxidative polymerization, [28] and a wide variety of polymerization techniques were designed to procure PANi nanofibers [29–31]. One of the conventional, simple, and comparatively less expensive methods is interfacial polymerization. Sivakumar et al. [32] studied and reported the electrochemical properties of PANi nanofibers synthesized by means of interfacial polymerization. Though the cyclic stability of supercapacitors was very poor, a high specific capacitance of 554.0 F/g at a current of 1.00 A/g was sensed. Li et al. [33] reviewed hypothetical and experimental capacitances of PANi in acidic

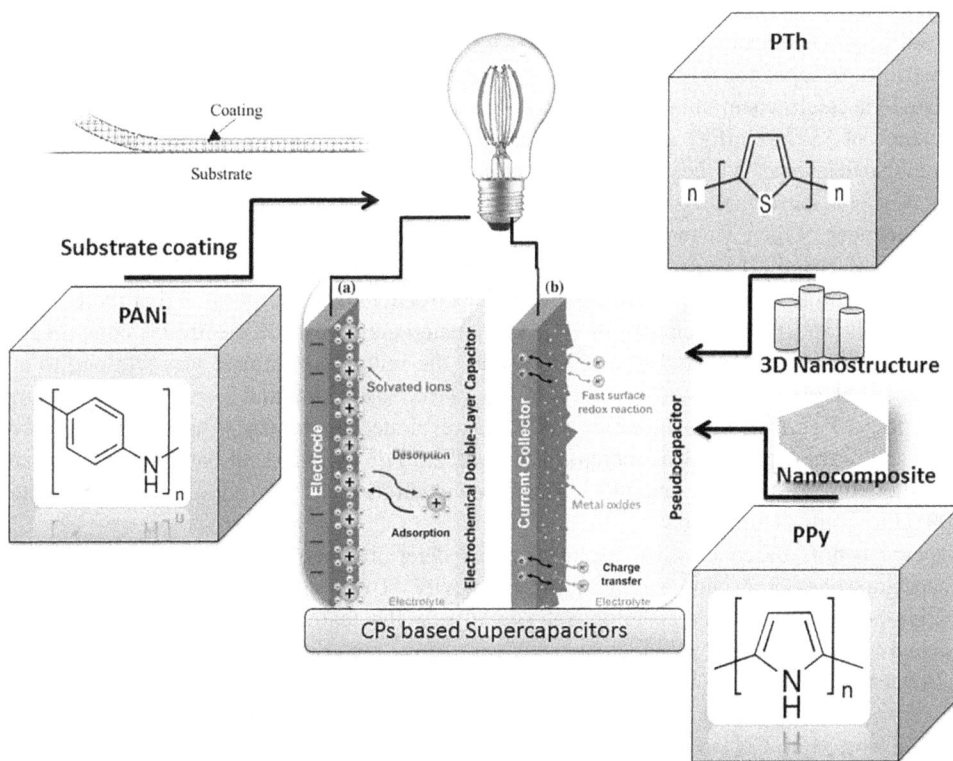

FIGURE 13.1   The schematic representation and design of CP-based supercapacitors.

medium. The theoretical value of capacitance was evaluated to be 2000 F/g which was greater than experimental value. Conductance of PANi and diffusion of counter ions establish a great contribution towards the capacitance. Even a little amount of PANi influences the specific capacitance.

PPy is another important CP, displays large number of advantages including ease of synthesis, moderately good capacitance, and prominent cycling stability. PPy films were prepared using interfacial polymerization employing surfactants by Yang et al. [34] and it possess a small pore size and exhibit noticeable electrochemical properties than those synthesized without the aid of surfactant. With the support of a reactive self-degradable template, Li and Yang [35] developed a characteristic PPy film through chemical oxidation technique. At current density of 0.2 A/g, PPy film displayed outstanding electrochemical properties with a specific capacitance of 576.0 F/g. Xu et al. [36] prepared conducting cotton materials treated with nanorods of PPy via, an in situ polymerization technique utilizing a template of $FeCl_3$-methyl orange complex. Although these engineered textiles display a low cycling stability, these finds application as electrodes of supercapacitors and presents exceptional specific capacity and energy density. The microstructure as well as the performance of PPy based electrodes depends on several factors like mode of preparation, substrate, dopant, the template employed etc. The electrochemical performance of these electrodes can be greatly modified by balancing these factors. Otherwise, these PPy based electrodes creates different practical problems and scholars experience a great technical difficulty to meet application requirements. Serious investigations in the field of PPy based supercapacitor electrodes is befalling effectively.

Polythiophene and derivatives serves as excellent materials for supercapacitor applications due to excellent electric conductivity, good environment stability, and longer absorption [37–39]. Researchers have examined the electrochemical properties of pure PTh [40] and several synthesized techniques were formulated to improve their performance. Laforgue et al. [41] utilized a chemical method for synthesizing PTh and it showed exceptional specific capacitance and remarkable cycle stability. At current density of 2.50 $mAcm^{-2}$, the supercapacitor electrodes based on PTh exhibited a specific capacitance value of 260 Fg^{-1}. Ambade et al. [42] accomplished a solid-state symmetric supercapacitor with elastic features by incorporating PTh/TiO_2 electrodes which was fabricated via electrochemical process. The solid-state supercapacitor showed remarkable specific capacitance of 1357.31 mFg^{-1} and excellent cyclic stability. PTh nanoparticles and PTh/tartaric acid nanoparticles was synthesized by Gnanakan et al. [43–44] where tartaric acid is sourced as dopant. Cationic surfactant-assisted polymerization method was utilized for preparation. The specific capacitance of the PTh and PTh-tartaric acid-based nanoparticles were respectively 134.0 F/g and 156.0 F/g. Nejati [45] employed a method of oxidative chemical vapor deposition for preparing a non-substituted PTh film. Investigation of electrochemical test revealed that there is a 50% increase in the specific capacitance of PTh film coated active carbon electrodes compared with activated carbon electrodes and after 5000 cycles, the initial capacitance was retained to about 90%. Patil et al. [46–47] framed a method of ionic layer adsorption and reaction for the synthesis of amorphous PTh thin films in which $FeCl_3$ was exploited as an oxidizing agent. The specific capacitance of the electrode was increased up to 252.0 Fg^{-1}. A chemical bath deposition method was further employed for preparing PTh film-based electrodes and maximum specific capacitance of 300.0 Fg^{-1} could be attained.

Supercapacitors based on CPs displayed many deteriorations in properties like low electrical capacitance, poor cycling strength, the perplexity of doping/de-doping, poor conductivity, etc. High-performance supercapacitors rely greatly on the features of electrodes. Experts have fabricated binary and ternary nanocomposites with active materials like CNTs, graphenes, fibers, etc., to boost the electrical and chemical properties and stabilities of CP-based supercapacitors. In next section, negotiations on current advancement in the elastic supercapacitors centered on binary nanocomposites like CNTs/CPs, graphene/CPs and ternary nanocomposites CNTs/graphene/CPs will be discussed.

13.3   CNT-BASED POLYMER SUPERCAPACITORS

CNTs based thin film polymer supercapacitors are widely used in flexible electronics. CPs suffers from less energy density owing to the small surface area of it. CNT films' energy density can be enhanced by adding CPs. The most popularly applied one is PANi due to its flexible oxidation state and large pseudocapacitance. Basically, it is associated with its doping-de doping characteristics [48–50].

Now way days, various approaches are being used for the synthesis of PANi-CNT film for super-capacitor applications [51–54]. Several templates were used for chemical polymerization of PANi, such as CNT based Bucky papers and CNT networks. Various solid-state paper like supercapacitors with superior specific capacitance have been successfully fabricated from these templates [55]. The conductivity attained is less than 150 S/cm and energy density is lesser than 2.2 kW k/g for the freely standing PANi/CNT flexible films. Based on this Niu et al. [51] developed another strategy known as "skeleton/skin" strategy. Using this strategy Single-walled CNT film was used as skeletal framework and PANi was employed as skin to prepare PANi/CNT hybrid flexile film. The CVD technique was applied for electrochemical deposition of PANi on the bundles of CNT. They form an interlinked and continuous matrix of SWCNT films based on CNT/PANi strategy. This strategy confirmed effective electron transports covering a huge space. The specific conductance of 1138 Scm^{-1} was attained (30 times more than the published ones for conventional PANi/SWCNT hybrid films) and also exhibitted large energy density of 131 Wh kg^{-1} and power density of 62.5 kW kg^{-1} [52].

The associated CNT arrays are established to make easy transport of ions with special CNT architectures in CP/CNT hybrid films [56, 57]. PANi/CNT flexible films were prepared [53] and studied by electro-depositing PANi into voids within an aligned Multi-walled CNT (MWCNT) array [58]. The developed hybrid PANi/MWCNT film could be bend more than hundred time. Because of synergism in connections between PANi and MWCNT, the PANi-MWCNT hybrid films exhibit large capacitance and good cycling stability [53]. This research opened up the possibility for the strategic planning and designing of flexible and stretchable supercapacitors [54, 59].

CNTs have a lot of advantages for the development of polymer supercapacitors; they are employed as modifier for CPs with an aim of enhancing cycle stability and conductance. Making CPs/CNTs flexible films, can join the pseudocapacitance of CPs and enhanced charge- storage ability of CNTs together, thereby enhancing the energy and power densities of supercapacitors through synergetic effects. Comparing with pristine CPs, electrochemical performance of CPs/CNTs films have large energy and power density. Recent developments in graphene-based hybrid supercapacitors (CPs/Graphene) are discussed in the coming section.

13.4   GRAPHENE-BASED POLYMER SUPERCAPACITORS

Graphene is a two-dimensional nanomaterial with large electric conductivity, mechanical flexibil-ity, and superior chemical stability, and is regarded as the most important candidate for flexible supercapacitors [60–62]. Achieving the high theoretical specific capacitance (2630 m^2/g) based on pure graphene is really difficult. This is due to the aggregation of individual sheets as a result the surface area is lost during their fabrication procedure. This can be reduced by the introduction of CNTs as spacers between graphene sheets to form composite film. This is called LBL electrostatic assemblies. The first CNT/reduced graphene oxide (CNT/rGO) hybrid film was prepared [63] with the intercalation of carbon nanotubes in between reduced graphene oxide sheets. The LBL assem-bly of poly(ethyleneimine) modified GO sheets is carried out by the intercalation of acid oxidized CNTs. The hybrid films prepared; retained the outstanding mechanical properties, higher surface area and showed high capacitance of 120 Fcm^{-2}. Like CNT- spacer technique, insertion of CPs into layers of graphene is a successful method to avoid the accumulation of graphene plates in order to enhance the electrochemical functions. Researchers put more hard work for preparing CP/graphene hybrid films as electrodes [64–68].

Normal chemical reduction methods (using reducing agents) create some deterioration in composite structures of hybrid films. This defect can be overcome by adopting electrochemical reduction methods GO to rGO in the electrochemical polymerization of CPs and was seen as an efficient method to create high performance rGO/CP composites [66, 69–71]. With the technique, researchers synthesized polyaniline in a Graphene Oxide bath to design PANi-rGO composite as supercapacitor electrode. These developed films exhibited an improvement in conductivity around 30% and enhancement in capacitance by 15% owing to the existence of rGO. After 10000 charging and discharging cycles, the capacitance of 15 mF/cm^2 sustained upon the tunable PANI-rGO electrode [72].

Creation of 3-D graphene structures is one main method to avoid the accumulation of graphene nanosheet. Lately, the use of three dimensional porous rGO framework [67] was reported with a large capacitance of 311.95 F/cm^3 [73]. The prepared PANi-rGO composite film showed a specific capacitance of 401.5 Fg^{-1} [74]. The large volumetric energy density of 6.80 m W/cm^3 and a high cycle stability were obtained for a device based flexible supercapacitor [75], prepared by Li and co-workers. The schematic representation of carbon nanomaterial-based supercapacitors is shown in Figure 13.2.

In this part we discuss the effective way to synthesize Graphene/CPs composites to improve electrochemical effectiveness of CPs supported electrode. Compared with different structures, the free-standing three-dimensional structure has several benefits, so appropriate 3D structures are being constructed by using new techniques. But, since the capacitance of graphene is low, enhancement on the capacitance of ternary hybrid is reduced in the field of supercapacitors.

The binary CPs/CNT and CPs/Graphene structures has been employed extensively in supercapacitor materials, but their instrumentalities still couldn't reach high levels. Traits like conductivity, specific capacitance and cycling stability need further improvement in their levels [76]. In the

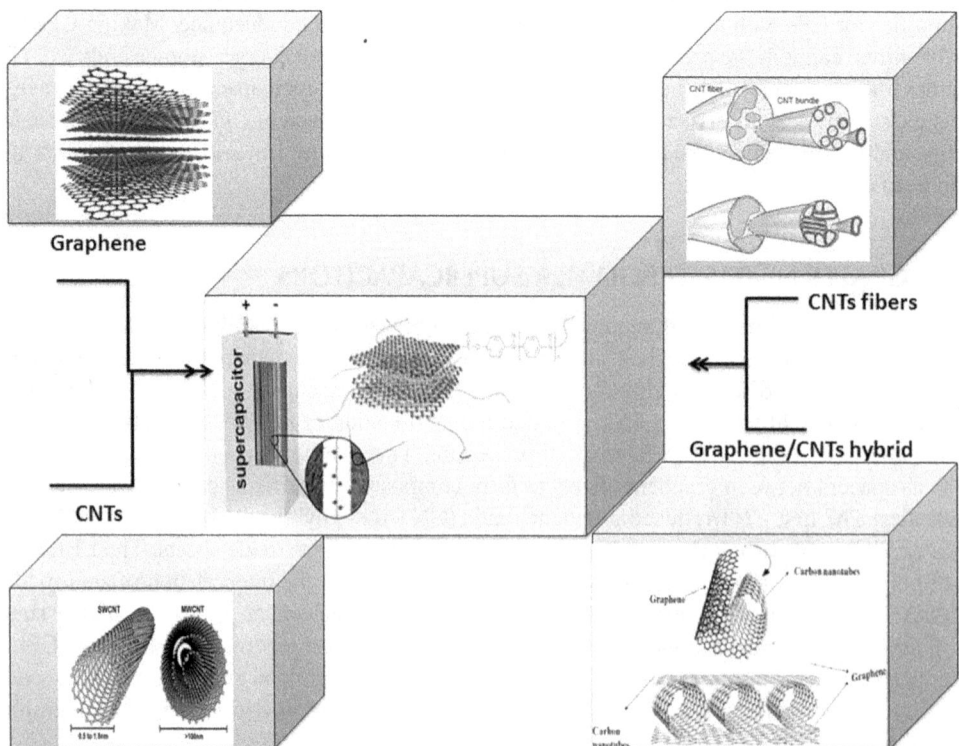

FIGURE 13.2 Schematic of carbon nanomaterial-based supercapacitors.

coming sections, recent progress of CPs/Graphene/CNT composites in supercapacitor electrode materials for higher performance will be introduced.

13.5 CPS/GRAPHENE/CNT TERNARY SUPERCAPACITORS

Several numbers of scientists have tried to develop PANi base composites for improvement of electrochemical performances to meet practical purposes. This can be achieved by utmost interaction between each component by synergism by adjusting their microstructure.

Metal oxides and carbon materials are also important for supercapacitor electrode materials. Many research groups [77–78] have made an attempt to produce CP-metal-oxide-carbon ternary systems with various type of CPs, metal oxide and carbon material. Research group of Chen [77] designed a polymer supercapacitor electrode where $NiMoO_4$ nanowires array alone with consistently dispersed proportions acting as the major pseudocapacitively active material and coat them on highly conducting carbon substrate. The developed electrodes displayed superior electrochemical properties like improved rate capabilities and cycling stabilities.

Metallic nanoparticles like gold, silver etc. are ornamented on the surface of carbon nanomaterials for enhancing the features like conductivity (both thermal and electric), mechanical strength, and cyclic life of CPs [79–81]. Researchers including Dhibar [82] developed nanocomposites where silver nano particles are ornamented on PANi-GN through low-cost in situ polymerization. The developed system shows capacitance of 591.0 F/g and power density around 20.240 Wh/kg. In addition, the electrode can attain retention of 96% even after 1500 cycles because of the high stability. Researchers [83] developed PANi-rGO-Au ternary systems coated on a shiny glass like carbon electrode by Cyclic Volta metric technique. The ternary system showed superior and stabilized capacitance compared to the pristine PANi electrode.

As mentioned earlier, PPy/CNT or PPy/metal oxide binary electrode supercapacitor showed some limitations in their electrochemical performance including cycling stability, specific capacitance, cycle life etc. This can be overcome by synergism in ternary systems between CPs, carbon materials as well as metal oxides are of great interest in scientific community. A lot of work is still performed for the development of PPy-based ternary systems. Scientists [84] have developed 3D CNT-MnO_2 nanostructures in PPy. In this system MnO_2 is placed over Carbon Nano Tubes consistently and a tough pseudocapacitive shell is formed with large surface area. The developed electrode showed better specific capacitance and current density. In addition, the capacitor showed high cycling and bending stabilities.

For preventing degradation in the capacitance of CNT-MnO_2 based electrodes, research group of Yun [85] developed PPy coated MnO_2 nanomaterials which are deposited on CNT fabric. Developed system showed superior energy capacity, cyclic reliability and bending flexibility and they can be used in several applications. Other important class of carbon materials in EDLC field is GN and its derivatives; they are widely used for developing different ternary-systems. Chen et al. [86] developed new PPy-polyoxometalate-rGO nano ternary hybrids by a one-pot strategic method. The nano ternary composites displayed superior specific capacity, and high-rate stability, good flexibleness, and mechanical strength. Scientists [87] used ultrasonic irradiation to develop PPy-GN-MnO2 ternary hybrid composites were p-toluene sulfonic acid (PTSA) was the doping agent. This kind of ternary composite showed a capacitance value of 258 Fg^{-1} and the current density was 1.0 Ag^{-1}. Ng's research team [88] developed stretchable PPy-GO-MnOx supercapacitors with large cycle stability. TiO_2 [89], ZnO [72] are also studied with PPy, GN and their derivatives. Such composite materials showed superior features and are possible to be employed in several supercapacitor applications.

The CPs based ternary composites exhibited superior electro chemical performance like cyclic stability and other superior characteristics due to its synergism between the components. Fiber-shaped polymer supercapacitors are getting attention owing to their high performance in portable and wearable electronics. These composite systems showed excellent flexibility, low volume, and good deformability over traditional 2D planar structured supercapacitors. In the coming section, we focused on the recent advancement in fiber shaped polymer supercapacitors.

13.6 FIBER-BASED SUPERCAPACITORS

Fiber-shaped asymmetric supercapacitors (FASCs) with one-dimensional (1D) structure are paid wide attention in future transportable and practical electronics. In recent times, great attempts have been broadly used to fabricate FASCs with superior performance and excellent mechanical properties. Fiber-shaped electrodes like CNT fibers, graphene fibers (GF) etc. have been discussed in this section. The schematic illustration of asymmetric supercapacitors and different carbon nanomaterials used as electrode is shown in Figure 10.3.

CNT fibers have many superior characteristics like excellent mechanical properties and good electrical and thermal conductivities. In addition, they show remarkable structural flexibleness and higher surface area, which make them a competing option to the coming generation devices [74–91]. By spinning CNT dispersions into a PVA coagulation bath, CNT fiber can be prepared [92]. Four major techniques have been used for producing Carbon Nano Tube fibers. They are spinning from CNT solution [93–94], spinning of CNT array over a substrate [95–97], spinning from a CNT aero gel [98], and twisting or rolling from a CNT film [99]. CNT fibers are widely studied due to their application in FASCs owing to the excellent physical and mechanical characteristics [100–102].

Scientists developed FASCs by making CNT-ZnO-NWs-MnO$_2$ as the positive part of the electrode and CNT as the negative part of the electrode [102]. Developed supercapacitor exhibited large specific capacitance of 31.150 mF/cm^2. The developed system based on these devices in series or parallel showed excellent electrochemical performance. Peng's research group developed FASCs with improved volumetric energy density by means of CNT fibers as flexible and conducting substrate [103]. Because of the magical properties such as conductive nature, lightweight, and flexibility of CNT fibers, the CNT based FASCs could be weaved into flexible textiles showing good mechanical strength.

Other type supercapacitors designed in which CNT fiber was the negative electrode and CNT fiber coated with MnO$_2$ -polymer composite was the positive electrode [104]. These FASCs model displayed superior energy density, flexibility, and good cycle stability, when compared to conventional asymmetric and symmetric supercapacitors.

MoS$_2$ has received considerable attention as energy storing material. MoS$_2$ can competently store charges through pseudocapacitance due to the faradaic charge-transfer process on the metal (Mo) center. It is slow procedure. The development of a double layer at the electrode materials (electrode-electrolyte) interface is a fast process [105]. As a result, MoS$_2$ is treated as a favorable electrode material for developing asymmetric supercapacitors.

The research group of Chen et al. incorporated MoS$_2$ with rGO into aligned multiwalled CNT film to construct rGO decorated MoS$_2$ with CNT and CNT/reduced GO fibers [106]. The ideal arrangement of individual CNTs sheet provided the fine electrical properties and mechanical properties of the prepared fibers. Modified FASCs is made-up by using MoS$_2$-rGO/CNT (positive) and rGO/CNT (negative) as electrodes. This system displayed a good capacitivity of 5.20 F/cm^3 and cycling constancy. These studies revealed the advantages of nanotube fibers in wearable electronics for future applications. Generally, the prepared FASCs are high price tag and it is needed for several complex processes, which create hard in commercialization.

Graphene based fiber is a new trend in carbon fiber for supercapacitors applications. They paid significant attention because of its thermal and electrical conducting properties, good mechanical properties, and good flexible nature. Various methods are used to fabricate GF architectures, which includes solution spinning techniques, laser ablation technique and dimensionally confined hydrothermal process or soft chemical synthesis [107–109].

Zhu et al. developed hierarchical nanowire of MnO$_2$/Graphene fibers via the method of solution spinning by means of MnO$_2$ nano-wire and GrO [110]. From this method 100 meters lengthy MnO$_2$/GrO fibers is acquired. The developed design of approach is widely used for other graphene based nano hybrid fiber fabrication is intended for the future energy storage space applications.

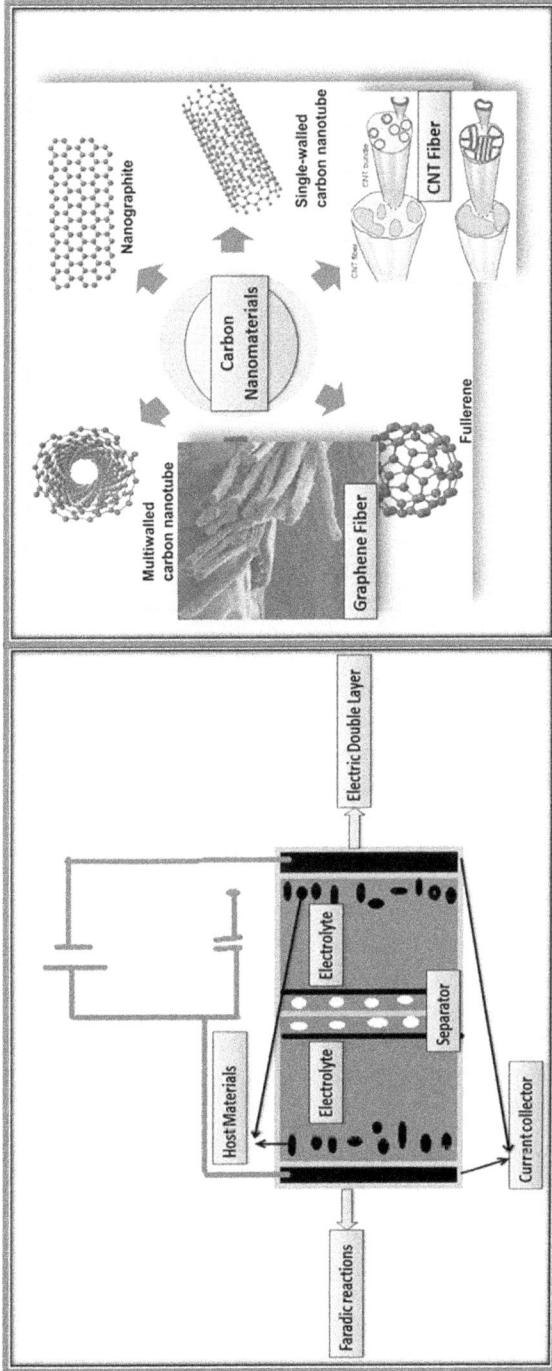

FIGURE 13.3 Asymmetric supercapacitors and different carbon nanomaterials used as electrodes.

FASCs have more and more interest of scientists due to their flexible nature and woven ability. Major dispute in this regard is how to improve their energy-density. Research group of Gao et. al. developed another category of FASCs by two diverse types of fiber-based graphene electrodes [111]. The core sheath morphology of MnO_2-graphene fibers was made by the deposition of a MnO_2 sheath (flower like) on the solution spinning graphene fiber.

Asymmetric supercapacitor was fabricated with graphene/MnO_2 fibers (positive) and graphene/CNT (negative) electrode respectively. This system can be operated by a potential of 1.60 V by means of an area energy-density of 11.90 mWh/cm^2. This supercapacitor system displayed superior cycle steadiness with 93% retention even after cycles of 8000. The research lab of Cai et. al. published that the graphene fiber can be synthesized by solution spinning which act as a reacting chamber [112]. $NiCo_2S_4$ nanoparticles covered through the typical solvothermal method of decomposition, the fabricated GF/$NiCo_2S_4$ fiber exhibited a high mechanical and conducting property. The graphene fiber/$NiCo_2S_4$ electrode exhibited a good capacitance of 388.00 F/cm^3 in an assembly of three electrode cell at 2.0 mV/s. In addition, developed system shows good flexibility at different bending angles and zero deterioration in their performance electrochemically.

Fabricated supercapacitors were rush into a textile and series of supercapacitors connected be able to power a LED and can be used in wearable applications. The volumetric energy density is high for the developed supercapacitors based on GF/$NiCo_2S_4$. The FASCs design and fabrication depend on graphene fiber can be treated as the excellent contestant for the upcoming flexible electronics.

13.7 CHALLENGES AND FUTURE PERSPECTIVES

This chapter extensively investigates the number of key findings and recent advances on developing high performance energy storage devices mainly based on CPs. The major objective is to present an outline of different fabrication techniques followed by inclusion of CPs as a component in supercapacitors to direct upcoming investigation on CPs. CPs have several advantages like ease of preparation, flexible nature, high energy storage etc. to resolve exceptions in the advancement of supercapacitors. Based on the charge storage response, comparison between the fabricated systems is established in terms of precise capacitance and cycling steadiness. Pristine CP electrodes display many troubles like inferior energy-density, time rate of energy transfer (power density), and reduced cycling life. The electrochemical properties of CP-based supercapacitors are enhanced by improving their crystallinity, optimum structure on a microscopic scale, and surface topography or morphology carried out by different polymerization techniques, concentration of doping agents, their oxidative intensity, nature of surfactants used, and their concentration, etc. Other important properties taken into consideration include stability towards elevated temperatures, processing nature, and mechanical strength in order to use for a reasonable purpose. Figure 13.4 shows important application areas for polymer-based solid-state supercapacitors.

Fabrication of CP-based composites with a number of active materials like CNTs, graphene, their fibers, and metal oxides has been investigated in recent years. Their electrochemical performances have been enhanced by additive effects (synergism) and have made a lot of progress in hybrid nanocomposite electrodes. Even though their electrochemical properties are enhanced, they cannot meet all the supercapacitor practical applications. As a result, researchers have developed nanocomposites with three different types of resources such as conducting polymers, oxides of nanometals and nano carbon materials, and new types of pseudo or fake capacitive resources. Fabricated nanocomposites displayed better performance over binary systems. In recent years, lot of research been undertaken for the development of next-generation supercapacitors.

In this chapter, recent progress in conducting polymer-based supercapacitors by means of electrodes between from one-dimensional fibers, especially CNT and graphene fibers, through two-dimensional films, has also been reviewed. CPs cannot be recommended for attractive doping-de-doping characteristics. They possess interesting morphological structure and conducting

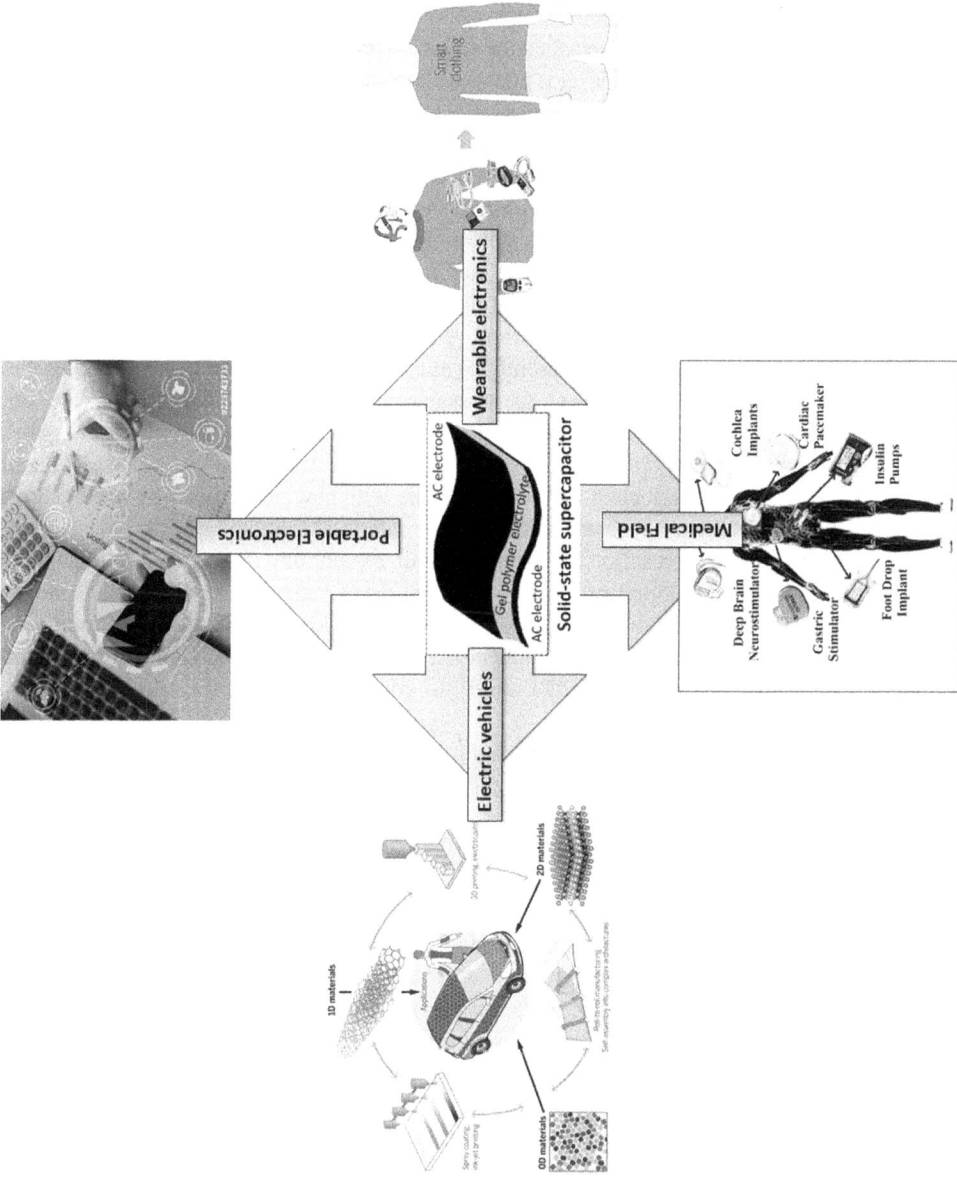

FIGURE 13.4 Important application areas of polymer-based solid-state supercapacitors.

properties, but as well explain some structural difference throughout the charge-discharge process. Different strategies include electro deposition process, self-crosslinking process, macrocyclic assembling, and electro spinning process which have been reported by several researchers to fabricate highly advanced conducting polymer-based supercapacitors with good flexibility and an exceptional stability in their structure.

Growth in the development of polymer-based capacitors mainly depends on their extraordinary properties in order to achieve practical application criteria. This can be overcome to a certain extent by using specially fabricated three-dimensional CP hydrogels by means of their porous structures. Another challenging matter is in the direction of controlled the structure-property bonding for various materials. In addition, the control of doping agent's level is necessary to enhance the conducting property and the capacity energy storing. A promising doping approach is to incorporate immobilized dopants into two dimensional materials that permit the inclusion of small ions for adjusting the change in the volume. Other main challenge is to develop supercapacitors based on different microscopic techniques which will support the development of new class of CPs composites. As a final point in this regard, computational modeling and different learning techniques are anticipated to develop high performance supercapacitors to direct/encourage the development of conducting polymer-based nanocomposites. Research and development in this field should launch opportunities for fabricating the flexible supercapacitor technologies into commercialization, which will improve our life style in near future.

Several three-dimensional macroscopic assemblies and architectures of electrode materials have immense effect on the configurations of energy storage materials. CPs nanocomposites are able to be developed into dissimilar nanostructures from one dimensional to three dimensional, which expedite the understanding of non-conventional supercapacitor in diverse configurations like hierarchic sandwich nanostructure, interlocked and fiber type and with new properties like stretchability, compressibility, electrochromism, and self-healing at the nanoscale. Supercapacitors of this type have superior strength and quick reaction to exterior stimuli, assembling those best components for portable electronics. In the developed process of 3D printing of supercapacitors, polyaniline and PEDOT:PSS are more striking for their good processability. In addition, in the case of multifunctional supercapacitors, polyaniline and polythiophene derivatives are excellent conducting polymers because of their extraordinary properties like ability to color changing in the special oxidation-reduction states.

Recent advancements in the preparation procedure and performance utility of perpendicular CPs nanostructures and their composite for supercapacitors electrodes are reported. Well-arranged CPs nano arrays are able to be fabricated starting its monomer by using chemical/electrochemical polymerization technique. By adjusting the reaction dynamics, CPs nano structures are able to be fabricated lying on different surface of the template. No template is used for these fabrication purposes. The capacitance of this category for nanostructures of PANi and PPy nanostructure is little better than conventional structures as electrodes for supercapacitors due to the high surface activity in electrolytes for electrochemical reactions. Besides, this exhibits higher charge capability credited to the well-arranged structure reducing the ion diffusion resistance and ion diffusion path. In the direction of the superior electrochemical property of nano structures in a supercapacitor, and the nanocomposites of PANi nano arrays and rGO are obtain by means of polymerization process within the medium itself. The developed composite has exhibited synergism of the CPs and carbon nanomaterials. It showed superior capacitance than pristine PANi and graphene, and this nanocomposite demonstrated improved cyclic life stability. The outstanding feature is credited to the well-arranged nanocomposition in addition to the established the excellent interfacial interface among the conducting polymer along with rGO. Fabricated composite with dissimilar elastic substrates of CNTs OR graphene, we get even thread-like supercapacitors. In addition, nanoscale/multifunctional supercapacitors are fabricated based on CPs like PANi nano structures. The ideal stretchy strategy is simple to fabricate and operate the high capacitivity of CPs. The assimilation of novel working in one supercapacitor can auxiliary develops the practical application areas of supercapacitors.

Self-powering energy storage devices are an important factor for many wireless products/applications. The major benefit is that it be capable of be developed as a system with free of maintenance. Particularly, such systems are new in a huge amount and this system is hard to access, working may offer an immense improvement by enable an elevated scale of consistency and dropping continuation expenses. Generally, it is able to be done by two major strategies: the first one is the energy to be capable of be occupied by the storage space system exceed the whole energy desirable during the life cycle of the device and the second one is the device itself is set with an energy harvesting mode. The two parameters, energy provide and energy storage must be designed and measured while designing a self-powering device.

The fast progress in the development of self-powered energy systems has illustrated their potential towards practical applications in the future. However, many issues still exist and more efforts are required for the commercial development of self-powered energy systems.

- In order to overcome the limitation of energy density, energy storage devices occupy a large space and contributes to the weight of portable devices, which contradicts the miniaturization of wearables [113].
- The development of these ESDs is still in their early stages and so the practical energy storage devices developed at present are very expensive.
- Safety is of major concern as the energy storage devices pose potential risks of toxicity and flammability.
- Synthetic materials like CNT, graphene, Metal Organic Frameworks (MOFs), metal oxide nanoparticles are employed for the development of energy storage devices. Several problems are encountered with these materials like complex preparation process, low yield and inconsistency.

Supercapacitors make their attention in the fields of science due to their excellent power density and wide choice of in-service temperatures and very fast time of charge. The main disadvantage on comparing with the batteries is the low energy density of the device. Analyzing the energy formula, $E = \frac{1}{2}CV^2$, it is clear that there are three main strategies are available to improve the energy density.

I. Fine-tuning of the properties of electrode surface.
II. Boosting the voltage by the selection of proper electrode and electrolyte along with the stable electrochemical performance.
III. Increase the capacitance by assembling a hybrid system (i.e., by assembling a faradaic electrode with a non-faradaic one) [114].

Owing to increase the capacitance, a proper way is to increase the surface area. But this trend is not so good that it doesn't guarantee the performance of the device. In most of the cases micropores and mesopores materials are the best choice. However, materials with pore diameter very low (less than 0.4 nm) could not add much to the value of capacitance thereby energy density. In recent times, Ruoff's research group developed graphene with surface area of 2400.00 m²/g and a pore diameter of 0.6 to 5 nm result power density of 20.0 Wh/kg [115].

Recent breakthrough to achieve high surface area is the use of carbon materials such as high porous carbon, carbon nanosheets, holey graphene frame work. Incorporation of these materials provide a noticeable capacitance value to the material. Along with the increase in surface area, modification of surface by attaching surface functionalities is also another alternative. The heteroatoms like oxygen, nitrogen etc. can influence the capacitance by improving the wettability of electrode surface. This leads to provide additional capacitance follow-on from faradaic-redox mechanism. The working voltage also provides an improvement in the power density of the supercapacitor. Generally, for aqueous electrolyte the working voltage is nearly 0.1 V where as if an organic electrolyte is using this can be reaches to a value 0.2 V. The chief voltage devices were normally realize in ionic liquid systems, in which the voltage can reach more than 3 V, which is comparable with Lithium ion based devices. Because of their excellent voltage (>3.0 V) and excellent performances, ionic liquids are smart candidates for future applications as electrolytes. Although the supercapacitor is an efficient energy storage device there needs to be more progress made to enhance its energy density.

The electrochemical properties of superconductors can be improved by selecting appropriate electrode materials [116]. The electrode materials of superconductors should endow with high temperature stability, high electrical conductance, high specific surface area, decomposition resistance, appropriate surface wettability and suitable chemical stability. They should also be economical and ecologically benevolent. The electrodes with smaller pores exhibit high capacitivity and results in the reduction of power density.

Considerable amount of research has been devoted in current decades for improving the electrochemical properties of the supercapacitors during the growth of novel electrode materials. As electrode materials determine the electrical properties of supercapacitors, selection of electrodes if supercapacitors is of vital significance [117]. But the obstacles faced by the supercapacitors include high production cost and self-discharge, low voltage per cell and energy density. These challenges can be tackled by the development of novel electrode materials. Carbonaceous materials, conducting polymers and metal oxides are popularly used as electrode materials nowadays. Due to high surface area, low cost, accessibility and established electrode production technologies, carbonaceous materials in their various forms are the most used electrode materials in the fabrication of supercapacitors. These are promising electrodes which offer high specific surface area, thermal and chemical stability, and low electrical resistance. However, large scale applications are restriction as the low energy density, which arises from their surface or quasi-surface energy storage, cannot be tackled. Metal oxides provide high specific capacitance and low resistance, which make it easier to construct high energy supercapacitors. These materials favor diffusion of ions onto the bulk of material. Metal oxides such as RuO_2, V_2O_5 and MnO_2 provide an exceptional improvement as electrode materials, both as cathode and anode. However, one of the drawbacks, which weaken their potential, is the low electrical conductivity of metal oxides. Conducting polymers store and release charge using reduction-oxidation process [118]. They exhibit high specific capacitance, facile processability, favorable flexibility, and advanced Ed. But the major obstacles faced by these electrode materials are swelling and shrinking during charging and discharging and this results in shorter lifetime and will restrict the cyclic performance of the CPs. The discovery of graphene as electrode material for supercapacitors has opened a lot of research opportunities being carried out. Besides the selection of suitable electrode materials, the electrochemical properties can be improved through the proper combination of the anode and cathode, particularly in the case of hybrid devices.

A performance of supercapacitors can be described by a number of key parameters, such as capacitance, power density, energy density, operating voltage, resistance, and time stability. The major trends in improving supercapacitor performance are to ensure high capacitance and to allow the electrolyte to easily pass through the electrode and to provide a high SSA and pore size distribution. A large portion of current research is focused on the factors that influence specific capacitance and series resistance in electrode materials. The early studies focused primarily on carbonaceous materials. More initiatives have recently been introduced to better understand the applications of graphene. Activated carbon is the best choice for increasing surface area. However, the evolution of carbon nanotubes (CNTs) is playing a significant role due to their performance. in addition to cycling stability. The utilization of pseudocapacitance that is by using conducting polymer along with metal oxide are creating further boosting to energy and power density of supercapacitors thereby paving a way for forthcoming generation of supercapacitors. The commonly used metal oxides are RuO_2 and IrO_2 along with MWCNTs attained much attention. But the cost and toxicity of these materials reduced the choice. On considering the cost and performance certain oxides and hydroxides of metals like vanadium, manganese, cobalt, nickel is also extensively studied. Recently ceramic metal oxides used as the electrode materials, their output is still needing modification. The development of novel electrode materials making the supercapacitor research area as a hotspot in research community. The next advancement to enhance the performance is the choice of electrolyte used. Owing to the desirable advantages like safety, economical and simple production procedures, hydrogel redox-active electrolytes are potential candidates for high energy-density electrochemical supercapacitors. Ionic liquid-based electrolytes are regarded as the better choice for the supercapacitor electrolyte. Through the theoretical and experimental investigations, the compatibility between electrode and electrolyte can be improved. The application of supercapacitors in stretchable or wearable devices taking the demand for the advancements into a great need. So that the research society is constantly searching for the new methods and techniques to get better output.

The development on the FASCs rapidly increases because of its fabrication and design and their performance electrochemically. But extra strategies are still needed for additional advancements in their electrochemical performance. More attention on designing electrode and electrolyte by means of good ionic conduction, and safety precautions are essential. Furthermore, FASCs be able to be reactive to changes in configurational integrity, strength properties, elctrochromic and thermal properties, self-healing activity, etc. And they need substantial enhancements which will boost up in the future applications. In reality, excellent properties of FASCs are reliant on employing costly electrode materials and complicated manufacturing techniques. The enhancements of FASCs in cost wise and performance wise are essential in future prospects. The systematic investigation of the charge-discharge and leakage of current troubles of the supercapacitors is essential. In the FASCs systems, the mechanism of self-discharge is not known till now. The detailed mechanism is essential to meet the necessities for future energy storage. The used eco-friendly materials and an eco- friendly approach are more preferred for the future developments of polymer supercapacitors. The detailed understanding of the mechanism of polymer supercapacitors and their correlation with theoretical calculations/data needs substantial improvements for future applications.

Conducting and non-conducting polymer-based supercapacitors exhibit almost similar patterns. Conducting polymers show electrochemical properties by participating directly, whereas non-conducting polymers do so in a more indirect manner in electrochemical reactions. They either reinforce the active materials or improve additional properties like stretchability and flexibility. As a result, materials made from non-conducting polymers have improved electrochemical properties and versatility. The unaltered device electrochemical performance, which can impart good function even under external distortions, is another advantage of these functions. The use of polymeric elastomers in supercapacitor electrode materials improves the existing electrochemical active material by adding new features and functionality. The use of several polymeric elastomers in supercapacitor electrode materials enhances the existing electrochemical active material by adding new features and functionality.

Polyurethane (PU) is a commonly used elastomer for supercapacitors, in addition to various silicon rubbers such as Polydimethylsiloxane (PDMS). Characteristic advancement in supercapacitors can be achieved by fine-tuning the properties of these elastomers. Non-elastomeric polymers like polyvinyl alcohol as well as natural polymer like cellulose, are used to make the electrode material for supercapacitors. There have been numerous reports on textile-based supercapacitors in the field of electrochemical supercapacitors. Aside from textile, cellulose-based materials, which are natural polymers, are also often used for supercapacitor applications. These have been extensively used in electrochemical supercapacitors applications particularly as the paper electrode in energy storage equipment.

Polymers also found application as the binders in supercapacitors. Fluorinated polymers like polyvinylidene fluoride are the most commonly used binders for conventional electrodes. PVDF demands costly and dangerous organic solvents to interact with other materials. Although these polymers have reasonable chemical stability, they combine weakly due to its dependence on weak van der Waals forces. Polyurethanes, catechol-bearing polymers, and a combination of Carboxymethyl cellulose and Styrene butadiene rubber show promise as aqueous binders due to their improved properties. The electrolyte plays a vital role in the electrochemical properties as it links with the energy density of the device directly. There occur many advancements in the part of electrolyte. One of the main polymers used here is PVA based solid polymer electrolyte with different functionalities. The proper blending of the nonconducting polymer pave a huge ground to the existing supercapacitor researches.

Composite electrode materials made of non-conducting polymers are a popular choice for supercapacitor electrodes because they can provide not only electrochemical performance but also multifunctional properties. However, because the polymers used to make composites have limited electrochemical energy storage capabilities, the device's specific capacitance and energy density must be considered. When electrochemical active materials are incorporated, however, the pristine properties of the polymers used, such as mechanical elasticity, are reduced. It is critical to find a balance between composition and properties in order to solve these issues.

The ever-increasing demand for portable and wearable electric devices necessitates energy storage devices that are flexible and wearable compatible without compromising performance. This necessitates the electrode and electrolyte materials becoming robust and durable under mechanical deformations, which provides inherent benefits to polymers. As a result, non-conducting polymers for supercapacitors, whether as a substrate/matrix or active materials, have received a lot of attention.

13.8. CONCLUSION

Polymer supercapacitors are finding extensive use as important energy storage devices for modern life. This chapter tries to cover the current status of design, fabrication, and application of polymer supercapacitors. In this part, the major challenges are recognized and future trends are discussed. The structural design/fabrication plays a significant part in the enhancement in the property advancement of polymer supercapacitors. One-dimensional objects like CNTs and 2D objects such as graphene and their binary and ternary hybrids are important electrodes in polymer-based capacitors for high efficiency applications. Currently, no united assessment standard is used to characterize the flexibility property and their novel performance. Therefore, a comprehensively standardized metrology that precisely evaluates mechanical flexibility should be recognized and made clear to all researchers to facilitate evaluation/comparison between various flexible supercapacitors. Constant effort is continued in polymer-based supercapacitors area to make low cost and eco-friendly stretchable supercapacitors as competent in storing energy to meet the needs of modern society.

REFERENCES

[1] M. Winter, R.J. Brodd, What are batteries, fuel cells, and supercapacitors? *Chem. Rev.* 104 (2004) 4245–4270.

[2] A.S. Aricò, P. Bruce, B. Scrosati, J.M. Tarascon, W.V. Schalkwijk, Nanostructured materials for advanced energy conversion and storage devices, *Nat. Mater.* 4 (2005) 366–377.

[3] J.R. Miller, P. Simon, Electrochemical capacitors for energy management, *Science* 321 (2008) 651–652.

[4] K.S. Ryu, K.M. Kim, N.G. Park, Y.J. Park, S.H. Chang, Evaluation of activated carbon fiber applied in supercapacitor electrode, *J. Power Sources* 103 (2002) 305–309.

[5] D. Li, J.X. Huang, R.B. Kaner, Polyaniline nanofibers: A unique polymer nanostructure for versatile applications, *Acc. Chem. Res.* 42 (2008) 135–145.

[6] Q.H. Yang, Z.Z. Hou, T.Z. Huang, Self-assembled polypyrrole film by interfacial polymerization for supercapacitor applications, *J. Appl. Polym. Sci.* 132 (2015) 41615—(1–5).

[7] P. Simon, Y. Gogotsi, Materials for electrochemical capacitors, *Nat. Mater.* 7 (2008) 845–854.

[8] G.A. Snook, P. Kao, A.S. Best, Conducting-polymer-based supercapacitor devices and electrodes, *J. Power Sources* 196 (2011) 1–12.

[9] L. Nyholm, G. Nystrom, A. Mihranyan, M. Stromme, Toward flexible polymer and paper-based energy storage devices, *Adv. Mater.* 23 (2011) 3751–3769.

[10] B. Senthilkumar, P. Thenamirtham, R.K. Selvan, Conductivity study of thermally stabilized RuO_2/ polythiophene nanocomposites, *Appl. Surf. Sci.* 257 (2011) 9063–9067.

[11] C. Niu, E.K. Sichel, R. Hoch, D. Moy, H. Tennent, High power electrochemical capacitors based on carbon nanotube electrodes, *Appl. Phys. Lett.* 70 (1997) 1480.

[12] D.N. Futaba, K. Hata, T. Yamada, T. Hiraoka, Y. Hayamizu, Y. Kakudate, O. Tanaike, H. Hatori, M. Yumura, S. Iijima, Shape-engineerable and highly densely packed single-walled carbon nanotubes and their application as supercapacitor electrodes, *Nat. Mater* 5 (2006) 987.

[13] T. Chen, Z. Cai, Z. Yang, L. Li, X. Sun, T. Huang, A. Yu, H.G. Kia, H. Peng, Nitrogen- doped carbon nanotube composite fiber with a core-sheath structure for novel electrodes, *Adv. Mater* 23 (2011) 4620e4625.

[14] Y.P. Wu, E. Rahm, R. Holze, Effects of heteroatoms on electrochemical performance of electrode materials for lithium ion batteries, *Electrochim. Acta* 47 (2002) 3491–3507.

[15] A.K. Geim, K.S. Novoselov, The rise of graphene, *Nat. Mater.* 6 (2007) 183–191.

[16] J. Xia, F. Chen, J. Li, N. Tao, Measurement of the quantum capacitance of graphene, *Nat. Nanotechnol.* 4 (2009) 505–509.

[17] Q. Ke, J. Wang, Graphene-based materials for supercapacitor electrodes—a review, *J. Materiomics* 2 (1) (2016) 37–54.

[18] Y. Zhu, S. Murali, M.D. Stoller, A. Velamakanni, R.D. Piner, R.S. Ruoff, Microwave assisted exfoliation and reduction of graphite oxide for ultracapacitors, *Carbon* 48 (7) (2010) 2118–2122.

[19] C. Liu, Z. Yu, D. Neff, A. Zhamu, B.Z. Jang, Graphene-based supercapacitor with an ultrahigh energy density, *Nano Lett.* 10 (12) (2010) 4863–4868.

[20] K. Chen, S. Song, F. Liu, D. Xue, Structural design of graphene for use in electrochemical energy storage devices, *Chem. Soc. Rev.* 44 (17) (2015) 6230–6257.

[21] Y. Huang, H. Hu, Y. Huang, M. Zhu, W. Meng, C. Liu, Z. Pei, C. Hao, Z. Wang, C. Zhi, From industrially weavable and knittable highly conductive yarns to large wearable energy storage textiles, *ACS Nano* 9 (2015) 4766–4775.

[22] L. Kou, T. Huang, B. Zheng, Y. Han, X. Zhao, K. Gopalsamy, H. Sun, C. Gao, Coaxial wet-spun yarn supercapacitors for high-energy density and safe wearable electronics, *Nat. Commun.* 5 (2014) 3754.

[23] W. Zeng, L. Shu, Q. Li, S. Chen, F. Wang, X.-M. Tao, Fiber-based wearable electronics: A review of materials, fabrication, devices, and applications, *Adv. Mater* 26 (2014) 5310–5336.

[24] L. Chen, Y. Liu, Y. Zhao, N. Chen, L. Qu, Graphene-based fibers for supercapacitor applications, *Nanotech* 27 (2016) 032001.

[25] D. Yu, Q. Qian, L. Wei, W. Jiang, K. Goh, J. Wei, J. Zhang, Y. Chen, Emergence of fiber supercapacitors, *Chem. Soc. Rev.* 44 (2015) 647–662.

[26] K.S. Ryu, K.M. Kim, N.G. Park, Y.J. Park, S.H. Chang, Hybrid supercapacitor based on polyaniline doped with lithium salt and activated carbon electrodes, *J. Power Sources* 103 (2002) 305–309.

[27] D. Li, J.X. Huang, R.B. Kaner, Polyaniline nanofibers: A unique polymer nanostructure for versatile applications, *Acc. Chem. Res.* 42 (2008) 135–145.

[28] J.X. Huang, R.B. Kaner, Nanofiber formation in the chemical polymerization of aniline: A mechanistic study, *Angew. Chem.* 116 (2004) 5941–5945.

[29] N.R. Chiou, A.J. Epstein, Polyaniline nanofibers prepared by dilute polymerization, *Adv. Mater.* 17 (2005) 1679–1683.

[30] J.X. Huang, S. Virji, B.H. Weiller, R.B. Kaner, Polyaniline nanofibers: Facile synthesis and chemical sensors, *J. Am. Chem. Soc.* 125 (2003) 314–315.

[31] J.X. Huang, R.B. Kaner, A general chemical route to polyaniline nanofibers, *J. Am. Chem. Soc.* 126 (2004) 851–855.

[32] S. Sivakkumar, W.J. Kim, J.A. Choi, D.R. Macfarlane, M. Forsyth, D.W. Kim, Electroplating of polyaniline on carbon fiber cloth in a simple two electrode system: Application for the electrochemical filter in wastewater treatment, *J. Power Sources* 171 (2007) 1062–1068.

[33] H.L. Li, J.X. Wang, Q.X. Chu, Z. Wang, F.B. Zhang, S.C. Wang, Theoretical and experimental specific capacitance of polyaniline in sulfuric acid, *J. Power Sources* 190 (2009) 578–586.

[34] Q.H. Yang, Z.Z. Hou, T.Z. Huang, Self-assembled polypyrrole film by interfacial polymerization for supercapacitor applications, *J. Appl. Polym. Sci.* 132 (2015) 41615—(1–5).

[35] M. Li, L.L. Yang, Synthesis and characterization of polypyrrole/NiO doped Nanocomposites (NCs) for Dielectric studies, *J. Mater. Sci. Mater. Electron.* 26 (2015) 4875–4879.

[36] J. Xu, D.X. Wang, L.L. Fan, Y. Yuan, W. Wei, R.N. Liu, S.J. Gu, W.L. Xu, Alternate and random copolymers of bay substituted rylenebisimides for energy applications, *Org. Electron.* 26 (2015) 292–299.

[37] P. Simon, Y. Gogotsi, Materials for electrochemical capacitors, *Nat. Mater.* 7 (2008) 845–854.

[38] G.A. Snook, P. Kao, A.S. Best, Conducting-polymer-based supercapacitor devices and electrodes, *J. Power Sources* 196 (2011) 1–12.

[39] L. Nyholm, G. Nystrom, A. Mihranyan, M. Stromme, Toward flexible polymer and paper-based energy storage devices, *Adv. Mater.* 23 (2011) 3751–3769.

[40] B. Senthilkumar, P. Thenamirtham, R.K. Selvan, Conductivity study of thermally stabilized RuO_2/polythiophene, *Appl. Surf. Sci.* 257 (2011) 9063–9067.

[41] A. Laforgue, P. Simon, C. Sarrazin, J.F. Fauvarque, Polythiophene-based supercapacitors, *J. Power Sources* 80 (1999) 142–148.

[42] R.B. Ambade, S.B. Ambade, R.R. Salunkhe, V. Malgras, S.H. Jin, Y. Yamauchi, S.H. Lee, Flexible-wire shaped all-solid-state supercapacitors based on facile electropolymerization of polythiophene with ultrahigh energy density, *J. Mater. Chem. A* 4 (2016) 7406–7415.

[43] S.R.P. Gnanakan, M. Rajasekhar, A. Subramania, Synthesis of polythiophene nanoparticles by surfactant – Assisted dilute polymerization method for high performance redox supercapacitors, *Int. J. Electrochem. Sci.* 4 (2009) 1289–1301.

[44] S.R.P. Gnanakan, N. Murugananthem, A. Subramania, Organic acid doped polythiophene nanoparticles as electrode material for redox supercapacitors, *Polym. Adv. Technol.* 22 (2011) 788–793.

[45] S. Nejati, T.E. Minford, Y.Y. Smolin, K.K.S. Lau, Enhanced charge storage of ultrathin polythiophene films within porous nanostructures, *ACS Nano* 8 (2014) 5413–5422.

[46] S.B. Madasu, N.A. Vekariya, M.N. Kiran, B. Gupta, A. Islam, P.S. Douglas, K.R. Babu, Synthesis of compounds related to the anti-migraine drug eletriptan hydrobromide, *Synth. Met.* 62 (2012) 1400–1405.

[47] B. Patil, S.J. Patil, C.D. Lokhande, Synthesis and crystal structures of transition metal ion complexes of Di(2- thienyl)imide, *Electroanal* 26 (2014) 2023–2032.

[48] J. Jang, J. Bae, M. Choi, S.H. Yoon, Fabrication and characterization of polyaniline coated carbon nanofiber for supercapacitor, *Carbon* 43 (2005) 2730.

[49] J. Zhang, D. Shan, S.L. Mu, A rechargeable Zn-poly (aniline-co-m-aminophenol) battery, *J. Power Sources* 161 (2006) 685.

[50] M.Q. Wu, G.A. Snook, V. Gupta, M. Shaffer, D.J. Fray, G.Z. Chen, Electrochemical fabrication and capacitance of composite films of carbon nanotubes and polyaniline, *J. Mater. Chem.* 15 (2005) 2297.

[51] Z.Q. Niu, P.S. Luan, Q. Shao, H.B. Dong, J.Z. Li, J. Chen, D. Zhao, L. Cai, W.Y. Zhou, X.D. Chen, S.S. Xie, A "skeleton/skin" strategy for preparing ultrathin free-standing single-walled carbon nanotube/polyaniline films for high performance supercapacitor electrodes, *Energy Environ. Sci.* 5 (2012) 8726.

[52] L.N. Jin, F. Shao, C. Jin, J.N. Zhang, P. Liu, M.X. Guo, S.W. Bian, Hollow core-shell ZnO@ZIF-8 on carbon cloth for flexible supercapacitor with ultrahigh areal capacitance, *Electrochim. Acta* 249 (2017) 387.

[53] H. Lin, L. Li, J. Ren, Z. Cai, L. Qiu, Z. Yang, H. Peng, Conducting polymer composite film incorporated with aligned carbon nanotubes for transparent, flexible and efficient supercapacitor, *Sci. Rep.* 3 (2013) 1353.

[54] X. Chen, H. Lin, P. Chen, G. Guan, J. Deng, H. Peng, Smart, stretchable supercapacitors, *Adv. Mater.* 26 (2014) 4444.

[55] C.Z. Meng, C.H. Liu, L.Z. Chen, C.H. Hu, S.S. Fan, Highly flexible and all-solid-state paperlike polymer supercapacitors, *Nano Lett.* 10 (2010) 4025.

[56] J.L. Liu, J. Sun, L. Gao, Flexible single-walled carbon nanotubes/polyaniline composite films and their enhanced thermoelectric properties, *Nanoscale* 3 (2011) 3616.

[57] F. Huang, E. Vanhaecke, D. Chen, In situ polymerization and characterizations of polyaniline on MWCNT powders and aligned MWCNT films, *Catal. Today* 150 (2010) 71.

[58] M. Gao, S. Huang, L. Dai, G. Wallace, R. Gao, Z. Wang, Aligned coaxial nanowires of carbon nanotubes sheathed with conducting polymers, *Angew. Chem. Int. Ed.* 39 (2000) 3664.

[59] Y.P. Zhu, N. Li, T. Lv, Y. Yao, H.N. Peng, J. Shi, S.K. Cao, T. Chen, T. Ag-doped PEDOT:PSS/CNT composites for thin-film all-solid-state supercapacitors with a stretchability of 480%, *J. Mater. Chem. A* 6 (2018) 941–947.

[60] Y. Wang, Z. Shi, Y. Huang, Y. Ma, C. Wang, M. Chen, Y. Chen, Supercapacitor devices based on graphene materials, *J. Phys. Chem. C* 113 (2009) 13103.

[61] X. Cao, Y. Shi, W. Shi, G. Lu, X. Huang, Q. Yan, Q. Zhang, H. Zhang, Preparation of novel 3D graphene networks for supercapacitor applications, *Small* 7 (2011) 3163.

[62] B.G. Choi, M. Yang, W.H. Hong, J.W. Choi, Y.S. Huh, 3D Macroporous graphene frameworks for supercapacitors with high energy and power densities, *ACS Nano* 6 (2012) 4020–4028.

[63] D.S. Yu, L.M. Dai, Self-assembled graphene/carbon nanotube hybrid films for supercapacitors, *J. Phys. Chem. Lett.* 1 (2010) 467.

[64] Q. Wu, Y. Xu, Z. Yao, A. Liu, G. Shi, Supercapacitors based on flexible graphene/polyaniline nanofiber composite films, *ACS Nano* 4 (2010) 1963.

[65] L. Mao, M. Li, J. Xue, J. Wang. Bendable graphene/conducting polymer hybrid films for freestanding electrodes with high volumetric capacitances, *RSC Adv.* 6 (2016) 2951.

[66] T. Lindfors, R.M. Latonen, Improved charging/discharging behavior of electropolymerized nanostructured composite films of polyaniline and electrochemically reduced graphene oxide, *Carbon* 69 (2014) 122.

[67] X.D. Hong, B.B. Zhang, E. Murphy, J.L. Zou, F. Kim, Three-dimensional reduced graphene oxide/polyaniline nanocomposite film prepared by diffusion driven layer-by-layer assembly for high-performance supercapacitors, *J. Power Sources*, 343 (2017) 60–66.

[68] Y. Liu, B. Weng, J.M. Razal, Q. Xu, C. Zhao, Y. Hou, S. Seyedin, R. Jalili, G.G. Wallace, J. Chen, High-performance flexible all-solid-state supercapacitor from large free-standing graphene-PEDOT/PSS film, *Sci. Rep.* 5 (2015) 17045.

[69] T. Lindfors, A. Österholm, J. Kauppila, M. Pesonen, Electrochemical reduction of graphene oxide in electrically conducting poly(3,4-ethylenedioxythiophene) films, *Electrochim. Acta* 110 (2013) 428–436.

[70] T. Lindfors, A. Österholm, J. Kauppila, R.E. Gyurcsányi, Enhanced electron transfer in composite films of reduced graphene oxide and poly(N-methylaniline), *Carbon* 63 (2013) 588.

[71] Y. Tang, N. Wu, S. Luo, C. Liu, K. Wang, L. Chen, One-step electrodeposition to layer-by-layer graphene–conducting-polymer hybrid films, *Macromol. Rapid Commun.* 33 (2012) 1780.

[72] W.K. Chee, H.N. Lim, I. Harrison, K.F. Chong, Z. Zainal, C.H. Ng, N.M. Huang, Performance of flexible and binderless polypyrrole/graphene oxide/zinc oxide supercapacitor electrode in a symmetrical two-electrode configuration, *Electrochim. Acta* 157 (2015) 88–94.

[73] X.W. Yang, C. Cheng, Y.F. Wang, L. Qiu, D. Li, Liquid-mediated dense integration of graphene materials for compact capacitive energy storage, *Science* 341 (2013) 534.

[74] J. Di, X. Zhang, Z. Yong, Y. Zhang, D. Li, R. Li, Q. Li, Carbon-nanotube fibers for wearable devices and smart textiles, *Adv. Mater.* 47 (2016) 10529–10538.

[75] Z. Li, G. Ma, R. Ge, F. Qin, X. Dong, W. Meng, T. Liu, J. Tong, F. Jiang, Y. Zhou, K. Li, X. Min, K. Huo, Y. Zhou, Free-standing conducting polymer films for high-performance energy devices, *Angew. Chem. Int. Ed.* 55 (2016) 979–982.

[76] Y. Li, C.L. Zhu, T. Lu, Z.P. Guo, D. Zhang, J. Ma, S.M. Zhu, Simple fabrication of a Fe2O3/carbon composite for use in a high-performance lithium ion battery, *Carbon* 52 (2013) 565–573.

[77] Y.P. Chen, B.R. Liu, Q. Liu, J. Wang, J.Y. Liu, H.S. Zhang, S.X. Hu, X.Y. Jing, Flexible all-solid-state asymmetric supercapacitor assembled using coaxial NiMoO4 nanowire arrays with chemically integrated conductive coating, *Electrochim. Acta* 178 (2015) 429–438.

[78] M.M. Sk, C.Y. Yue, R.K. Jena, Non-covalent interactions and supercapacitance of pseudo-capacitive composite electrode materials (MWCNTCOOH/MnO2/PANI), *Synth. Met.* 208 (2015) 2–12.

[79] B.P. Grady, A. Paul, J.E. Peters, W.T. Ford, Glass transition behavior of single-walled carbon nanotube–polystyrene composites, *Macromolecules* 42 (2009) 6152–6158.

[80] K.I. Winey, T. Kashiwagi, M.F. Mu, Improving electrical conductivity and thermal properties of polymers by the addition of carbon nanotubes as fillers, *MRS Bull.* 32 (2007) 348–353.

[81] A. Ehsani, B. Jaleh, M. Nasrollahzadeh, Electrochemical properties and electrocatalytic activity of conducting polymer/copper nanoparticles supported on reduced graphene oxide composite, *J. Power Sources* 257 (2014) 300–307.

[82] S. Dhibar, C.K. Das, Electrochemical performances of silver nanoparticles decorated polyaniline/graphene nanocomposite in different electrolytes, *J. Alloy. Compd.* 653 (2015) 486–497.

[83] J.S. Shayeh, A. Ehsani, M.R. Ganjali, P. Norouzi, B. Jaleh, Conductive polymer/reduced graphene oxide/Au nano particles as efficient composite materials in electrochemical supercapacitors, *Appl. Surf. Sci.* 353(2015) 594–599.

[84] J.Y. Zhou, H. Zhao, X.M. Mu, J.Y. Chen, P. Zhang, Y.L. Wang, Y.M. He, Z.X. Zhang, X.J. Pan, E.Q. Xie, Importance of polypyrrole in constructing 3D hierarchical carbon nanotube@ MnO 2 perfect core–shell nanostructures for high-performance flexible supercapacitors, *Nanoscale* 7 (2015) 14697–14706.

[85] T.G. Yun, B.I. Hwang, D. Kim, S. Hyun, S.M. Han, Polypyrrole–MnO2-coated textile-based flexible-stretchable supercapacitor with high electrochemical and mechanical reliability, *ACS Appl. Mater. Interface* 7 (2015) 9228–9234.

[86] Y.Y. Chen, M. Han, Y.J. Tang, J.C. Bao, S.L. Li, Y.Q. Lan, Z.H. Dai, Polypyrrole–polyoxometalate/reduced graphene oxide ternary nanohybrids for flexible, all-solid-state supercapacitors, *Chem. Commun.* 51 (2015) 12377–12380.

[87] W.H. Sun, L.H. Chen, Y.B. Wang, Y.Q. Zhou, S.J. Meng, H.L. Li, Y.Q. Luo, Research progress on conducting polymer based supercapacitor electrode materials, *Synth. React. Inorg. Met.* 46 (2016) 437–444.

[88] C.H. Ng, H.N. Lim, Y.S. Lim, W.K. Chee, N.M. Huang, Fabrication of flexible polypyrrole/graphene oxide/manganese oxide supercapacitor, *Int. J. Energy Res.* 39 (2015) 344–355.

[89] L.L. Jiang, X. Lu, C.M. Xie, G.J. Wan, H.P. Zhang, Y.H. Tang, Flexible, free-standing TiO$_2$–graphene–polypyrrole composite films as electrodes for supercapacitors, *J. Phys. Chem. C* 119 (2015) 3903–3910.

[90] K. Koziol, J. Vilatela, A. Moisala, M. Motta, P. Cunniff, M. Sennett, A. Windle, High-performance carbon nanotube fiber, *Science* 318 (2007) 1892–1895.

[91] W. Lu, M. Zu, J.H. Byun, B.S. Kim, T.W. Chou, State of the art of carbon nanotube fibers: Opportunities and challenges, *Adv. Mater* 24 (2012) 1805–1833.

[92] B. Vigolo, A. Penicaud, C. Coulon, C. Sauder, R. Pailler, C. Journet, P. Bernier, P. Poulin, Macroscopic fibers and ribbons of oriented carbon nanotubes, *Science* 290 (2000) 1331–1334.

[93] A.B. Dalton, S. Collins, E. Munoz, J.M. Razal, V.H. Ebron, J.P. Ferraris, J.N. Coleman, B.G. Kim, R.H. Baughman, Super-tough carbon-nanotube fibres, *Nature* 423 (2003) 703.

[94] L.M. Ericson, H. Fan, H.Q. Peng, V.A. Davis, W. Zhou, J. Sulpizio, Y.H. Wang, R. Booker, J. Vavro, C. Guthy, A.N.G. Parra-Vasquez, M.J. Kim, S. Ramesh, R.K. Saini, C. Kittrell, G. Lavin, H. Schmidt, W.W. Adams, W.E. Billups, M. Pasquali, W.F. Hwang, R.H. Hauge, J.E. Fischer, R.H. Smalley, Macroscopic, neat, single-walled carbon nanotube fibers, *Science* 305 (2004) 1447.

[95] K.L. Jiang, Q.Q. Li, S.S. Fan, Nanotechnology: Spinning continuous carbon nanotube yarns, *Nature* 419 (2002) 801.

[96] M. Zhang, K.R. Atkinson, R.H. Baughman, Multifunctional carbon nanotube yarns by downsizing an ancient technology, *Science* 306 (2004) 1358.

[97] X.B. Zhang, K.L. Jiang, C. Teng, P. Liu, L. Zhang, J. Kong, T.H. Zhang, Q.Q. Li, S.S. Fan, Spinning and processing continuous yarns from 4-inch wafer scale super-aligned carbon nanotube arrays, *Adv. Mater* 18 (2006) 1505–1510.

[98] Y.L. Li, I.A. Kinloch, A.H. Windle, Direct spinning of carbon nanotube fibers from chemical vapor deposition synthesis, *Science* 304 (2004) 276.

[99] W.J. Ma, L.Q. Liu, R. Yang, T.H. Zhang, Z. Zhang, L. Song, Y. Ren, J. Shen, Z.Q. Niu, W.Y. Zhou, S.S. Xie, Monitoring a micromechanical process in macroscale carbon nanotube films and fibers, *Adv. Mater* 21 (2009) 603–608.

[100] J. Ren, L. Li, C. Chen, X. Chen, Z. Cai, L. Qiu, Y. Wang, X. Zhu, H. Peng, Twisting carbon nanotube fibers for both wire-shaped micro-supercapacitor and micro-battery, *Adv. Mater* 25 (2013) 1155–1159.

[101] Z. Yang, J. Deng, X. Chen, J. Ren, H. Peng, A highly stretchable, fiber-shaped supercapacitor, *Angew. Chem. Int. Ed.* 52 (2013) 13453–13457.

[102] Y. Li, X. Yan, X. Zheng, H. Si, M. Li, Y. Liu, Y. Sun, Y. Jiang, Y. Zhang, Fibershaped asymmetric supercapacitors with ultrahigh energy density for flexible/wearable energy storage, *J. Mater. Chem. A* 4 (2016) 17704–17710.

[103] X. Cheng, J. Zhang, J. Ren, N. Liu, P. Chen, Y. Zhang, J. Deng, Y. Wang, H. Peng, Design of a hierarchical ternary hybrid for a fiber-shaped asymmetric supercapacitor with high volumetric energy density, *J. Phys. Chem. C* 120 (2016) 9685–9691.

[104] L.J. Cao, S.B. Yang, W. Gao, Z. Liu, Y.J. Gong, L.L. Ma, G. Shi, S.D. Lei, Y.H. Zhang, S.T. Zhang, R. Vajtai, P.M. Ajayan, Direct laser-patterned micro-supercapacitors from paintable MoS2 films, *Small* 9 (2013) 2905–2910.

[105] G. Sun, X. Zhang, R. Lin, J. Yang, H. Zhang, P. Chen, Hybrid fibers made of molybdenum disulfide, reduced graphene oxide, and multi-walled carbon nanotubes for solid-state, flexible, asymmetric supercapacitors, *Angew. Chem.* 127 (2015) 4734–4739.

[106] Z. Xu, C. Gao, Graphene chiral liquid crystals and macroscopic assembled fibres, *Nat. Commun.* 2 (2011) 571.

[107] Z. Dong, C. Jiang, H. Cheng, Y. Zhao, G. Shi, L. Jiang, L. Qu, Facile fabrication of light, flexible and multifunctional graphene fibers, *Adv. Mater* 24 (2012) 1856–1861.

[108] Y. Hu, H. Cheng, F. Zhao, N. Chen, L. Jiang, Z.H. Feng, L.T. Qu, All-in-one graphene fiber supercapacitor, *Nanoscale* 6 (2014) 6448–6451.

[109] W. Ma, S. Chen, S. Yang, W. Chen, Y. Cheng, Y. Guo, S. Peng, S. Ramakrishna, M. Zhu, Hierarchical MnO2 nanowire/graphene hybrid fibers with excellent electrochemical performance for flexible solid-state supercapacitors, *J. Power Sources* 306 (2016) 481–488.

[110] B. Zheng, T. Huang, L. Kou, X. Zhao, K. Gopalsamy, C. Gao, Graphene fiberbased asymmetric micro-supercapacitors, *J. Mater. Chem. A* 2 (2014) 9736–9743.

[111] W. Cai, T. Lai, J. Lai, H. Xie, L. Ouyang, J. Ye, C. Yu, Transition metal sulfides grown on graphene fibers for wearable asymmetric supercapacitors with high volumetric capacitance and high energy density, *Sci. Rep.* 6 (2016) 26890.

[112] P. Xu, B. Wei, Z. Cao, J. Zheng, K. Gong, F. Li, J. Yu, Q. Li, W. Lu, J.H. Byun, B.S. Kim, Y. Yan, T.W. Chou, Stretchable wire-shaped asymmetric supercapacitors based on pristine and MnO_2 coated carbon nanotube fibers, *ACS Nano* 9 (2015) 6088–6096.

[113] F. Zabihi, M. Tebyetekerwa, Z. Xu, A. Ali, A.K. Kumi, H. Zhang, R. Jose, S. Ramakrishna and S. Yang, Perovskite solar cell-hybrid devices: Thermoelectrically, electrochemically, and piezoelectrically connected power packs, *J. Mater. Chem. A* 7 (2019) 26661–26692.

[114] Edurne Redondo et al., Enhancing supercapacitor energy density by mass-balancing of graphene composite electrode, *Electrochimia Acta*, 30 (2020), 36957

[115] Y. Zhu, S. Murali, M.D. Stoller, et al., Carbon-based supercapacitors produced by activation of graphene, *Science* 332 (2011) 1537–1541

[116] L. Lai, S. Yang, L. Wang, B.K. Teh, J. Zhong, H. Chou, L. Chen, W. Chen, Z. Shen, R.S. Ruo et al., Preparation of supercapacitor electrodes through selection of graphene surface functionalities, *ACS Nano* 6 (2012) 5941–5951.

[117] Y. Li, M.V. Zijll, S. Chiang, N. Pan, KOH modified graphene nanosheets for supercapacitor electrodes, *J. Power Sources* 196 (2011) 6003.

[118] P. Sharma, T.S. Bhatti, A review on electrochemical double-layer capacitors, *Energy Convers. Manag.* 51 (2010) 2901.

Index

Note: Page numbers in *italics* indicate a figure and page numbers in **bold** indicate a table on the corresponding page.